Advances in Epitaxy and Endotaxy
Selected chemical problems

AUTHORS

Chapter 0:   *V. Ruth*
Chapter 1:   *H. G. Schneider*
Chapter 2.1: *C. M. Jackson, J. P. Hirth*
              Columbus, Ohio, USA
Chapter 2.2: *W. J. Riedl*
              Warsaw, Poland
Chapter 2.3: *J. Noack*
              Berlin, GDR
Chapter 2.4: *Yu. D. Čistyakov, Yu. A. Baikov*
              Moscow, USSR
Chapter 3.1: *F. Bailly, L. Svob, G. Cohen-Solal*
              Bellevue, France
Chapter 4.1: *G. Kühn, A. Leonhardt*
              Leipzig, GDR
Chapter 4.2: *Yu. D. Čistyakov, Yu. A. Baikov*
              Moscow, USSR
Chapter 5.1: *J. Oudar*
              Paris, France
Chapter 5.2: *Ch. Weissmantel, G. Hecht, J. Herberger*
              Karl-Marx-Stadt, GDR

Advances in Epitaxy and Endotaxy

Selected chemical problems

Edited by

H. G. Schneider

Dr. rer. nat. habil.,

Professor of Technology of Electronic Components,
Technische Hochschule Karl-Marx-Stadt,
Department of Physics—Electronic Components,
German Democratic Republic

and

V. Ruth

Dr. rer. nat.,

Professor of Physics,
University of Oldenburg,
Federal Republic of Germany

with the cooperation of

T. Kormány

Dr. rer. nat.,

Department of Semiconductor Technology,
Research Institute for Telecommunication,
Budapest, Hungary

Elsevier Scientific Publishing Company
Amsterdam–Oxford–New York 1976

6425-8038

CHEMISTRY

The distribution of this book is being handled by the following publishers:

for the U.S.A. and Canada

American Elsevier Publishing Company, Inc.
52 Vanderbilt Avenue
New York, New York 10017

for East Europe, China, Democratic People's Republic of Korea, Cuba, the Democratic Republic of Vietnam and of Mongolia

Akadémiai Kiadó, The Publishing House of the Hungarian Academy of Sciences, Budapest (Hungary)

for all remaining areas

Elsevier Scientific Publishing Company
335 Jan van Galenstraat
P.O. Box. 211, Amsterdam, The Netherlands

Library of Congress Cataloging in Publication Data

Main entry under title:

Advances in epitaxy and endotaxy.

Includes bibliographical references and indexes.
1. Epitaxy 2. Thin films.
I. Schneider, Helmut Günther. II. Ruth, Volker. III. Kormány, T.
QD921.A3 1976 548'. 5 75-45005
ISBN 0-444-99845-4

Joint edition published by Elsevier Scientific Publishing Company, Amsterdam, The Netherlands and Akadémiai Kiadó, The Publishing House of the Hungarian Academy of Sciences, Budapest, Hungary

Printed in Hungary

ACKNOWLEDGMENTS

The editors and the authors wish to express their appreciation
- to G. Bognár and the management of the Research Institute for Telecommunication, Budapest, for their unselfish help in making it possible to publish this book in Hungary,
- to K. Drescher for reading the manuscript of Chapter 1 and for many useful and interesting discussions on this subject,
- to the many companies that cooperatively sponsored the fundamental program on CVD at Battelle, Columbus, for their financial support of portions of the research described in Chapter 2.1, and to C. A. Alexander, who performed the mass spectrometric study of chromous iodide, R. D. Gretz, J. M. Blocher, Jr. and the late A. C. Loonam, who encouraged this work and also contributed to it,
- to H. Neels for encouragement and valuable discussions concerning the work described in Chapter 4.1,
- to Mrs. I. Vogel and R. Glauche for their careful work during the experiments described in Chapter 5.2,
- and to the many other unnamed colleagues in many countries who contributed to and promoted the realization of this book.

H. G. Schneider translated the manuscript of Chapter 3.1 from French into English and of Chapter 4.1 from German into English.

V. Ruth prepared the last linguistic revision and the present English version of the text of this book. During linguistic revision, however, priority has been given to matching the text as closely as possible to the authors' depositions rather than to linguistic smoothness.

5

PREFACE

It is well known that often in the past a whole human generation had to elapse before results of basic research were applied in the interest of society. Nowadays this space of time has been reduced so much that it seems to be practically nonexistent. Sophisticated experimental research apparatus is sometimes converted without delay into powerful production tools and new methods of investigation immediately incorporated in quality control procedures. Thus, engineering and fundamental research have reached the stage where they can enter a "real time" dialogue, the one putting questions, the other answering them, the roles sometimes being exchanged. This lively dialogue is nowhere so fruitful as in the fields of solid state science and technology.

Fundamental research concerned with a quantitative description of the physical and chemical phenomena in semiconducting, dielectric and metallic materials has promoted considerable advances in the technology of solid state devices. Device technology, in turn, has provided some of the impetus and motivation as well as the framework for the formulation of many research problems. Current extensive basic and applied investigations on epitaxial and endotaxial intergrowth processes — a major technique in the microelectronics industry — provide an example of such feedback cycle. They represent a logical extension of continuing studies of the formation, structure and properties of thin films and they are also concerned with the search for better, cheaper and more reliable solid state devices and components than those in present use.

The main purpose of this monograph is to present the state of research in the field of epitaxy and endotaxy up to December 1973. The growing interest shown in epitaxial and endotaxial processes and the need to describe the theoretical background and experimental results concerning the formation and properties of these films justify the publication of a new book on "advances in epitaxy and endotaxy".

In the tradition of former volumes, significant new contributions by internationally accepted authors promote the solution of chemical problems in epitaxy and endotaxy and show the different methods of approach. Chapter 1 emphasizes the technological significance of epitaxy and endotaxy for the production of solid state devices and IC's. Chapter 2 indicates several methods of theoretical treatment for chemical vapor deposition: the heterogeneous nucleation theory, the thermodynamic approach, as well as aspects of defect growth. Chapter 3 deals with evaporation and interdiffusion phenomena under isothermal conditions. Chapter 4 takes into account the formation of epitaxial semiconductor films from the melt. In Chapter 5 the authors set out the theory of chemoepitaxial and chemoendotaxial layer growth on metals, supported by experimental results.

I hope that these new aspects of important theoretical and experimental results will assist scientists and engineers to widen their knowledge of epitaxial-endotaxial intergrowth processes and help them to perceive the relationship existing between theory and product research.

Prof. D. Sc. Iván Péter Valkó

Technical University, Budapest, Hungary

CONTENTS

0. BRIEF INTRODUCTION TO THE DIFFERENT ASPECTS OF CHEMICAL PROBLEMS OF EPITAXY AND ENDOTAXY CONSIDERED IN THIS BOOK

V. Ruth

University of Oldenburg, Oldenburg, Federal Republic of Germany

A general introduction to the taxonomy of epitaxy and endotaxy and the definitions of different kinds of epitaxy and endotaxy have been given in the previously published "Advances in Epitaxy and Endotaxy — Physical Problems of Epitaxy" [1]. Whereas, in "Physical Problems of Epitaxy", different aspects of the physics of epitaxial processes are considered, selected chemical aspects of epitaxy and endotaxy are emphasized in the present book.

The phenomena of epitaxy and endotaxy have attained appreciable interest due to their significance in semiconductor technology. The *first chapter* of this book is, therefore, devoted to the technological aspect of epitaxy and endotaxy, as far as chemical aspects are involved. *Schneider* discusses the most likely technological developments in the field of microelectronics (production of information processing components). The understanding of epitaxial and endotaxial intergrowth processes is already now a necessary requirement for the design of microelectronic devices. These processes, and the deposition and precipitation techniques based upon them, may become even more important in the future. The fundamental definitions of the different intergrowth processes are given according to the mechanisms involved. The technological necessity of homoepitaxial layer combinations of semiconducting materials is outlined for the examples of Si/Si and GaAs/GaAs. The applications, advantages, disandvantages, and details for the realization of homoepitaxial layers (epitaxial planar technique) are compared for these materials. A similar discussion is presented for the heteroepitaxial layer combinations GaAs/$A^{III}B^{V}$, HgTe/CdTe, and Si/insulator. Sapphire and MgAl-spinel are considered as suitable insulator substrates for Si deposits.

The undesired effect of endotaxial precipitations in silicon often occurs during doping with boron and phosphorus. This effect may only be prevented if details of the mechanism are known. Possible mechanisms are discussed.

Chemoepitaxial and chemoendotaxial layer formation by oxidation is important in numerous fields which involve thin film techniques. Here, the oxidation of tantalum is considered as a model process.

Eutectic and eutectoid hetero-crystals, epitaxially or endotaxially intergrown with each other, could be employed in order to achieve a higher reliability of electronic compounds. Such crystals with epitaxial or endotaxial interfaces can be obtained in a reproducible way by means of oriented solidification or transformation.

Surface layers with epitaxial layer/substrate interfaces can be achieved under quite

15

different conditions; *e.g.* films may be grown by deposition from a supersaturated vapor phase, by solidification from a melt, during exposure of a substrate to a reactive vapor phase, or by electrolytic deposition. Here, however, only the first three possibilities are considered. The respective chemical reactions may occur at different stages of film growth, depending on the chemical properties of the reactants. The combination of these two aspects, the medium from which the deposited film is grown and the kind of reaction involved, leads to different general problems of epitaxy and endotaxy which are discussed in the remaining four chapters.

Deposition processes from the vapor phase including chemical reactions in the vapor phase and/or at the substrate surface are known as Chemical Vapor Deposition processes (CVD) which are discussed in the *second chapter*.

During vapor deposition onto a substrate, epitaxial interfaces may be formed during the nucleation process itself, the subsequent growth process, or during an eventually occurring final process of rearrangement of deposits (agglomeration). *Jackson* and *Hirth* restrict their discussion to the case of epitaxial nucleation. After a brief review of the classical theory of heterogeneous nucleation, its assumptions, and the nucleation rate equations as obtained for different nucleus models, alternative rate equations are given applicable for different cases with respect to a combination of different conditions: (i) whether impingent and desorption fluxes are equal or not; (ii) whether metastable equilibrium between adsorbed monomers and critical nuclei prevails or not; (iii) whether translational or rotational degrees of freedom in the nuclei or adsorbed monomers are activated or not; and (iv) whether the sub-critical nuclei grow by addition of monomers via surface diffusion or direct impingement. The nucleation rate equations derived for the case of Physical Vapor Deposition (PVD) have to be modified for the case of chemical vapor deposition only inasmuch as the nucleation rate is affected by the chemical reactions involved. The effective supersaturation of the material being nucleated cannot, generally, be measured and must be calculated from the known flux of the impingent reactant. For this calculation, however, the mechanism of the chemical reaction involved has to be known. Furthermore, the impingent reactant and gaseous products of the chemical reaction may function as surface impurities and affect nucleation kinetics. Hence, in principle, the same nucleation rate equations may, according to *Jackson* and *Hirth*, be employed for both PVD and CVD.

Jackson and *Hirth* assume that for epitaxial nucleation the classical capillarity model generally applies rather than the *Walton–Rhodin* model of a critical nucleus consisting of only a few atoms. They argue that the former model is favored at high substrate temperatures and low supersaturations while the latter one should apply at low temperatures and high supersaturations, conditions corresponding to the observation of epitaxial and non-epitaxial interfaces, respectively. These arguments are supported by observations of deposits for the case of CVD of chromium from a chromous iodide beam effusing from a Knudsen cell under various conditions. At a substrate temperature range of about 840 to 940°C epitaxial deposits have been observed, while at the lower temperature range of about 650 to 740°C randomly oriented deposits occurred. While the assumption of the capillarity model for epitaxial nucleation seems to be justified, the experimental findings reveal no hint concerning the applicable nucleus model.

Riedl discusses the theory of CVD kinetics at relatively high pressures (approx. 1 atm) when viscous flow conditions prevail. In this case, the mass transport of the reactant from the gas phase towards the substrate surface, or of gaseous reaction products from the surface into the gas phase, may become rate controlling, depending on the prevailing flow conditions. In any case, when viscous flow conditions hold, diffusion through the interface adjacent laminar gas layers is involved. The influence of flow dynamics (laminar or turbulent flow), temperature, and a correlation of the diffusivities of the different species upon the mass transfer coefficient is discussed. *Riedl* defines the supersaturation ratio of the externally exerted partial vapor pressure p_i to the corresponding equilibrium partial vapor pressure over the deposit, p_{i0} of the deposited material as its activity at metastable equilibrium at the gas/solid interface. The solid deposit is assumed to be an ideal crystal. An equation relating this activity to the partial pressures involved, the enthalpy of reaction, and the equilibrium constant is derived for the case where only one product of the chemical reaction occurring during CVD is solid, while all other reactants and products are gaseous. This relation implies that molecular flow conditions and ideal gas behavior of all gaseous constituents are prevailing. Furthermore, the corresponding chemical reactions have to be known.

The theory of *Burton*, *Cabrera* and *Frank* on the growth of crystal films is extended to multi-component systems. A set of equations of metastable equilibria, of transport, and of material balances is developed enabling a description of the deposition of pure elemental crystal films in CVD flow systems. Previous deposition experiments for Si films by means of the CVD method are reviewed. Discrepancies between the results of different authors are attributed to the sensitivity of nucleation and growth mechanisms and the pertinent rates on temperature, flow intensity, and geometry whenever viscous flow conditions prevail, which should apply for the experiments reviewed.

A system of equations is developed from which the pertinent thermodynamic data, determining Si deposition rates, may be obtained for given supersaturations and $SiCl_4$ partial pressures in a gas stream consisting of a $SiCl_4/H_2$ mixture. The mechanism may comprise 6 different chemical reactions with 9 different components. The calculations carried out are based on reliable thermodynamic data for the system $Si-H-Cl$. A complete analysis is undertaken by means of computer techniques including a consideration of the appropriate mass transfer coefficients for CVD of Si from flowing $SiCl_4/H_2$ mixtures. However, because of the lack of sufficiently reliable experimental data, the analysis has still to include many simplifying estimations and assumptions.

Noack describes the usual CVD methods for GaP epitaxial layers. Differences in deposition temperatures, possibilities of influencing the Ga : P ratio, and hints for deposition mechanisms are mentioned. Of importance regarding contamination and graded junctions is the back etching of the substrate. For various usual substrate materials etching methods are given. Special attention is given to defects occurring in epitaxial and heteroepitaxial GaP films. Cracks occurring in these films grown on Si substrates at high temperatures are correlated to substance combinations with great differences in the coefficients of thermal expansion. In order to prevent cracking in this system, low deposition temperatures on small areas of deposition have to be applied.

Hillocks and pitting are attributed to dust particles or Ga droplets on the surface of the substrate and to etching by HCl excess or impurity segregations, respectively. Twins are only considered for substrate materials with non-polar structure. Stacking faults are found in the case of several substrates, even with homoepitaxial depositions. However, twins and stacking faults are correlated neither to substance combinations nor to deposition reactions. It is possible to obtain GaP films free of twins and stacking faults on all substrates mentioned.

Point defects have an important influence on the optoelectronic behavior. Contaminations like Ge from Ge substrates or Si from the quartz apparatus cause losses in quantum yields. Stoichiometric deviations in the form of Ga vacancies are detected with concentrations strongly dependent on the preparation methods. The importance of the origin of recombination centers is investigated by several authors.

Orientation defects as detected by broadening the X-ray reflexion width for GaP films on Ge and Al_2O_3 substrates are due to the departure of subgrains as shown by X-ray topographs. This departure is due to the chosen substance combinations. The densities of dislocations detected on GaP homoepitaxial junctions deposited by several methods lie in the range of 10^4 to 10^5 EPD cm^{-2} which is in the limit of the substrates used. A correlation is difficult because there are no investigations with highly perfect substrates. Misfit dislocations in heteroepitaxial layers as detected by transmission electron microscopy exhibit high dislocation densities in the range predicted from theory. The junctions considered are GaP on Ge and homostructures. Back etching, solid solution transition layers, and intermediate layers are discussed. Stress may be introduced into epitaxial layers by the misfit not balanced by misfit dislocations, or differences in the coefficients of thermal expansion. Stress results in the occurrence of bending and optical anisotropy as described for GaP on Ge. As demonstrated, certain kinds of defects such as hillocks and stacking faults are defects which may be prevented; some, such as stress, are due to certain substance combinations, whereas others are complex, such as dislocations which are affected by growth parameters, misfit, stress etc. The interaction of defects is important for the electronic behavior and it seems, considering recent papers, that point defects, especially, could explain differences in the CVD and LPE material.

In the fourth section of this chapter *Čistyakov* and *Baikov* discuss the model of thin liquid films on heated surfaces of solid materials during epitaxial crystal growth from the vapor phase. The conditions for the existence of such films and their thicknesses are derived on the basis of the van der Waals interaction theory by means of classical thermodynamics. The estimated thickness is in good agreement with experimental values, determined for the case of Si homoepitaxy. The appearance of protrusions on top of the quasi-liquid films at some critical film thickness is explained theoretically by a consideration of the appropriate surface energies.

In the *third chapter* a special type of vapor deposition, the formation of epitaxial semiconductor films in connection with volume interdiffusion under isothermal conditions, is discussed. The authors of this chapter, *Bailly*, *Svob*, and *Cohen-Solal*, treat the problem for the technologically important example of the growth of a HgTe film ($A^{II}B^{VI}$ compound) on a CdTe substrate when exposed to mercury vapor. Previous investigations have revealed that the film growth and the corresponding interdiffusion processes sensitively depend upon temperature, geometry of the arrangement, the excess vapor pressure, and the duration of treatment. The latter has indicated that the rate of equilibration is relatively slow.

In the first section of this chapter the simpler problem of the kinetics of equilibration between solid HgTe and a surrounding atmosphere of mercury vapor is considered, in particular in dependence upon the initial externally exerted excess mercury vapor pressure. The experimental results are theoretically interpreted.

In the second section of this chapter the kinetics of layer growth of HgTe on a CdTe substrate is described as obtained under various conditions, particularly at different excess mercury vapor pressures. The results are interpreted in terms of a complex model comprising volume interdiffusion between the growing HgTe layer and the CdTe substrate giving rise to a Kirkendall effect and the associated formation of vacancies in the Kirkendall interface, diffusion of these excess vacancies towards the layer surface, the partial annihilation of these vacancies at the surface by mercury atoms (incorporation of mercury atoms from the gas phase), and the equilibration process in the gas phase (the establishing of an actual mercury vapor pressure dependent upon the initial externally exerted excess mercury vapor pressure).

The *fourth chapter* is devoted to the formation of semiconductor films by crystal growth from the melt. In the first section of this chapter the problems of growing $A^{III}B^V$ compound layers by means of the Liquid Phase Epitaxy method (LPE) are thoroughly discussed by *Kühn* and *Leonhardt*. The combinations of different $A^{III}B^V$ films forming *pn* junctions have gained great interest for the production of light-emitting semiconductor elements. Since the structure of the deposit/substrate interface determines decisively the electrical properties of the junction, the lattice constants and expansion coefficients of layer and substrate have to be equal within a small range of tolerance. Combinations of different $A^{III}B^V$ compounds are, therefore, favorable. There is a wide range of possible combinations, since the solubility of $A^{III}B^V$ compounds in each other is appreciable. In most cases, the substrate/deposit combination contains three different elements from the third and fifth groups of the periodic table of the elements. The knowledge of the respective ternary phase diagrams is indispensible for the appropriate choice of the materials, the employed epitaxy technique, and the conditions being maintained during the growth process (*e.g.* temperature). Most important are ternary systems of the type $A^{III}B^{III}C^V$. Because of experimental difficulties (the necessity of high temperatures, slow equilibration rate) experimental investigations of these ternary phase diagrams have to be backed by semi-empirical evaluations. The most commonly applied experimental techniques are reviewed. Of the different semi-empirical approximations of ternary phase diagrams considered, most are based on extrapolations from known binary phase diagrams and some idealizing assumptions about the solid and liquid solutions involved, (*e.g.* ideal solution, regular solution, quasi-regular solution etc.), which allow an estimation to be made of the appropriate activity coefficients.

The different LPE techniques are distinguished by the methods of establishing a supersaturation of the melt and the ways of bringing the substrate into contact with the melt. The different techniques are discussed in detail. The relations between phase diagrams, growth parameters, and resulting layer characteristics are considered particularly for the systems Ga–Al–As, Ga–In–As, and Ga–As–X (X is the doping element).

2*

In the second section of this chapter *Čistyakov* and *Baikov* propose a model in order to explain the kinetics of phase transitions during crystal growth from a 50% binary alloy melt. The model is characterized by a roughness of the melt/crystal interface which is related to surface steps which are often observed at the surface of crystalline phases. Based on a simplified model of the surface roughness, the conditions for the appearance and the effectiveness of surface protrusions on the kinetics of crystal growth are derived leading to critical interface temperatures which define the ranges of different observable growth kinetics. Completely irregular crystals should be formed in the temperature range between the two critical interface temperatures T_{cr} and T_k (with $T_{cr} < T_k$), while regular growth should occur between T_k and the phase equilibrium temperature. The values of the critical temperatures depend on both the specific properties of the melt and the model assumed for the crystallization mechanism.

When a metallic surface is exposed to a reactive atmosphere, chemical compounds may be formed covering the metallic surface as a solid layer (*e.g.* oxides, sulphides, carbides, or chlorides). The formation of these surface layers and their crystallographic structures and orientations do not depend only on the materials involved but also on the crystallography of the metallic surface. Hence, the formed surface compounds may be considered as epitaxial or endotaxial layers. The formation of these chemoepitaxial or chemoendotaxial layers is considered in the *fifth chapter*. In the first section *Oudar* gives an extensive summary about recent experimental work and its interpretation concerning the early stages of layer formation for the cases of sulphides, oxides, and carbides. Layer formation is considered to consist of the steps adsorption, nucleation, growth of nuclei, and uniform layer growth. In particular, the epitaxy of the initial adsorption layers, alterations of the crystalline lattice of the matrix metal by solving adsorbed species, and surface structure changes due to the adsorbed layers, are discussed. Furthermore, the resulting epitaxial relationships between metal substrates and surface compound layers are considered along with the corresponding interface models, as they have been proposed in more recent literature especially by *v. d. Merwe*. Of special interest are the conditions and mechanisms of the transition from adsorption layers to nucleation of epitaxial deposits. Experimental results have indicated the existence of transition layers during the formation of epitaxial deposits. An extensive discussion about this problem is given by *Mayer* in the above mentioned "Advances in Epitaxy and Endotaxy — Physical Problems of Epitaxy" [2]. Numerous observations indicate the decisive role of structural differences between metallic substrates and surface compound deposits at the interface for the growth rate of nuclei and of films. Different growth models have been developed.

The second section of this chapter deals with another aspect of chemoepitaxial and endotaxial layer growth. *Weissmantel*, *Hecht*, and *Herberger* discuss the kinetics of the growth of film thickness. The time-dependence of the film thickness can be described by means of different empirical laws: a "square root law", an "inverse logarithmic law", a "logarithmic law", and laws according to which the film thickness is proportional to different powers $1/n$ of the time, with $n \geq 2$. The different laws apply to different stages of layer growth, different conditions, and different chemical reactions involved. Combinations of these laws which had been experimentally verified should apply to overlapping stages of layer growth. The rather complex kinetics of layer growth is due to different mechanisms ascribed to different stages of growth, like adsorption, nucleation, nuclei growth, recrystallization, coalescence of separate crystals, and continuous layer growth. Different elementary mechanisms may overlap in certain stages of layer growth, leading to complex growth laws.

In order to detect the different mechanisms prevailing at the different stages of growth kinetics, *Weissmantel et al.* have carried out extensive morphology (electron microscopy) and structure investigations as well as a continuous measurement of layer thickness for the cases of oxidation and sulphidation of predeposited Zn and Cd films. It was possible to attribute the different stages of growth kinetics to different observed mechanisms and to propose a general model consisting of consecutive and overlapping elementary mechanisms for the formation of tarnish layers on metals.

This book does not reflect a uniform conception of epitaxy and endotaxy, but the most substantial problems of this field will be elaborated from the point of view of certain chemical aspects. Different views of specific problems will also be considered within this book and reference is given to previous publications of the editors [3], [4].

REFERENCES

[1] *Schneider, H. G:* in "Advances in Epitaxy and Endotaxy" (Physical Problems of Epitaxy) (Eds.: H. G. Schneider and V. Ruth), VEB Deutscher Verlag für Grundstoffindustrie, Leipzig (1971), p. 13.
[2] *Mayer, H.:* in Reference [1], p. 63.
[3] *Schneider, H. G.* (Ed.): Epitaxie—Endotaxie, VEB Deutscher Verlag für Grundstoffindustrie, Leipzig (1969).
[4] *Schneider, H. G., Ruth, V.* (Eds.): Advances in Epitaxy and Endotaxy (Physical Problems of Epitaxy), VEB Deutscher Verlag für Grundstoffindustrie, Leipzig (1971).

1. TECHNOLOGICAL SIGNIFICANCE OF EPITAXIAL AND ENDOTAXIAL INTERGROWTH PROCESSES FOR THE PRODUCTION OF DISCRETE ELECTRONIC COMPONENTS AND INTEGRATED MICROELECTRONIC CIRCUITS

H. G. Schneider

Technische Hochschule, Karl-Marx-Stadt, Department of Physics — Electronic Components
German Democratic Republic

1.1.   PROBLEMS OF RELIABILITY

The tendencies of technical development in modern industrial states have, for some time, been characterized by intensive efforts to rapidly develop electronics, and, in particular, to develop and improve the electronic components required, which should, more comprehensively, be called information processing components. This tendency is most likely to exist for some time in the future.

These trends concentrate especially on the microminiaturization of such components since modern electronic instruments have to contain increasingly more components (about $10^6 \ldots 10^7$ in large-size equipment) in order to meet the functional requirements. In respect to their functions, these components have now almost reached the limits imposed by physical laws.

Therefore, the main interest is now focussed on lifetime and reliability, which have to be improved, in addition to the increasing technological requirements.

In principle, an electronic plant is already considered to have failed when one of its components has totally failed or when a decisive parameter deviates more than the permissible range of tolerance and a redundant circuit is not available.

If it can be expected with a probability p_e that a component is still working after a certain defined time, then this probability is, for an electronic plant with n components, p_e^n. If, for instance, 1 or 2 components out of 1 000 fail within this defined time, then $p_e = 0.999$ to 0.998. However, $p_e^n = 0.999 \ldots 0998^{10\,0000}$ which is practically zero. This value is very insufficient. To ensure sufficient reliability of such a plant, p_e^n should be as close to 1 as possible in order that q_e, the failure rate, may approach zero. Therefore, efforts should be made to achieve values of about 10^{-9} to 10^{-12} for q_e [1]–[3].

Considerations like these inevitably lead to the conclusion that all available scientific and technological knowledge and experience should be applied in order to manufacture such reliable microminiaturized components. One way to meet these requirements is the increasing tendency of integration in microelectronics. However, the production of such integrated circuits requires an extremely complicated, heterogeneous technology. Moreover, according to the present conceptions, this technology which has been established all over the world with high investment costs partially consists of less compatible steps. Silicon serves as the base material for this technology. The principles of this technology have become known under the name "Epitaxy planar technique". This technique will remain the most important technique for the production of integrated circuits of microelectronics and of the preponderant parts of discrete, active, electronic semiconductor components in the near future.

It can be foreseen, however, that development will not cease here. The search for simpler technologies for the production of more reliable components and circuits, while still utilizing, as far as possible, the technological investments and experiences including those made in the field of materials science, is steadily continuing everywhere.

A solution will be achieved only by utilizing, for the design of components and circuits, a more complex preparation of materials in the stage of crystal growth by using known and new principles in order to avoid too many, and non-compatible, single steps of the subsequently applied technology. This would mean a shift towards the field of materials science and would also lead to an increasing entanglement of materials science and technology.

The fact that epitaxy plays a decisive role here should be mentioned, without, however, indulging in too much speculation.

1.2.   EPITAXIAL AND ENDOTAXIAL INTERGROWTH MECHANISMS

The significance of epitaxial and endotaxial intergrowth processes for the technology of producing electronic components − not only in the field of semiconductor techniques − is here demonstrated by means of some selected examples. The appropriate systematic approach relating to the intergrowth processes will be used in the following discussion to allow a better understanding.

Table 1. Designation of the various intergrowth processes

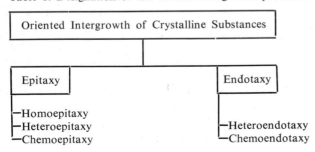

A general explanation of these terms can be found elsewhere [1], [4]−[6]. These terms will be referred to in connection with the technological problems which are discussed below. However, the intergrowth mechanisms corresponding to the terms given above will be explained again with the help of the examples cited. All five different mechanisms of intergrowth are definitely necessary in some specific way to prepare the basic structures for the production of electronic components. Occasionally, they give rise to disturbing side effects. In any way, all these intergrowth processes exhibit more-or-less weighty chemical or physico-chemical problems. Priority is given in the present book to topical aspects to these problems. This either refers to deposition processes such as chemical vapor deposition of single-component or multi-component semiconductor substances; formation of solid solutions or chemical compounds via diffusion processes; chemically reactive deposition or precipitation growth; intergrowth processes from the melt; or to the discussion of fundamental chemical or physico-chemical problems. The physico-chemical problems encompass also the problems of bonds between partner substances also in the case of growths due to physical separation of the deposited layer, *i.e.* of crystal growth in the course of intergrowth. These are only a few essential aspects. In any case, it is necessary to know exactly such intergrowth mechanisms to enable the technological process of preparation to be optimized, and to eliminate disturbing effects.

1.3. EPITAXIAL AND ENDOTAXIAL INTERGROWTH PROCESSES IN THE SEMICONDUCTOR TECHNOLOGY

1.3.1. HOMOEPITAXY

1.3.1.1. Si–Si

Homoepitaxy of Si on Si is one of the substantial bases of the Si epitaxy planar technique which will now be dealt with briefly for a better understanding of homoepitaxy [2], [7]–[12]. However, this technology has become also increasingly important for other semiconductor materials. This applies especially to the $A^{III}B^{V}$ compound semiconductors [13] as discussed in other sections of this book, as well as to germanium (see below).

By using the epitaxy planar technique, it is possible to produce nearly all of the modern active components and microelectronic circuits; also those used in power electronics such as junction power rectifiers and thyristors. The particularly favorable possibilities for the realization of the principles of this technology by applying the base material silicon (thermal oxidation to produce non-porous, detachable SiO_2 surface layers) resulted in the monopoly that silicon has had, and will continue to have, in semiconductor techniques in the near future. This tendency has become generally accepted, although Ge has, to some extent, more favorable electronic properties, *e.g.* a higher mobility of charge carriers. However, Ge is found less frequently than Si. Due to its larger band gap, Si, on the other hand, has enabled the semiconductor technique to be used for power electronics. The quantitatively predominant application of Si is a wise compromise in material-scientific, technological, physical, and economic respects.

The planar technique permits especially the production of more stable, and more reliable components and switching circuits, and, along with the advantages of production technology, proved to be extremely economic. The technical characteristics obtained for the components due to the geometrical conditions and material properties can be optimized so that they are largely restricted by the crystal parameters only. The preparation of surface layers provides a surface protection which has a positive effect on the increase of reliability and which facilitates sealing. The planar

Table 2. The most important steps of Si epitaxial planar technology according to data taken from [8], [11]. As to the Table, great importance was attached to show that, apart from thermal oxidation still preferably used to produce SiO_2 surface layers, more possibilities of producing surface layers at lower process temperatures are investigated. Various such methods are already being used in industry. This search also refers to Ge and the $A^{III}B^V$ compound semiconductors (GaAs and its solid solution variants with Al and P) in order to make them applicable to the planar technology and, thus, to produce reliable and stable components for correspondingly suited applications

Process Steps	Influenced Parameters	Controllable Range	Characterization of the Process			
					see Table 3	
			Layer	Process	Temperature of Reaction of Substrate	Average deposition Rate
E P I T A X Y	$X_E + W + X_C$	from $\approx 1\ \mu m$				
Masking and Process Engraving	S_1, S_2 (see Fig. 1)	from $\approx 1\ \mu m$	SiO_2	Thermal oxidation in O_2 – dry in N_2 – wet in O_2 – wet	1 200°C	0.15 $\mu m\ h^{-1}$ 0.4 $\mu m\ h^{-1}$ 1 $\mu m\ h^{-1}$
				Anodic oxidation	20 . . . 30°C	
				Pyrolysis of Si compounds such as $Si(OCH_3)_4$, $Si(OC_2H_5)_4$ etc.	500°C 800°C	0.4 $\mu m\ h^{-1}$
				Decomposition of Si compounds in gas discharge	⫫ 200°C	
				Hydrolysis of Si halides	1 200°C ⫫ 800°C	8 $\mu m\ h^{-1}$ 2 $\mu m\ h^{-1}$
				Reactive sputtering of Si in O_2	⫫ 200°C	2 $\mu m\ h^{-1}$
			Si_3N_4	Pyrolysis of Si compounds, such as $SiCl_4$, SiH_4 and others, with N_2 or NH_3	800°C 1 300°C	1.2 μm . . . . . 2.5 μm

Reaction of Si compounds with NH_3 in the gas discharge	VII	300°C	
Reactive sputtering of Si in N_2 or Ar/N_2	VII	300°C	$0.12 \ldots 1$ μm h^{-1}

Diffusion (Doping)

W (see Fig. 1)

from 1 μm

Diffusion in the open system
- Diffusion of B or P *from gaseous sources* into the carrier gas flow at temperatures of about 1 200°C, e.g. PH_3 intermediate gas mixture as the original source, P_2O_5 as local source
- Indiffusion *from solid sources*, such as P_2O_5 as the original source, evaporation temperature 215 ... 300°C. N_2 is mostly used as the carrier gas.
- Indiffusion *from liquid sources* such as $POCl_3$ (or PCl_3 or PBr_3) as original sources, P_2O_5 as the local source. Deposition takes place at ambient temperature.

Diffusion in the closed system
For instance, the base material and contaminant are placed in a quartz tube closed by melting. This process is now of little significance for element semiconductors. It is especially employed for compound semiconductors such as GaAs (different gas pressures, see below).

Vacuum diffusion
The apparatus used correspond to those employed for diffusion in the closed system. Such diffusion is restricted to special cases.

Paint-on diffusion
The base material is coated with the original source. The entire arrangement is heated in an inert atmosphere. Such diffusion is mainly used for the diffusion of Au into Si.

Diffusion from doped covering layers
Intermediate oxide glass phases are formed on the semiconductor surface. Such diffusion is advantageous for compound semiconductors (see above). If SiO_2 is used as a covering layer there is the danger of oxygen being incorporated as a low-positioned donor.

27

technique is characterized by a high degree of coordination of the successive steps thus providing the fundamental condition for the monolithic integration technology. Moreover, due to its flexibility, this technology makes possible the production of a wide range of semiconductors and integrated components of this kind by means of standard equipment.

In Fig. 1 a classical example of the basic structure of a simple Si planar transistor is shown, whereas in Fig. 2 that of a Si–pn-field effect planar transistor is given.

Tables 2 and 3 briefly, and generally, describe the basic steps of planar technology (especially with respect to the use of silicon as the base material) and the most important processes used for the deposition of epitaxial silicon layers on silicon substrates.

Further details on Si epitaxial planar techniques are omitted here since this lies outside the purpose of this paper. However, one problem encountered in all one-sided transistors should be emphasized. In practice, it is not possible to produce larger silicon slices less than 75 to 100 μm thick without great losses due to slice failure. Also, the application of a higher impedance layer with a thickness of 70 μm and more as a collector, contacting, in transistors, with a base width of a few μm cannot solve this problem in a satisfactory way since the saturation voltage $U_{CE\,sat.}$ and the storage time t_s attain values so unfavorable that such transistors cannot be employed in digital techniques.

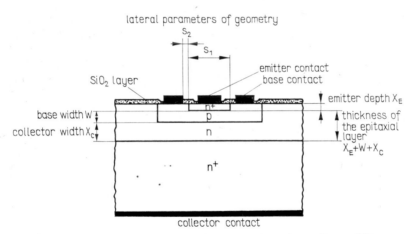

Fig. 1. Basic structure of a simple Si planar transistor (according to [8])

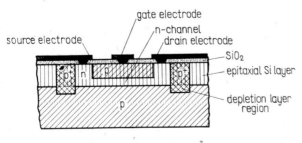

Fig. 2. Typical structure of a pn-field-effect transistor (n channel) in Si epitaxy planar technique (according to [2])

28

Table 3. Various deposition methods for silicon in Si epitaxial planar technology (empirical formulae given without taking into account side-reactions) using data of [2], [8], [11], [12]

High-Temperature Epitaxy Process	Low-Temperature Epitaxy Process
Reduction of Si Cl_4 by H_2 at 1 200. . . 1 300°C *(employed preferably in industry at first)* $Si Cl_4 + 2 H_2 \rightleftharpoons Si + 4 HCl$ (Typical rate of deposition 0.5 . . . 1 μm min^{-1})	Pyrolysis of silane at 950°C $SiH_2 \rightleftharpoons Si + 2 H_2$
Reduction of trichlorosilane by H_2 at 1 150 . . . . 1 250°C $SiHCl_3 + H_2 \rightleftharpoons Si + 3 HCl$	Reduction of Si_2Cl_6 at 650°C while under ultraviolet radiation
Range of temperatures 1 150 . . . . 1 300°C	(The two-step process is not taken into consideration) *Range of temperatures* 650 . . . 950°C

Note:

Si is deposited from a 2nd and 3rd source depending on the requirement, along with a desired amount of acceptor or donor atoms (B or P). The deposition of epitaxial layers may also be effected selectively at different points of the substrate crystal.

The device advantage of the epitaxial planar technique which must be stressed here is that the silicon, as the effective semiconductor material in the transistor, can, by means of this method, be deposited as a very thin, monocrystalline layer onto an extremely low-impedance, sufficiently thick silicon slice (200 μm) which serves only as a substrate and contact area. Mostly, a {111} substrate surface is used, in special cases with an intended misorientation of 3° towards ⟨110⟩. This monocrystalline silicon layer (10 to 15 μm thick) enables the above required values to be achieved. It must be added that the preparation of monolithic, integrated solid state circuits is achieved in an analogous way. As a base, a substrate is used into which many components can be incorporated simultaneously by means of diffusion. The preparation of the above-mentioned required basic structure is the same as in the preparation of the respective transistor (see Fig. 3).

In this connection, it is not necessary to discuss the production of corresponding planar diodes.

Methods different from those applied for the production of conventional discrete components have to be utilized for the production of other circuit elements, for example, passive circuit elements as resistors and capacitors. Thus, for instance, semiconductor resistors can be produced by modifying impurity concentrations in silicon by means of adequate diffusion. Semiconductor capacitors can be varied by using depletion layers of transistors as corresponding switching elements or by building up metal–SiO_2–Si (MOS) structures serving as capacitors.

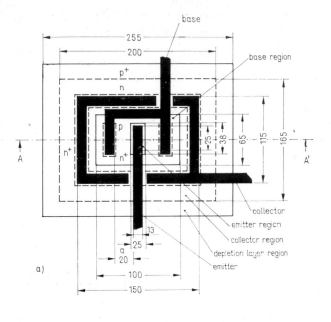

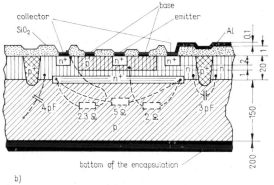

Fig. 3. Typical structure of Si planar transistors in integrated circuits (all dimensions given in μm) a) elevation; b) section A–A' (according to [2])

The epitaxy process, however, is also of great importance for the formation of *pn* junctions for high currents and voltages since the time required for preparation can thus be reduced to 1/20 to 1/25 of the time required for the formation of structures by diffusion. Furthermore, reaction vessels with cold walls can be used and the incorporation of undesired impurities can be diminished. In the case of large area, and thicker depositions special care has to be taken to ensure that the crystal is perfect over the entire volume of the epitaxial layer. It is especially the critical region in the *pn* junction which must be free of inhomogeneities to avoid local excesses of field strength. Only in this way can rectifiers with controlled avalanche breakdown be obtained. To achieve this, the epitaxial process is particularly well suited. This is due to the fact that in the case of epitaxial intergrowth in the interfaces (*i.e.* here *pn* junctions) endeavours are made to reach the lowest possible free interface energy yielding particularly stable conditions. This also applies to the preparation of the

30

four-layer structure of thyristors with three successive *pn* junctions. In addition, the lateral geometry requirements are particularly high in this case.

After this characterization of the technological significance of Si/Si homoepitaxy, the basic definition of this term is stated again: "Homoepitaxy" is the process of the growth of a crystalline substance on a substrate crystal, or on a monocrystalline substrate layer with corresponding crystallographic orientations, when both crystals do not differ from each other in respect to their crystal symmetry dependent properties and when they do not differ at all, or differ only slightly (in the case of semiconductor substances having different doping concentrations and equal or different types of conduction) in respect to their chemical compositions [1], [4], [5]. This definition is assumed to apply regardless of the separation process of the deposit.

1.3.1.2. GaAs–GaAs

The conditions become more complicated when the technology of $A^{III}B^V$ compound semiconductor components is regarded. The most important of these components are also based on the principle of homoepitaxial deposition of GaAs on GaAs substrate crystals in the form of chemical vapor deposition (gaseous phase epitaxy) and in the form of liquid epitaxy (Table 4).

In principle, almost all of the components prepared by the silicon technology and all of the circuit types prepared by the epitaxial planar technique can also be produced on the basis of the $A^{III}B^V$ compound semiconductors, especially on the basis of GaAs and its solid solutions [13].

However, the question arises whether such a procedure is useful. An attempt is made to find out whether it is useful or not and, if not, to state the reasons for such a negation (see above) and to discuss the applications for which this group of substances should be preferred.

There is, indeed, no absolute necessity to generally replace silicon by an $A^{III}B^V$ compound semiconductor, as the latter has no *basically* different, or better electronic properties than the element semiconductors Ge and Si. The differences in question are only minor (Table 5). This also applies in spite of the fact that GaAs is mostly considered as a base material which, in accordance with its intended application, is modified by adding Al or P to form solid solutions of the $Ga_{1-x}Al_xAs$, or $GaAs_{1-x}P_x$, types. The result of these additions is a certain variation of physical properties. There is no remarkable difference as to the other $A^{III}B^V$ compound semiconductors, which, due to their properties, could be included in the compounds which are of interest regarding electronic devices. Details relating to these problems. are given in [14].

On the other hand, the growing of high-purity and structurally perfect silicon single crystals, even those with a great diameter, is by and large mastered. Nevertheless, the growing of such GaAs crystals is more complicated than that of an element crystal and, in fact, still entails considerable technological difficulties.

Highly perfect GaAs crystals are required as substrate materials for the production of GaAs components. Difficulties in growing these crystals are, among others, due

Table 4. Most important stages of gaseous phase and liquid epitaxy of GaAs (applicable to other binary and ternary $A^{III}B^{V}$ compound semiconductors) using data from [12] and [13]

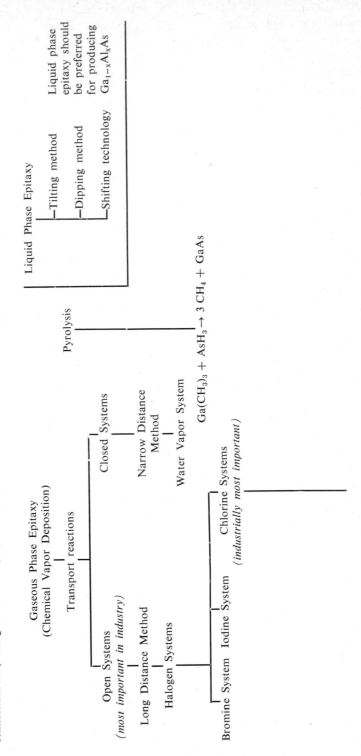

Gaseous Phase Epitaxy
(Chemical Vapor Deposition)

Transport reactions
— Open Systems
  (*most important in industry*)
  — Long Distance Method
  — Halogen Systems
    — Bromine System
    — Iodine System
    — Chlorine Systems
      (*industrially most important*)
  — Closed Systems
    — Narrow Distance Method
    — Water Vapor System

Pyrolysis

$$Ga(CH_3)_3 + AsH_3 \rightarrow 3\ CH_4 + GaAs$$

Liquid Phase Epitaxy
— Tilting method
— Dipping method
— Shifting technology

Liquid phase epitaxy should be preferred for producing $Ga_{1-x}Al_xAs$

Ga–AsH$_3$–HCl–H$_2$ GaAs–HCl–H$_2$ GaCl$_3$–As–H$_2$ Ga–As–HCl–H$_2$

Arsine and phosphine processes are becoming more and more important in producing GaAs$_{1-x}$P$_x$ layers.

Arsenic Trichloride Method

GaCl$_2$–AsCl$_3$–H$_2$ Ga–AsCl$_3$–H$_2$

Source reaction:

$$6\,H_2 + 4\,AsCl_3 \rightleftarrows 12\,HCl + As_4$$
$$6\,Ga + 6\,HCl \rightleftarrows 6\,GaCl + 3\,H_2$$

(between 800°C and 900°C)

Disproportioning after transport in the carrier gas flow *at lower temperatures in the substrate zone:*

$$6\,GaCl \longrightarrow 2\,GaCl_3 + 4\,Ga$$

Epitaxial deposition:

$$4\,Ga + As_4 \longrightarrow 4\,GaAs$$

Relative mean rates of growth:

(GaAs – substrate orientation)

(110)	(111)	($\overline{111}$)	(100)
0.4	1	0.1...0.2	0.2...0.4

(industrially most important)

Table 5. Important physical data of Si, Ge, and GaAs. These data apply at 300 K. Any deviations are specially marked (according to [13])

Properties	Ge	Si	GaAs	Unit
Band gap ΔE_{min}				
0 K, opt.	0.75	1.15	1.52	[eV]
300 K, opt.	0.67	1.11	1.35	[eV]
300 K, electr.	0.78	1.21	1.44	[eV]
Effective mass of				
electrons m_n	0.55 m_0	1.1 m_0	0.067 m_0	
Holes m_p	0.37 m_0	0.59 m_0	0.5 m_0	
	(4 K)	(4 K)		
Mobility of electrons μ_n	3 900	1 900	8 800	[cm²/Vs]
Holes μ_p	1 800	480	450	[cm²/Vs]
Intrinsic resistance ρ_i	46	$2.3 \cdot 10^5$	$> 10^8$	[Ω cm]
Dielectric constant ε_r	16	12	11.5	
Lattice constant a_0	5.657	5.431	5.654	[Å]
Density d	5.328	2.329	5.316	[g cm^{-3}]
Debye temperature Θ_D	406	689	355	[K]
Expansion coefficient β	$6.1 \cdot 10^{-6}$	$2.4 \cdot 10^{-6}$	$5.8 \cdot 10^{-6}$	[°C^{-1}]
Thermal conductivity λ	0.61	1.41	0.46	[W/cm°C]
Melting point T_S	936	1412	1238	[°C]
Vapor pressure at the				
melting point	$8.4 \cdot 10^{-7}$	$5.6 \cdot 10^{-4}$	740	[torr]

to the different vapor pressures of the two components at the melting point of the compound (arsenic is exceptionally volatile).

Therefore, the application of GaAs components will essentially be restricted to special fields where particularly favorable conditions exist due to the material properties. Such fields are, for instance, hyper-frequency diodes, electron transfer components and components of opto-electronics, which shall not be discussed here. These are discussed in references [13] and [14].

It is of basic interest here that the principles of epitaxial planar technology are applicable to the production of components having *pn* junctions as the basic structure, and GaAs as a base material. This means that, as in the case of Si, a thin, monocrystalline layer of GaAs (homoepitaxy), or of the two above-mentioned ternary solid solutions, is deposited onto a monocrystalline substrate (GaAs in our case) by means of gaseous phase epitaxy (the most important method used in industry), or by liquid epitaxy. The subsequent procedure of producing component structures is similar to the one employed in Si planar epitaxy technology (selective diffusion) or corresponds to the mesa-technique also employed for Si (lateral limitation by etching processes), especially when the liquid epitaxy process is applied. An interesting application of homoepitaxial deposition of GaAs is the one used for producing Gunn diodes. In this case, both epitaxy processes mentioned above are used.

The gaseous phase epitaxy serves to produce elements for higher frequencies of the S band, whereas liquid epitaxy is used for elements of the lower part of the frequency spectrum (S and L bands), for which higher demands concerning purity are made.

The gaseous phase epitaxy of GaAs enables the production of elements with a maximum output in pulsed service of 240 W in the S band and of 350 W in the C band. In the X band, elements for c.w. service furnish 2 W. For the L band, liquid epitaxy produces elements which are able to deliver a maximum output exceeding 500 W [15], [16].

A list of specific technological data, together with the corresponding scientific fundamentals, is given in another section of this book on GaP.

1.3.2.   **HETEROEPITAXY**

1.3.2.1.   **GaAs-$A^{III}B^{V}$ solid solutions and Ge**

For opto-electronic applications, the technique of epitaxial CVD (Chemical Vapor Deposition) of the two ternary compounds (already mentioned above) having Al, or P, as the third component on a GaAs substrate is used more and more because injection luminescence components emitting in the visible spectral range (red light) may be obtained by this method. This involves already, however, a transition to *heteroepitaxy*, and this hampers the growth of such layers due to the different lattice parameters of the substrate and the deposited layer which, however, can be reasonably controlled up to a deviation of 5%. Thus, there are relatively favorable conditions for the combination of Ga $P_{0.4}As_{0.6}$ ($a_0 = 5.57$ Å) and GaAs ($A_0 = 5.65$ Å). Moreover, it should be possible also to grow $In_{0.4}Ga_{0.6}P$ ($a_0 = 5.61$ Å). If the *pn* junction is placed into the solid solution layer by subsequent Zn diffusion, *i.e.*, if the *pn* junction is not formed by heterotransition, an emission could be expected at 1.95 eV in the yellow-green spectral range [14].

The heteroepitaxy of Ge/GaAs or GaAs/Ge is practised, for example, in order to provide either high-impedance GaAs as a substrate material and surface protection for integrated Ge components, or low-impedance Ge as a substrate material with power supply for GaAs components. (Ge as a substrate exhibits higher mechanical stability) [13].

In both cases, however, a relatively wide junction zone is formed during the intergrowth process due to interdiffusion of the three components. The Ge/GaAs phase boundary range is severely disturbed by stacking faults and dislocation concentrations. Finally, referring to the difficulties in growing monocrystalline GaAs it must be pointed out that homoepitaxial deposition of GaAs, especially that obtained in accordance with the liquid method, is often used to obtain particularly perfect and high-purity GaAs monocrystalline materials that could not be produced by any other method.

1.3.2.2.   **HgTe – CdTe**

At the preparation of heteroepitaxial intergrowths of $A^{II}B^{VI}$ compounds (HgTe on CdTe, parallelism of the (111) planes) by vapor deposition the interdiffusion for producing a continuous series of solid solutions of the general form $Cd_xHg_{1-x}Te$

under isothermal conditions is of special significance. The width of the zone of solid solutions of the general form $Cd_xHg_{1-x}Te$ can be controlled and varied. The band gap of the ternary substance continuously changes with changes in its composition between $\Delta E \approx 1.5$ eV for CdTe and $\Delta E \approx 0$ eV for HgTe; the composition being defined by the interdiffusion between the deposit and the substrate. In this way, we succeed in producing basic structures for highly sensitive infrared detectors on the basis of epitaxial $Cd_xHg_{1-x}Te$ diodes [18], [19].

Rapidly operating and highly sensitive detectors of this kind which are obtained by means of controlled diffusion of Hg into p-$Cd_xHg_{1-x}Te$ and which exist in the form of epitaxial layers work in the temperature range between 77 K and 130 K. The spectral sensitivity per quantum at 77 K has a uniform distribution over a wide wavelength range from 3 to 14 μm. The quantum yield is 13%, and the time constant is less than 1 ns. The detectivity is between $3 \cdot 10^9$ and 10^{10} cm $Hz^{1/2}W^{-1}$.

1.3.2.3.   Si-insulator crystal

After this digression of the discussion from Si to the $A^{III}B^V$ and $A^{II}B^{VI}$ compound semiconductors, motivated by the transition from homoepitaxy to heteroepitaxy, let us return to Si, but now to *heteroepitaxy of Si on insulator* substrates. Sapphire (α-Al_2O_3) and magnesium-aluminium spinel are suitable substrate crystals for this purpose. The spinel to be used for this application has the composition $MgO \cdot (1.5$ to $2.5) Al_2O_3$ [20], [21]. The idea of dispensing with the Si substrate, and choosing a monocrystalline insulator substrate instead, is based on the expectation that a better mutual insulation of the switching elements in integrated circuits, an improvement of important electronic parameters by an appreciably reduced capacity, and the suppression of parasitic capacities, may be achieved [21]. These expected, and already partly attained, improvements are confronted, however, with problems that cannot be disregarded:

(a) the problem of growing perfect insulator crystals,
(b) the much more complicated preparation of suitable monocrystalline silicon layers under the conditions of heteroepitaxy.

Among all the possible variations, the deposition of Si by means of pyrolysis of silane on spinel having the above-mentioned composition has proved, internationally, to be the most favorable process, whereas sapphire as a substrate, and other depositing methods, have had to be abandoned [22]–[24]. In this way, diodes, MOS transistors, and, especially, MOS circuits with good electronic properties and technological simplifications, can be produced. Information on the use of the insulator crystal-semiconductor interface as the active part of possible new components is not yet available [21].

Figure 4 depicts the principal structure of an MOS transistor prepared according to this technique. The progress achieved in realizing bipolar structural components is, by far, much less satisfactory as compared to the fairly favorable results obtained for unipolar components. This is due mainly to the higher density of irregularities in the silicon layers which arise from the difficulties of growing adequately perfect

36

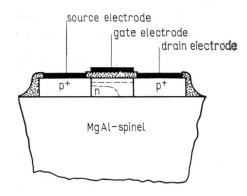

source electrode
gate electrode
drain electrode

p+   n   p+

MgAl–spinel

Fig. 4. Deep depletion MCS transistor on MgA spinel (according to [21])

insulator crystals under economically acceptable conditions. Furthermore, related to this, the major difficulties in heteroepitaxy to produce sufficiently perfect mono-crystalline layers of Si as compared to the causally more favorable conditions of homoepitaxy impede the realization of bipolar structural components. The latter could, for instance, be evidenced in the case of GaAs. Apart from the fact that spinel can readily be grown in a more perfect manner than sapphire and, hence, also proves to be a better substrate, the differences of crystal geometry between Si and sapphire are greater than those between Si and spinel (Table 6) which both form cubic crystals. However, in the case of heteroepitaxy the binding conditions between the substrate

Table 6. Possible orientational relations of sapphire — Si and MgAl spinel — silicon (according to [22]), limited to the indication of plane parallelism

Sapphire		Silicon	MgAl spinel	Silicon
(0112),	(1012)	(001)	(100)	(100)
(1124),	(1120)	(111)	(111)	(111)
(0001)				
(1120)		(110)	(110)	(110)

At least *another* 13 different crystallographic planes of sapphire permit epitaxial deposition of silicon

Structural data:
sapphire; rhombohedral

$a = 4.758$ Å

$c = 12.991$ Å

$c/a = 2.73$

Silicon (see Table 7)

Structural data:
$MgAl_2O_4$ (stoichiometric) spinel, cubic

$a = 8.0808$ Å

Spinel and silicon are of the same structural type:

$O_h^7 - Fd\ 3m$

Note: In part, the data given as to the sapphire — silicon orientational conditions are very different.

37

and the deposit and – especially in the example discussed here – the coordination relations due to different lattice parameters are more complex than in homoepitaxy. As a result, homoepitaxy, which represents the simplest and clearest case of all epitaxial and endotaxial relations, should be technologically preferred because of the better chance of observing the condition of reproducibility.

Substantially, no methods other than those already described will be used for the preparation of components of this technique. Merely slight modifications are feasible. As a result, it can be stated that the preparation of components by heteroepitaxial deposition of Si on insulator crystals, at present, is likely to be neither an economically, nor a technically competitive technique in its comprehensive sense when compared to the Si epitaxial planar technique, where homoepitaxial deposition of Si on Si substrate crystals is carried out. For the time being, heteroepitaxial deposition of Si onto insulator crystals can, therefore, be used only for special purposes of industrial production in the above-mentioned variants.

1.3.3.   EPITAXY ON AMORPHOUS INTERLAYERS

A possible modification of the process discussed in the preceding section which, conceivably, might partially exclude the problems described while preserving the advantages could be the successful deposition of epitaxial Si layers on Si single crystal slices provided with thin, amorphous* insulator interlayers. This is a special case of homoepitaxy where the orientation is somehow transmitted from the substrate crystal through the thin insulating layer. The fundamental possibility of transmitting such "information" has already been found in other substance combinations such as $PbS - C - NaCl$, $CdS - C - NaCl$, and similar layer systems [25]–[30].

In these investigations the thickness of the amorphous layer was between 100 Å and 700 Å, and, in single cases, to a maximum of 1 500 Å.

At the moment, however, this effect has been observed only up to an interlayer thickness of about 400 Å.** Such a layer must reveal electret effects. The presence of water vapor seems to promote the formation of a layer with these characteristics. For the semiconductor technique, the question arises how thick such a layer *should* be in order to have a sufficiently high insulation effect, and how thick it *may* be without losing its ability to transmit the "orientation information". This question has a certain bearing on the selection of materials and technological details.

To conclude this paragraph dealing with some significant examples of heteroepitaxy and its technological importance for the production of components, we give the more general definition of the term "heteroepitaxy":

"*Heteroepitaxy*" is the oriented growth process of a crystalline substance on another, foreign, crystalline substrate. In principle, both substances differ in their chemical composition and/or their crystal structure. The process of intergrowth may be linked

---

* The term "amorphous" is used here in the sense of a still-existing short-range order of about the same degree as found in glasses.

** According to a personal communication from *M. Krohn* relative to the special system NaCl–Se–anthrachinone. This effect of the possible interlayer thickness depends on the material combination.

38

with interdiffusion processes and may result in the formation of solid solutions in intermediate layers. In principle, chemical reactions between the two partners of intergrowth should be excluded. The problem of the formation of chemically reactive, intermediate layers due to impurities will be considered below.

In the treatment which follows, we shall not strictly follow the sequence given in Table 1 on the side of epitaxy. Considering the main concern of this chapter and in the interest of consolidating our considerations it is more advisable, first, to deal with heteroendotaxial and chemoendotaxial precipitations before discussing details of chemoepitaxial and chemoendotaxial processes of layer formation. Subsequently, the growth of multi-lamellar, and multi-bacular heterocrystals as basic structures of electronic components by means of oriented eutectic solidification and eutectoid transformation will be discussed. In this connection the significance of heteroepitaxial and chemoepitaxial, as well as heteroendotaxial and chemoendotaxial, intergrowth processes, which play a decisive role here, will be emphasized.

1.3.4. CHEMOENDOTAXIAL AND HETEROENDOTAXIAL PRECIPITATIONS DISCUSSED FOR THE EXAMPLE OF PRECIPITATIONS IN SILICON

If, under certain specific conditions (mainly determined by temperature and real structure), a chemical component solved in a host crystal attains the respective limit of solubility — either locally or in general — precipitation of this component occurs either without being accompanied by a chemical reaction or in connection with a chemical reaction taking place between the two substances to form a new, third, substance. The former process is defined as heteroendotaxy, when both substances, the host and the deposit, intergrow upon each other and, hence, obeying some kind of interrelationship between their crystallographic orientations. The latter is defined as chemoendotaxy when this new third substance's crystallographic orientation is somehow related to the host lattice.

The cases to be discussed below refer mostly to local oversaturations which are obtained by a particularly high supply of the respective component through regions which are favored by the real structure. On the other hand, however, oversaturation in respect of the corresponding limit of solubility is no absolute necessity to initiate precipitation when the presence of one or more components and their reactive participation influence the conditions favoring a reaction, thus enabling an earlier precipitation reaction, with a modified reaction product due to the participation of the additional components.

These precipitations lead to an increased concentration of the foreign element, either in chemically pure form or within the newly formed compound. This is due to diffusion processes occurring near the phase boundary of the resulting precipitates against the concentration gradient [31], [32]. This negative diffusion (also called up-hill diffusion) is caused by affinity relations which become determining during the formation of precipitates (reactive or non-reactive), thus overcompensating the efficiency of the concentration gradient [1]. Hence, a decrease of free energy is obtained. This is the thermodynamic condition for the accomplishment of the overall process. The structural change takes place at the phase boundary by means of a rapid interrelated displacement of the atoms existing there at the initial stage of precipitation. The energy threshold for nucleation and subsequent growth contains numerous energy contributions and can be determined only as a complex. The magnitude of the concentration increment at the interface is determined by the increment of the activity coefficient of the diffusing component in conjunction with the affinity relations already mentioned [33]. Precipitations and diffusion processes in primarily existing grain boundaries of the host material are not discussed here, as monocrystallinity of the host material can be assumed under the given circumstances.

Precipitations in high-purity semiconductor crystals (elementary semiconductors or compound semiconductors), or in epitaxial semiconductor layers, are undesired phenomena which should be avoided. In order to avoid them it is necessary to know their cause. Therefore, a basic knowledge is required, not only of the corresponding interdependencies as already briefly summarized, but also of the properties and the resulting behavior of particular chemical elements.

It will be useful here to deal, in more detail, with the occurrence of precipitates (probably silicon borides of various types) in conjunction with boron doping (acceptor doping to produce p-silicon) for both the compact monocrystalline silicon substrate and the epitaxial layer. When considering all the diffusion processes involved, it is necessary to assume, for the usually high concentrations and concentration gradients, a diffusion coefficient which depends on both the concentration and the real structure, whereas, for a simplified diffusion model, the diffusion coefficient may often be assumed to be constant. In this connection, the significance of the distribution of dislocations, for instance, should be emphasized. The limit of solubility of boron in silicon is, at $1\,250°C$, about $5 \cdot 10^{20}$ atoms cm^{-3} [34]. If this solubility value, or an analogous value applicable to other specific conditions (see above), is exceeded, precipitation occurs. Then it can be observed that, apart from the primary formation of various kinds of precipitates, a growth of larger precipitates at the expense of smaller ones occurs due to the causes mentioned above. In this case, it is very likely that, initially, diffusion processes described by positive diffusion coefficients are involved, which, after the dissolution of metastable, possibly low valence borides, cause boron also to diffuse towards regions of low boron content surrounding the above-mentioned larger and more stable, possibly higher valence boride precipitates. The growth of the larger boride precipitates obeys the general laws discussed: up-hill diffusion of boron in front of the phase boundary where its activity coefficient rises abruptly. In the course of this overall process the metastable smaller precipitates are consumed and, hence, the source of boron diffusion towards the growing precipitates is exhausted. Any information, based on theoretical or experimental investigations, on the effect of these facts on the diffusion process between the two types of precipitates is not yet available.

However, more information about this problem is urgently needed. The same applies to the precise identification and quantitative, as well as analytical characterization of the boride precipitates, taking into account the problems arising for the epitaxial planar technology of silicon.

An analogous case is represented by the precipitation of $\alpha''\text{-}Fe_{16}N_2$ accompanied by simultaneous growth of the stable $\gamma'\text{-}Fe_4N$ phase according to this mechanism [35]. The metastable $\alpha''\text{-}Fe_{16}N_2$ cell consisting of eight deformed, body-centered tetragonal subcells of martensitic structure represents an intermediate structure between body-centered cubic α-iron in which nitrogen has been interstitially incorporated, and γ'-Fe_4N composed of a face-centered cubic iron host lattice with the nitrogen atom centrally embedded on the 1/2 1/2 1/2 site.

Orientation relations of such precipitates with silicon are given in [36] without any structural and chemical designation of the boride precipitates as in [34]. Rod-shaped particles up to 1 μm in length and about 0.05 μm in diameter are embedded parallel with the $\langle 211 \rangle_{Si}$ direction and slab-shaped precipitates of similar sizes parallel with the $\{111\}_{Si}$ planes.

Figure 5 gives a survey of the best known silicon borides and Table 7 the corresponding structural data as far as these are known.

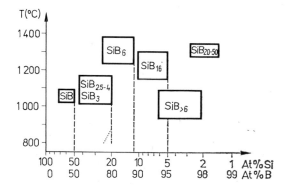

Fig. 5. Schematic view of formation zones of various silicon borides (according to [37])

The assumption that the frequently observed precipitates might be pure borides is certainly not justified in each case. There is still a residual content of carbon and oxygen, among others, in "pure transistor silicon", even after 2 to 10 passes of the floating zone melting technique in high vacuum [12]. These contaminations are likely to influence the boron solubility in the sense already discussed in connection with the precipitation tendency. Therefore, it is reasonable to take into consideration also the presence of silicon carbide borides and silicon oxide borides having a different content of carbon and oxygen. Such compounds are known also in other connections. Thus, for instance, a SiC_2B_{12} compound was described in [39] and a $Si_3C_2B_{20}$ compound in [38], the structures of these do not significantly differ from those of the corresponding pure borides.

Also, the assumption that pure boron is precipitated cannot be excluded [42]. This assumption is based on the fact that the specific volumes of the borides are higher than those of silicon at normal temperatures. The orientation relations obtained for pure boron precipitates correspond to those already mentioned for the borides. Consequently, precipitates of this kind are either typically chemoendotaxial or, in the case of pure boron precipitates, heteroendotaxial.

Similar problems arise for the case of phosphorus doping (donor doping to produce n-silicon). It should be mentioned, without going into details, that both oriented silicon phosphide precipitates and (inactive) pure phosphorus or phosphorus-boron ones have been detected [43]–[58].

Table 7. Structural data for Si, silicon borides and B (according to data from [12], [37]–[42])

Substance	Structural Formula	Structural Designation
Si		cubic; diamond lattice $a = 5.43072 \pm 0.00001$ Å
SiB		
$SiB_{2.5-4}$		rhombohedral
SiB_3 ($SiB_{2.89}$)	$Si_2(B_{10.5}Si_{1.5})$	rhombohedral $a = 5.52$ Å, $\alpha = 69.1°$
SiB_{3-4}	$Si_2(B_{12-x}Si_x)$	rhombohedral $a = 14.39$ Å, $B = 18.27$ Å $c = 9.89$ Å
SiB_6	Si_8B_{67}	orthorhombic $a = 14.346$ Å, $b = 18.226$ Å $c = 9.848$ Å
$SiB_{>6}$		amorphous (in the temperature range given in Fig. 5)
SiB_{14}	$Si(B_{84})$ $(B_7Si_3)_2$	β-rhombohedral, isotopic with B
SiB_{16}		hexagonal, $a = 6.42$ Å $c = 7.99$ Å
SiB_{20-50}		(β-rhombohedral)
B		a) β-rhombohedral $a = 10.145$ Å, $\alpha = 65.28°$ (crystallization from the melt) b) α-rhombohedral (pyrolithically deposited below 1 200°C c) tetragonal $a = 8.57$ Å $c = 8.13$ Å (by decomposition of BBr_3 at 1 300 – 1 450°C by vapor-deposition on Si at 900°C)

Silicon phosphide is orthorhombic ($a = 6.90$, $b = 9.40$, $c = 7.68$ Å). Rod-shaped SiP precipitates can be embedded parallel with $\{111\}$ $\langle 110 \rangle_{Si}$ when $\{110\}$-Si slices are used as the substrate material [50], [51]. They are preferably arranged at the Si-SiO$_2$ interface. In the case of a normal real structure, $\{111\}$-Si specimens exhibit smaller precipitates with the same orientation. If, however, during the preparation

42

of thick SiO_2 layers on such Si specimens, dislocations are induced due to the strain of the Si crystal thereby caused, larger precipitates (increasing up to μm size) can be detected along the glide plane track in the $\langle 110 \rangle$ direction. A report on spheroidally shaped precipitates is given in [58]. Because of their good coherence with the matrix such precipitates are considered to be composed of pure P or of material rich in P. Other cases, for instance, chemoendotaxial precipitations in high-purity silicon crystals which have not been treated after being grown can be observed after heat treatment at a temperature ranging from 700°C to 1000°C and a critical cooling rate of 10°C s^{-1}. The precipitates in question are of the β-CuSi and, more rarely, of the α-Fe$_3$Si type [59], [60]. A copper concentration of only $3 \cdot 10^{16}$ atoms cm^{-3} is sufficient to enable β-CuSi to be precipitated. The solubility of Cu in Si is about 10^{18} atoms cm^{-3} at 1 300°C [12]. According to another report [61], it is 10^{16} to $3 \cdot 10^{19}$ atoms cm^{-3} at a temperature ranging from 700°C to 1 000°C. In the so-called pure transistor silicon, the content of Cu is commonly lower: 10^{14} to 10^{12} atoms cm^{-3} so that, as a rule, such precipitations cannot be expected though the solubility decreases rapidly as the temperature decreases. More details on this subject are not given here as they have only a minor practical importance in comparison to the problems of precipitations connected with B and P doping. Of particular interest in this case is also the reference made in [59] that minor SiO_2 precipitates are promoting nucleation. These SiO_2 inclusions play a similar part in the precipitation processes linked with B and P doping.

The specific problems arising in the discussion of doping diffusion in compound semiconductors, concerning the formation of undesired precipitations are not discussed in this paper. The discussion of the considerable technological, and basic difficulties concerning the diffusion technique would lie outside the scope of this paper.

Therefore, only silicon as a base material has been discussed because silicon has, by far, the greatest practical and economic importance in the field of semiconductor techniques (see above), and, hence, the phenomena linked with it are of major interest. Heteroendotaxial and chemoendotaxial precipitations in metals and alloys will be mentioned in the following paragraph as far as they are of interest since they are an accompanying phenomenon of the considered layer growth processes.

1.4. THE SIGNIFICANCE OF EPITAXIAL AND ENDOTAXIAL INTERGROWTH PROCESSES FOR THE PRODUCTION OF NON-SEMICONDUCTOR COMPONENTS OF THE THIN FILM AND THICK FILM TECHNIQUES

Regularly oriented intergrowth processes of single crystalline substances are of no importance for the production of all non-semiconductor types of components such as resistors, capacitors, coils, vacuum electronic components, light sources, electromechanical components, and other components of conventional types.

Concerning the problem of distinguishing between thin film and thick film techniques the following definition is given: The limit between a thin layer and a thick one is assumed to be at about 10 000 Å according to [62]. This limit will be slightly above,

Table 8. Chemical processes to produce metal, alloy and insulator layers to be applied in micro-
to data from [62]–[67])

Process	Substances to be deposited or to be produced, and specification of methods	Deposition rates
I Electrolytic Deposition	*from an aqueous medium* *Metals:* Al, Ag, Au, Cd, Cr, Cu, Fe, Co, Ni, Pb, Pt, Rh, Sn, Zn and others (totalling about 33) *Alloys:* more than 100 combinations of binary alloys, about 15 ternary alloys (specially referred to in [64]) *from a non-aqueous medium* Al, Cr, Ti, platinum metals (from molten salts)	up to 10^4 Å s^{-1} with an average of about $10^2 \ldots 10^3$ Å s^{-1} $\dfrac{G}{A} = JtE\,\alpha$ [g cm^{-2}] G = Weight A = Unit area J = Current density t = Time E = Electro-chemical equivalent α = Electrical efficiency factor *e.g.* 10 Å Ag s^{-1} at 1 mA cm^{-2} or 10 Å Ta s^{-1} at 2 mA cm^{-2} etc. in accordance with $\dfrac{d}{t} = \dfrac{JE\,\alpha}{p}$ [cm s^{-1}] d = Layer thickness p = Layer density
II Chemical Reduction (Currentless Deposition)	*Non-catalytic Reactions:* Ag (Deposition of very thick layers possible from argentic nitrate solutions, formaldehyde serving as reducing substance) *Catalytic Reactions* Ni, platinum metals (in case of Ni, reduction of $NiCl_2$ by means of potassium hypophosphite, the metal substrate being the catalyst: Ni, Al, Co, Fe) Catalytic reactions with activators Cu, Ni (activator: $PdCl_2$) Catalytic reactions with activators and sensibilizers Ni (sensibilizer: $SnCl_2$) Ni—Co on glass	on an average, at about 10 Å s^{-1}

44

electronics and to produce discrete passive electronic components on a layer basis (according

Parameters determinant for deposition	Advantages of the process	Disadvantages of the process	Applications of the substances mentioned
Current density	— Simple apparatus and equipment may be used. — Epitaxial deposition possible. — Low temperatures are predominant. — Layer thickness is well controlled.	— Metallic substrates are necessary. — Problematical as to compatibility in component fabrication processes.	— Electrically conductive layers — Magnetic layers — Resistive layers
Temperature of solution, pH-value of the solution	Simple apparatus and equipment may be used. — Insulator, semiconductor and metal substrates may be used. — Epitaxial deposition is possible.	— The number of substances which may be deposited is limited. — Problematical as to compatibility in component fabrication processes.	— Electrically conductive layers — Resistive layers — (Magnetic layers)

45

Table 8 (Cont.)

Process	Substances to be deposited or to be produced, and specification of methods	Deposition rates
III Chemical vapor deposition (as far as deposition of semiconductor substances is concerned, see Tables 3 and 4) The description of reaction formulae must be dispensed with as these are too extensive	*Polymerization* inorganic and organic polymers (Polymerization of monomeric vapor by electronic beams, ultraviolet rays or glow discharge) *Disproportioning* $Al(AlJ_3)$, $C(CO)$ *Reduction* $Al(AlCl_3)$, $Ti(TiBr_3)$, $Sn(SnCl_4)$, $Ta(TaCl_5)$, $Nb(NbCl_5)$, $Cr(CrCl_2)$ *Decomposition* (by pyrolysis or glow discharge) $Ti(TiJ_4)$, Pb (organic plumbic compounds), $Mo(MoCl_5)$ $Fe[Fe(CO_5)]$, $Ni[Ni(CO_4)]$ C (toluenes), $MnO_2[Mn(NO_3)_2]$ BN (boron trichloroborazole) *Oxidation* $Al_2O_3(AlCl_3)$, $TiO_2(TiCl_4)$ $Ta_2O_5(TaCl_5)$, $SnO_2(SnCl_4)$ *Nitridation* $TiN(TiCl_4)$, $TaN(TaCl_5)$ $SiC(SiCl_4 + CH_4)$	on an average, at about $1-10^3$ Å s^{-1}
IV Anodic oxidation	Oxides of: Al, Ta, Nb, Ti, Zr, (Si), and numerous other metals	on an average, about 10 Å s^{-1} 3–50 (Å V^{-1})
V Thermal oxidation	Oxides of: Al, Ta(Si) and numerous other metals (A variant is plasma oxidation to avoid high substrate temper-	

46

Table 8 (Cont.)

Parameters determinant for deposition	Advantages of the process	Disadvantages of the process	Applications of the substances mentioned
— High substrate temperatures ranging up to > 1 000°C — Low pressure required in the deposition system	— Insulator layers, dielectrics — Semiconductor layers in capacitors — Electrically conductive layers — Superconductive layers — Magnetic layers — Resistive layers	Pressure, temperature Effect of rays	— Deposited substances are highly pure — Substances are well suited for epitaxial deposition — Insulator, semiconductor and metal substrates may be used — Substances are well compatible in component production processes
Current density	— Simple apparatus and equipment — Dense layers can be produced — Amorphous layers can be produced — Epitaxial and endotaxial layers can be produced — Good control of layer thickness.	— Limited thickness of layers (up to 1.5 μm) — Metal substrates presupposed — Problematical as to compatibility in component fabrication processes	— Insulator layers, dielectrics — Sealing layers
Pressure, Temperature	— Simple equipment and conditions — Good compatibility in component fabrication processes.	— Limited layer growth (3 000 . . . 4 000 Å) — Only a restricted number of sub-	— Insulator layers, dielectrics — Sealing layers

47

Table 8 (Cont.)

Process	Substances to be deposited or to be produced, and specification of methods	Deposition rates
	atures and to obtain more rapid growth) Apart from oxidation, it is possible to produce nitrides, sulphides, carbides, etc. This merely presupposes the feeding of corresponding gases to the metal substrate	
VI *Screen printing* of prefabricated pastes for thick-film technique* *Process steps* — Printing — Drying — Burning (600—1 000°C) *This process is mentioned here, although it is not a chemical process, but the preparation of pastes is decisive for layer properties and of a true chemical nature	Pastes are composed of a special glass matrix (boron—lead glasses), organic plasticizers, solvents and ingredients determining their functions. These are in *conductive pastes:* Ag, Au, Pd, Pt, Pd/Ag, Pd/Au, Pd/Pt *resistive pastes:* Oxides of Ru, Ir, In, Pd, Te, Sn in part containing a portion of precious metals *dielectric pastes:* Titanates, zirconates, oxides of Ba, Bi, Pb, partially doped with rare earth metal oxides (lanthanides) *insulating pastes:* Oxides of different nature having low dielectric constants and lower electrical conductivity *finished pastes:* based on spinels	Average thickness of layer: 20 μm

or below that value, depending upon the material and the properties that are decisive in special cases. Also, for the time being, the practical importance of epitaxy and endotaxy, in the present case, is by far lower than it is for semiconductor technology. Homoepitaxy, generally, is of no interest. Direct investigations regarding the heteroepitaxial deposition of metal and metal alloy layers are presently carried out on model systems (for instance, with halides serving as a substrate, as far as the physical depo-

Table 8 (Cont.)

Parameters determinant for deposition	Advantages of the process	Disadvantages of the process	Applications of the substances mentioned
	— Epitaxial and endo-taxial layers can be produced	stances are suitable for coherent layer growth — Metal substrate presupposed	
Printing technology Sintering temperature	— Relatively simple equipment (low cost) — Very well suited for hybridization — Variability of circuits	— High sintering temperatures (up to 1 000°C)	— Electrically conductive layers — Resistive layers — Insulator layers
			— High reliability — Robust design
	— Protective glazing — Inductances		

sition of layers is concerned). The results are, however, of fundamental significance for practical application in the preparation of polycrystalline layers and layer systems to produce discrete, passive components (such as metal film resistors) and complex components (networks of passive components: RC networks) on polycrystalline, insulating substrates (mostly ceramic materials). It seems to be ensured that, in many cases, an interaction exists between the substrate and the film (which, possibly, is not provoked intentionally) as in epitaxial intergrowth processes, but here limited to the smaller regions of crystallites. It is, however, not necessary to use monocrystalline substances under the conditions given, since the perturbing effect of major lattice imperfections, such as grain boundaries, does not have such immense consequences for the electric properties as in the case of semiconductors. This, however, implies that such grain boundaries are not enriched with contaminations, neither dissolved atoms and molecules nor precipitations. This means, however, that a relatively high purity of the materials is required which has a certain economic bearing on the technology as well.

In addition, the preparation of monocrystalline epitaxial layers would affect this factor to a still greater extent.

The chemical processes suited for the deposition of metal, metal alloy, and insulator films, for use in microelectronics and for the production of discrete electronic components based on films are summarized in Table 8.

For completeness, this table also contains the thick film technique. It is obvious that the influence of substrates on film thicknesses of about 20 μm is, in any case, lower than it is on thinner films. Consequently, the above-mentioned assumption that epitaxy model investigations of thin films are applicable does not apply, or, at most, only to a certain extent.

In the following, only comments regarding problems of chemoepitaxial and chemoendotaxial layer growth will be given, whereas the numerous problems regarding processes of layer deposition and preparation, a survey of which is given in Table 8, will not be referred to in detail. Partly, these are only of minor importance for microelectronics. (This does not apply, of course, to the chemical vapor deposition of epitaxial semiconductor layers which have already been discussed in the preceding paragraphs and will be discussed in more detail in the relevant sections of this book.)

1.5. CHEMOEPITAXIAL AND CHEMOENDOTAXIAL LAYER FORMATION

This kind of layer formation is achieved by the methods listed in the horizontal columns IV and V of Table 8.

For a discussion of such processes of layer formation, among others thermal oxidation processes of Ta and Al are a good example since, under certain circumstances, they approach the respective model conceptions (Fig. 6). The restriction to semiconductor technology, on the one hand, and the production of non-semiconductor components by thin film and thick film techniques, on the other hand, cannot be maintained here as these layer formation processes are significant for both groups, *i.e.* in the figurative sense as accounted for in the preceding paragraph.

The three very important layer growth processes are, in this connection, the thermal oxidation of Si to form a surface layer in the planar process (see Table 2); the anodic, and the thermal oxidation of Al and Ta in the thin film technique to compensate the electric properties of metal films (also done by means of electron beam or laser beam treatment); or to build up oxide layers as dielectrics in thin film capacitors [68].

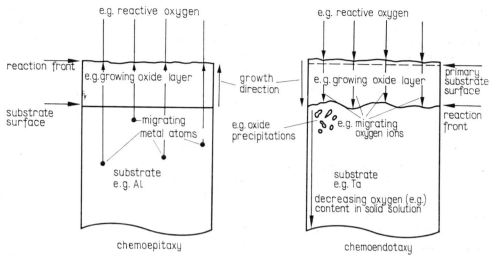

Fig. 6. Schematic view of chemoepitaxial and chemoendotaxial growth of layers demonstrated by the example of oxide film growth on Al and Ta

Table 9 gives a survey of materials and material combinations for thin film resistors and capacitors. Basic results of investigations on the solid–gas reaction and the processes of layer formation which are involved are given with the help of suitable model systems in Chapter 5 of this book. In this paragraph, priority will be given to the chemoendotaxial oxidation of tantalum because this problem is of particular interest for the modern thin film technique and particularly interesting problems arise in this process. These problems are also useful for aiding an understanding of thermal oxidation of silicon and the oxidation of aluminium.

In order to avoid any consideration of additional problems encountered in anodic oxidation, the model investigations with respect to the tantalum oxidation were carried out thermally.

1.5.1.   **THERMAL OXIDATION OF TANTALUM: CHEMOENDOTAXY**

Oxygen is largely soluble in tantalum (up to nearly 7 At% at 1 700°C). Tantalum tends easily to form oxides.

The reactive formation of oxide layers on tantalum is a typical representative of chemoendotaxial layer formation (see Fig. 6).

The mechanism and kinetics of the thermal oxidation of tantalum will be summarized now [1], [72].

As the basic material in the experiments, compact, high-purity tantalum single-crystal bars were used. The total content of contaminants was about 50 ppm consisting of: 10 ppm of O; 1 ppm of N; 0.2 ppm of H; 20 ppm of C; 10 ppm of Fe; < 0.5 ppm of Cu; and < 0.1 ppm of Nb (in at ppm), monocrystalline Ta deposited on W by

Table 9. Materials and material combinations for thin-film resistors and capacitors (according to data from [62]

Components	Materials
Resistors	*Alloy layers* NiCr, Ni(P) NiCr(Si) *Metal layers* $Ta(+N_2)$ *increasingly important for RC networks* (TaN) Cr, W, Mo, Re *Cermet layers* $Cr-SiO$ *(most significant in this group)* $Au-Ta_2O_5$, $Pt-Ta_2O_5$, $Au-WO_3$, $PtWO_3$ Cr_3Si, $TaSi_2$ (both plus Al_2O_3) *Semiconductor layers* C, SnO_2

Capacitors	*Dielectrics*	*Cover electrode materials*
	Ta_2O_5, $Mn_{2-x}-Ta_2O_5$	Au, Pd, Cu, Sb, Cd, Fe, In
	$SiO-Ta_2O_5$	Al, Ta
	Al_2O_3	$NiCr-Au$, $MnO_{2-x}-Au$
	SiO	*Base electrode materials*
	SiO_2	Ta, Al
	Parylene	*Substrate materials*
	TiO_2	Non-alkaline glasses, glazed and unglazed
	Titanates	ceramic bodies (polycrystalline)

Material combinations actually preferred:

Base electrode	—	Dielectric	—	Cover electrode
Ta		Ta_2O_5		Au
Ta		Ta_2O_5		$NiCr-Au$
Ta		Ta_2O_5		$MnO_{2-x}-Au$
Ta		$Ta_2O_5 + SiO$		$NiCr-Au$
Al		Al_2O_3		Al
Al		SiO		Al
Al		Parylene		Al

means of evaporation by electron bombardment heating [74] (500 to 1 000 Å thickness)

$$\{110\}\ \langle110\rangle_{bcc\ Ta}\ \ ||\ \ \{110\}\ \langle110\rangle_{bcc\ W}$$

and Ta deposited on NaCl in the same way [73]:

$$\{100\}\ \langle100\rangle_{bcc\ Ta}\ \ ||\ \ \{100\}\ \langle100\rangle_{NaCl}.$$

If the surface of the tantalum specimen is in thermal equilibrium with an oxygen atmosphere, the adsorption phenomena can be described as follows:

$$O_2 + Me \rightleftharpoons O_2 + Q_{phys.} \qquad \text{physisorption}$$
$$\underset{Me}{}$$

van der Waals forces

$$O_2 \underset{\text{on metal surfaces}}{\overset{\text{dissociation}}{\rightleftharpoons}} 2O + Q_1 \; \left. \begin{array}{c} \\ \\ \end{array} \right\} \qquad \text{chemisorption}$$

$$2 e^- + O \rightleftharpoons O^{2-} + Q_2$$

from the   physi-   chemi-
conduction   sorbed   sorbed
band of the   to Ta
Ta lattice

$$Q_1 + Q_2 = Q_{chem.}$$

(Q characterizes the released amounts of energy. $Q_{phys.}$ = some kcal mole^{-1} [80], $Q_{chem.}$ is above 900°C superior to $Q_{phys.}$ by 18 kcal mole^{-1} [77].)
The equilibrium of adsorption is initially determined by the gas pressure, the structure, and the temperature of the metal surface in the case of a pure tantalum surface and a gaseous oxygen phase. But it is also determined by the oxygen concentration in the tantalum because of the high solubility of oxygen in Ta. Quasistable physisorbed monolayers or multiple layers may be formed at very low temperatures, for example at 70 K. When the temperature increases, the degree of surface covering decreases since a part of the oxygen molecules is thermally activated to desorption. When the centers of gravity of the O_2 molecules and the centers of the metal atoms have approached each other sufficiently, a new chemisorption bond, likewise depending upon surface structure and surface orientation, emerges as a result of electronic interaction. The process of chemisorption can be considered in two steps. First, the oxygen molecule is dissociated. The required energy of 117 kcal mole^{-1} [88] is essentially provided from the metal lattice by thermal interaction. The subsequent ionization of oxygen confers on the metal a bond similar to the chemical bond. The energy thereby released corresponds to the amount of dissociation energy It is returned to the lattice. Likewise, chemisorbed oxygen may be desorbed in the case of elevated temperatures.
Oxygen may pass over from the chemisorbed layer into the tantalum lattice forming a solid solution. The solubility of oxygen in tantalum depends upon temperature (Fig. 7). At temperatures ranging from 700 to 1 400°C oxygen diffusion in tantalum occurs as genuine volume diffusion solely on interstitial paths, primarily through octahedron interstices of bcc tantalum. Uni-dimensional and two-dimensional lattice defects do not play any traceable, predominant part as diffusion channels in this temperature range [78], [79], [81], [87], [89] but contribute substantially to them at lower temperatures. The activation energy for the diffusion of oxygen in tantalum depends upon temperature and concentration (Table 10).

53

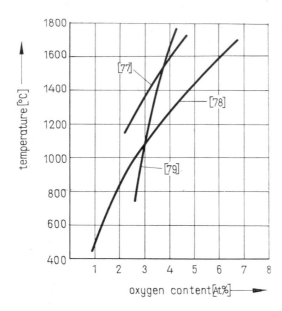

Fig. 7. Maximum solubility of oxygen in tantalum as a function of temperature

The oxygen concentration in the nonsaturated tantalum-oxygen solid solution drops appreciably with the distance from the margin towards the crystal interior [90]. The saturation concentration is rapidly attained in thin tantalum layers and in the marginal areas of compact tantalum as well.

The amount of oxygen supply from outside is larger than the amount that can be transported to the crystal interior in accordance with the reduced value of the diffusion coefficient.

Oxidation starts when the limit of solubility, valid under the given conditions, is exceeded, preferably in the marginal areas for the above reasons, and results in the formation of an oxide layer consuming the tantalum matrix while progressing from outside to inside (see Fig. 6). In front of the reaction interface, which shifts towards the interior, oxide precipitation occurs.

Hence, oxidation not only takes place at the metal-chemisorption layer interface, as is the case for Al at the oxide−oxygen atmosphere interface. The maximum oxygen solubility of Al is only about 30 ppm and, consequently, is about 3 orders of magnitude lower than that of tantalum [91].

Once a continuous, compact oxide film has developed, any further layer growth is determined only by the diffusion of anions, or anion vacancies, towards the reaction interface. For the growth of β-Ta_2O_5 layers, reliable results are available [77], [92]. A certain cation mobility was shown radiographically as well [93]. Therefore, oxidation occurs also within the oxide film.

The results available for the kinetics of tantalum oxidation are partly very contradictory. Table 11 contains a list of oxidation rate characteristics observed for various temperature ranges.

A comparison between the time dependencies found for different temperature ranges is hampered by the fact that the respective authors employed either the weight increase of the sample, the decrease of the oxygen partial pressure, the decrease of the

54

Table 10. Activation energies and solution heat in the oxidation of tantalum

Process or phenomenon	[kcal mole^{-1}]		References
Physical adsorption of O_2 on Ta	Activation energy: ≤ 10		[80]
Surface diffusion of oxygen on Ta	T		[77]
Chemisorption of oxygen on Ta	Activation energy: 18 kcal mole^{-1} greater than that of physical adsorption		[77]
Solution of oxygen in Ta at $T \leq 1\,400°C$	Solution heat: 6–8		[77]
Volume diffusion of oxygen in Ta or Ta$-$O solid solutions	Activation energy:		
<0.15 At% O (<400°C)		25.5	[81]
<0.6 At% O (100$-$300°C)		30.6	[82]
≈1 At% O (250$-$450°C)		27.0	[83]
1.07 At% O (≈500°C)		27.7	[84]
≈3 At% O (<1 200°C)		22.9	[85]
Formation of Ta$_4$O at $T<$ 600°C	Activation energy: *	60–65	[86]
		65–70	[87]
Formation of Ta$_2$O at $T>$ 800°C	Activation energy: *	14	[37]
Formation of Ta$_2$O$_5$	Activation energy: *		
at $T =$ 600°C		60	[86]
at $T >$ 800°C		27	[86]

(The rate of oxidation as a function of temperature can be demonstrated by m eans of an Arrhenius equation. The plotting of the oxidation rate against the reciprocal temperature T gives a straight line the slope of which is proportional to the activation energy referred to by the authors cited (here: activation energy*).

oxygen volume, or the increase in layer thickness with time as a measure of the oxidation rate without independently taking into account, except for the last case, the solubility of oxygen. For metals which, similar to tantalum, have a high oxygen solubility, it is solely the increase of thickness or mass of oxide layers with time which may be regarded as a quantitatively correct measure of the oxidation rate.

However, in spite of all apparent contradictions, it may be concluded from Table 11 that, in the temperature range up to 500°C, the reaction rate is determined by diffusion processes. The driving forces of diffusion may be attributed to electrical space charge fields (logarithmic or reciprocally logarithmic) at temperatures up to about 300°C and to the gradient of the chemical potential (parabolic) above 300°C.

Table 11. Time laws of tantalum oxidation in air, or in an oxygen atmosphere

Temperature range [°C]	Activation energy* [kcal · mole^{-1}]	Characteristic of time law	References	
50– 300	?	logarithmic[0), 00)]	[82]	
150– 300	27.4	reciprocally logarithmic [0), 00)]	[82]	[94]
200– 350	?	logarithmic [00)]	[95]	
250– 450	?	parabolic [00)]	[83]	
320– 350	27.2	parabolic [00)] or cubic?	[86]	[96]
400– 530	?	parabolic/linear [+)]	[97]	
500– 600	60	linear [+), 00)]	[85]	[86]
500– 650	60	linear [+)]	[86]	
500–1 000	?	linear [+)]	[86]	[98]
600– 860	?	linear [+)]	[86]	[99]
700–1 420	?	linear with slight deviations above 1 100°C [+), 00)]	[86]	[90]
800–1 000	25–27	linear [+)]	[86]	
800–1 250	12–25	linear [+)]	[77]	[85]
1 300–1 800	34–40	first linear, then quasi-parabolic [00)]	[77]	

(Activation energy *) see Table 9.
[0)] Determined by measuring the oxide film thickness.
[00)] Determined by measuring the increase in the weight of the sample.
[+)] Determined by measuring the change of oxygen partial pressure or vo   um .

In the temperature range between 500 and 1 300°C the increase of the oxide film thickness is linearly dependent upon time. Cracking in the layer occurs. This cracking accelerates the oxygen migration towards the reaction interface where a linear oxidation rate was found to apply.

The dependence of the oxidation rate k^+ upon temperature can be described, at least partially, by the following equation:

$$k^+ = k_0 e^{-\frac{\Delta E}{kT}}$$

where k_0 is a constant and k is Boltzmann's constant.

By plotting the logarithm of the oxidation rate versus the reciprocal temperature (Arrhenius plot) a straight line is obtained. From the slope of this line the energy ΔE can be evaluated. ΔE serves as a measure of the temperature dependence of the reaction rate. If, for example, the diffusion of anion vacancies through the already existing tantalum oxide layer at the corresponding temperature range is the slowest and, hence, the rate controlling step of the layer growth mechanism then ΔE designates the activation energy of anion vacancy diffusion.

Table 12 shows several values of ΔE evaluated under different conditions. The phases given in column I were found radiographically.

This list, despite its briefness, reflects the problematic nature and complexity of thermal tantalum oxidation.

Table 12. ΔE values for the formation of some tantalum oxides as obtained under different conditions

Phase	Oxidation temperature [°C]	Oxygen pressure [torr]	Time characteristic	ΔE [kcal mole^{-1}]	References
Ta$_4$O	500 — 600	760	parabolic	60 — 65	[86]
Ta$_2$O	800 — 900	10^{-2}	linear	14	[87] [90]
β-Ta$_2$O$_5$	600 — 650	760	linear	60	[86]
	800 — 1 000	760	linear	25 — 27	[86]
α-Ta$_2$O$_5$	1 300 — 1 400	> 10^{-2}	quasi-parabolic	35 — 40	[77]

1.5.1.1.   Tantalum oxides

The complexity of the oxidation process also accounts for the plurality of oxides that are stable, or only metastable, in different ranges (Table 13).

The tantalum oxides with a low oxygen content including the sub-oxides have only a narrow concentration range of stable existence (Fig. 8).

The transformation of these phases to phases of a higher oxygen content is irreversible. For this transformation, the difference of the corresponding formation energies has to be supplied.

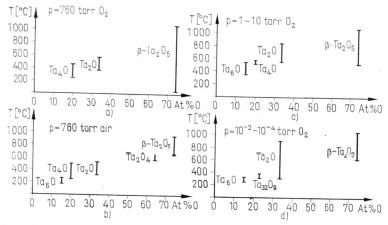

Fig. 8. Ranges of tantalum oxide existence according to data from [77], [86], [102], [104], [108], [110], [118], [128], [130], [131], [132]

Table 13. Structural data and conditions of preparation for different

Phase	Structure data (Lattice constants in Å)
Ta	bcc at 18°C; $a = 3.3029$ at 2 222°C $a = 3.3581$
Ta_6O	tetragonal $a = 3.36$; $c = 3.25$; $c/a = 0.97$
Ta_4O	orthorhombic $a = 3.61$; $b = 3.27$; $c = 3.20$
	orthorhombic $a = 7.22$; $b = 3.27$; $c = 3.20$
	orthorhombic $a = 7.194–7.238$; $b = 3.266–3.273$; $c = 3.204–3.216$;
$Ta_{32}O_9$ $Ta_{3.56}O)$	cubic $a^+ = 13.4$
Ta_2O	tetragonal $a = 6.675$; $c = 4.721$ $c/a = 0.707$
	tetragonal $a = 6.680$; $c = 4.758$; $c/a = 0.71$
	orthorhombic $a = 5.29$; $b = 4.92$; $c = 3.05$
	cubic $a^+ = 6.68$
(TaO_u)	Content of oxygen lower than for TaO
TaO	cubic $a = 4.422 \ldots 4.439$

tantalum oxides and suboxides

Conditions of Preparation	References
	[100]
Oxidation (330°C) of very fine-grained tantalum powder obtained by dehydrogenation at 1 100°C	[94] [101] [102] [104]
Oxidation (in air at 300 . . . 330°C) of tantalum wire, sheets and powder	[101]
Oxidation of tantalum sheets (10^{-1} torr O_2, up to 500°C)	[102]
Oxidation of tantalum anhydride (at 600°C)	[94] [102] [103] [105] [106] [107]
Oxidation of tantalum single crystal samples ($5 \cdot 10^{-4}$ torr O_2 at 300°C)	[106]
Oxidation of tantalum powder (1 torr O_2, 500 . . . 600°C) or disproportioning of $TaO_y(Ta_4O)$ in a high vacuum above 500°C	[108]
Oxidation of tantalum powder and sheets (1 atm O_2 at 400°C)	[102]
Oxidation of tantalum wire ($2 \cdot 10^{-2}$ torr O_2 at 900°C	[90]
Oxidation of tantalum single crystal samples ($5 \cdot 10^{-4}$ torr O_2 at 300 . . . 450°C)	[77] [103] [104] [105] [109] [110]
Oxidation of tantalum slices ($2 \cdot 10^{-2}$ torr O_2 at 1 700°C)	[77]
Ta carbides or nitrides oxidized for a period ranging from 1 hour to 2 days in the presence of hydrogen at 600 . . . 900°C	[103] [105] [111] [112]

Phase	Structure data (Lattice constants in Å)
Ta_2O_3	
TaO_2	tetragonal $a = 4.709$; $c = 3.065$; $c/a = 0.65$
$TaO_{2-2.5}$	complicated structure
Ta_2O_5 (β-phase)	orthorhombic $a^+ = 6.20$; $b^+ = 3.66$; $c^+ = 3.89$
	orthorhombic $a = 7.771$; $b = 7.664$; $c = 12.67$
	orthorhombic $a = 6.20$; $b = 69.69$; $c = 3.90$
	orthorhombic $a = 7.761$; $b = 12.38$; $c = 12.57$
	hexagonal $a = 7.228$; $c = 11.63$
	orthorhombic $a = 7.74$; $b = 6.47$; $c = 7.47$
	orthorhombic $a = 6.18$; $b = 43.93$; $c = 3.89$ $a = 12.38$; $b = 7.32$; $c = 7.79$
	orthorhombic $a^+ = 6.18$; $b^+ = 3.66$; $c^+ = 3.88$
Ta_2O_5 (α-phase)	monoclinic or triclinic

Conditions of Preparation	References
	[103] [113]
TaO oxidized for two hours in the presence of ammonia at 1 100°C; or Ta oxidized in air at 600°C	[103] [114] [115]
Powdered Ta_2O_5 reduced with carbon for two hours in vacuum at 1 500°C	[103] [105] [111]
Annealing of amorphous anodic Ta_2O_5 for 8 hours at 1 100°C	[113]
Annealing of Ta_2O_5 (in air at 920°C) obtained by chemical decomposition of K_2TaF_7 in H_2SO_4	[115]
Annealing of amorphous Ta_2O_5 at 750°C	[116]
Annealing (in air at 1 200°C) of Ta_2O_5 obtained by chemical decomposition of K_2TaF_7 in H_2SO_4	[115]
Annealing of Ta_2O_5 (in air at 800°C) obtained from a mixture of Nb_2O_5 and Ta_2O_5	[115]
	[117]
Hydrolysis of Pt wire at 1 100 ... 1 400°C: $2 TaCl_5 + 5 H_2O \rightarrow Ta_2O_5 + 10 HCl$	[118]
Annealing of Ta at 1 200°C	[86] [87] [94] [102] [103] [105] [108] [119] [120] [121] [122] [123] [124] [125] [126]
Oxidation of Ta slices (10^{-2} torr O_2 at 1 300°C)	[77]
	[127]

Phase	Structure data (Lattice constants in Å)
	hexagonal $a = 6.17; \; c = 11.7$
	tetragonal $a^+ = 3.80; \; c^+ = 35.67; \; c^+/a^+ = 9.39$
Ta_2O_5 (amorphous phase)	amorphous
	amorphous

The lattice constant marked $^+$ can be regarded as the best ensured value which could also be affirmed by own analyses.

By increasing the temperature this process is activated. In this way, the rearrangement of ions is favored and the oxygen absorption enhanced. For the transformation sequence of the phases the following scheme is obtained:

$Ta-O_{solid\ solution} \rightarrow Ta_6O \rightarrow Ta_4O \rightarrow Ta_2O \rightarrow TaO \rightarrow TaO_2 \rightarrow \beta\text{-}Ta_2O_5 \rightarrow \alpha\text{-}Ta_2O_5.$

(Amorphous oxide, Ta_2O_3, and Ta_2O_4 are omitted.)

$\alpha\text{-}Ta_2O_5$ is decomposed at temperatures above 1 600°C, TaO and TaO_2 being vaporized [102], [111], [133]. During this process, another sub-oxide with a lower oxygen content than TaO had been observed (see Table 13).
Tantalum pentoxide exists in 2 modifications, the so-called β-phase below 1 340 $\pm$ $\pm$ 20°C and the α-phase above this temperature.
Tantalum pentoxide is thermodynamically the most stable oxide phase of tantalum. Further oxide phases are given in different connections in Tables 12 and 13, and in Fig. 8.
$Ta_{32}O_9$ has a composition similar to that of Ta_4O. Different structures are given for these oxides. Nevertheless, $Ta_{32}O_9$ should be regarded as having a Ta_4O structure with a deficiency of tantalum or an excess of oxygen.
It can be seen from Table 13 that some of the structural data given for the tantalum oxides in the literature differ appreciably from each other. This may be due to considerable differences in the preparation conditions, to different contamination effects, and to the fact that most tantalum−oxygen compounds cannot be prepared, or separated in the amounts required for a careful structure analysis. These circumstances hamper, especially, attempts to establish the empirical formulae.
The sub-oxide and lower oxide phases exhibit an oxygen content lower than that of Ta_2O_5. It decreases from 72 At% for Ta_2O_5 to about 14 At% for Ta_6O (Fig. 9). Sub-oxide phases are preferably found in the interior of the tantalum matrix, and on the surface of tantalum. For an explanation of their primary formation at lower temperatures, where stable existence of $\beta\text{-}Ta_2O_5$ occurs, one could compare the specific

Conditions of Preparation	References
	[116]
Hydrolysis on Pt wire at 1 300 . . . 1 500°C: $2\,TaCl_5 + 5\,H_2O \rightarrow Ta_2O_5 + 10\,HCl$	[94] [105] [126] [127] [128]
Anodic oxidation with different electrolytes	[129]
Oxidation of Ta slices (pure O_2 atmosphere at 50 . . . 300°C)	[82]

interface energies of the sub-oxide and of the tantalum matrix, on the one hand, with those of tantalum pentoxide and of tantalum on the other hand. However, due to the lack of data it is impossible to obtain a quantitative conclusion.

For example, the intergrowth of Ta_2O with tantalum occurs always at {320} planes of the tantalum matrix in the interior of the oxidation zone. Obviously, the free interface energy is particularly small here because of a high correspondence of the unit cells. According to [104], the oxygen ions of Ta_2O are always situated in the centers of the joining lines between two adjacent tantalum atoms which have a distance of 2.90 Å. Since the expected diameter of the single oxygen ion O^{-2} is 2.60 Å, the electron configuration of oxygen in this oxide cannot be purely heteropolar. Probably,

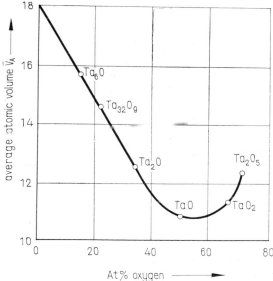

Fig. 9. Mean atomic volume V_A (volume of elementary lattice cells Å³/total number of atoms contained therein) of tantalum oxides and sub-oxides as a function of oxygen content

the sub-oxides Ta_6O, Ta_4O, and Ta_2O have to be regarded as tantalum-oxygen alloys with metallic bonding characteristics [110].

The activation energy* required for the formation of Ta_2O (see Table 10) is, by far, lower than that for the formation of Ta_2O_5. This is another reason why the primary formation of the sub-oxide is conceivable.

Similar arguments will certainly apply for the explanation of particular observations made during the formation of different boron, boride, phosphorus, and phosphide precipitates in Si and of the nitride precipitates in α-Fe.

1.5.1.2.   Orientation relations between tantalum and tantalum oxides

These simplest orientation relations hold for the suboxides, *i.e.* intergrowth always occurs along each other corresponding lattice planes of the sub-oxides and the tantalum matrix.

It is difficult to interpret the intergrowth of Ta_2O_5 with Ta since the former never intergrows directly with Ta, but only by forming intermediate layers of lower oxides, preferably the cubic Ta_2O.

A coherent growth of Ta_2O into the tantalum matrix is favored by the expansion of the tantalum elementary cell due to oxygen solubility (Fig. 10). Because of this adjustment, the formation of intermediate layers may always be assumed, particularly because a direct correspondence between the metallic Ta in both phases (tantalum matrix − tantalum sub-oxide) is most likely due to the bond characteristic of the sub-oxides.

The transition from Ta_2O (or a closely related sub-oxide) to β-Ta_2O_5 in the boundary region is conceivable because the uppermost tantalum atoms of a sub-oxide are re-

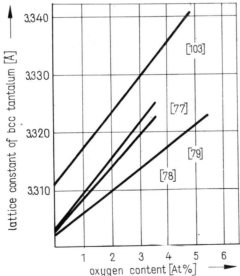

Fig. 10. Lattice constant of Ta as a function of oxygen content

leased from this bond, due to the reactive effect of oxygen during a critical stage of oxidation, to be coordinatively incorporated in the lattice structure of a first ionic low oxide, followed by a structural transition to the stable β-Ta_2O_5.

1.5.2. THERMAL OXIDATION OF Al, Si, AND Ge

Contrary to tantalum, the solubility of oxygen in Al, Si, and Ge is lower by some orders of magnitude. As mentioned above, the maximum oxygen solubility in Al is about 30 ppm. In silicon, this maximum value is closely below the melting point of silicon at about 300 ppm, *i.e.* one order of magnitude above that of Al and two orders of magnitude below that of Ta. The maximum solubility of oxygen in Ge corresponds to that of oxygen in Si. The oxidation of germanium is of no interest for the application in the planar technology because of the unfavorable properties of its oxides for this purpose (surface layer function).

The existence of oxides growing on the germanium surface, probably the growth of the stable, tetragonal GeO_2, has been unequivocally proved [12].

Preferred nucleation sites are vacancies, piercing points of dislocations, and other lattice defects. Moreover, a connection between oxide growth and the effect of doping atoms has been shown. From these considerations it may be concluded that, due to the effective range of excessive valence electrons and the depth of space charge regions, such an influence on oxide growth may not only originate from the semiconductor surface itself, but also from depths ranging from about 100 to 120 lattice planes below the surface.

The detection of such an oxide growth on the germanium surface indicates that a chemoepitaxial oxide growth is partially to be assumed. This will, however, not be the case if the phenomenon in question is merely an apparent growth caused by the greater specific volume of GeO_2 as compared with Ge. Further studies of the reaction mechanism in regard to diffusion through the growing oxide could lead to clear results. However, it is not necessary to undertake such efforts, because the formation of oxide films on Ge is of low practical significance. The above-mentioned results concerning oxide growth have been obtained during studies regarding the decoration of Ge surfaces. But they have not been considered with respect to the reaction mechanism.

The situation is quite different for silicon where the oxide film formation, and, hence, the influence exerted on the semiconductor surface, is of dominant significance in respect to planar technology.

According to a summary in [143] thermal oxidation of silicon is characterized by the fact that the oxygen-silicon reaction front is mainly situated in the region of the silicon-oxide (SiO_2) interface, *i.e.* it is linked with a preponderant diffusion of oxygen (anions, defect electrons) through the oxide. Consequently, the oxide film preferably grows into the silicon matrix which corresponds to chemoendotaxial layer growth with regard to the reaction mechanism.

Hence, chemoendotaxial layer growth is not only bound to high solubilities of the reactive medium in the substrate, *e.g.* oxygen during oxide layer formation (this is, for instance, the case for tantalum-oxygen). But chemoendotaxial layer growth is solely determined by the diffusion characteristics of the reaction partners in the film developing as a consequence of their reactive interaction, and it is only this criterion by which it can be distinguished from chemoepitaxial layer growth.

The degree of solubility is, however, among other factors which determine the extent of precipitates which may be formed during chemoendotaxial layer growth in front of the reaction interface.

During thermal silicon oxidation, phenomena of such kind as observed for Ta are not found to such a disturbing extent, although they are not missing.

Precipitates, which are formed to such a great extent as for Ta in front of the reaction interface, especially when the supply of oxygen exceeds the quantity required for reaction, decrease quantitatively with increasing thickness of the oxide film because of the continually increasing path of oxygen diffusion through the oxide film. Consequently, the precipitates are formed especially during the initial phases of growth and are partially re-incorporated into the growing layer while undergoing a structural change when sub-oxides or low oxides exist as precipitates, as has been shown in the case of tantalum oxidation.

Contrary to the examples referred to so far, thermal oxide layer formation (dry oxidation) on Al takes place predominantly according to purely chemoepitaxial principles (see Fig. 6) [4], [5], [69]. It may be considered to have been proved that, below 450°C, the cation transport controls the kinetics of oxide layer growth in this range. According to experience, chemoepitaxial layer growth is always linked with a very low or even negligible solubility of the reactive medium in the substrate material, *e.g.* the solubility of oxygen in Al in the present case. This also applies, for instance, to the chemoepitaxial formation of tungsten silicide layers on Si surfaces [70]. The maximum solubility of W in Si is less than 0.5 ppm. Likewise, chemoepitaxial copper, iron, molybdenum, and other metal transitional silicides or germanides are formed on Si and Ge, respectively [5], due to the similarly low solubility of these metals in Si and Ge.

In the case of purely chemoepitaxial layer growth, during which substrate atoms (cations, electrons) diffuse through the growing layer to the reaction interface, at the layer surface precipitations are not linked with layer growth.

It is of interest, however, that the oxygen solved in Al will not remain there in its elementary state, but exists nearly always in a bound state in an oxide because of its high affinity to Al. These oxide inclusions surely cannot be called true chemoendotaxial precipitates. Nevertheless true precipitates of this kind are conceivable. They would show that such precipitations are possible independent of the purely chemoepitaxial layer growth.

The knowledge of these interconnections and details of growth mechanisms is very significant from the point of view of the technological consequences relating to the control of desired material properties.

1.6. EPITAXIAL AND ENDOTAXIAL INTEGRROWTH BY ORIENTED EUTECTIC SOLIDIFICATION AND EUTECTOID TRANSFORMATION

In respect to the above-mentioned necessity of the need for a higher reliability of electronic components, eutectic or eutectoid binary and multiple heterocrystals obtained by oriented solidification or transformation are considered to be exceptionally interesting when epitaxial or endotaxial intergrowths result in regularly multilamellar, partially lamellar, or multibacular configurations which are reproducible over extended lengths and diameters [135].

The knowledge of intergrowth mechanisms leading to such structures is of decisive significance for the preparation of such heterocrystals constituting the basic structures of electronic components and microelectronic circuits. It is particularly the attainable high structural perfection of the crystal volumes of the individual phases and the phase boundaries between the various lamellae, on the one hand, the baculi (rods) and the matrix, on the other hand, which is important for such applications. A detailed, comprehensive description of growth rules, nucleation ideas and application aspects of eutectic and eutectoid heterocrystals is given in [135], [136]. Here, these connections shall, only briefly, be considered.

It is an obvious advantage of such structures that it is possible to prepare such configurations nearly simultaneously in a directed, reproducible way. This could not be

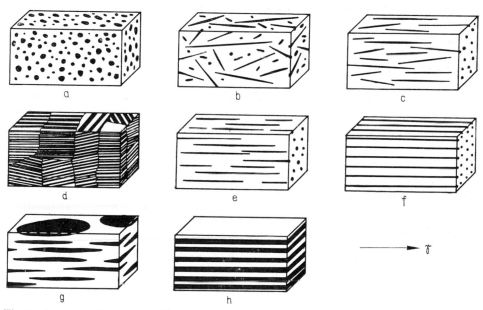

Fig. 11. Important basic types of eutectic structures: a) statistical globular structure; b) statistical bacular structure; c) textured bacular structure; d) textured lamellar structure; e) oriented bacular structure containing baculi of limited length and irregular positions; f) oriented perfect bacular structure; g) oriented partially lamellar structure (incorporation of spheroids); h) oriented perfect lamellar structure

achieved to the same degree of quality with other layer technologies. It is a fact, however, that some new, but promising, function models will have to be developed because of the modified geometrical conditions such as the stacking of monocrystalline layers (metals, semiconductors, dielectrics, etc.), periodical incorporation of baculi into a monocrystalline matrix, periodical incorporation of spheroids in lamellar systems, etc. (see Fig. 11).

Another positive aspect is the fact that, in eutectic and eutectoid structures, stable substances with particularly favorable physical properties can, possibly, be prepared easily (in a suitable combination with other substances) which have, under other conditions, to be considered unstable and, hence, cannot be prepared.

1.6.1.   EUTECTICS

Depending upon the substances involved and the conditions of solidification the products of eutectic crystallization exhibit only a more or less uniform structure. Two basically different kinds of configuration may be distinguished [137]:

a) configurations characterized by defined orientation relations between the solid phases, and

b) configurations not characterized by such relations, *i.e.* configurations constituting irregular intergrowth.

Therefore, all further discussions will be restricted to item *a)*.

When two substances co-crystallize they will always tend to attain a state of lowest possible free interface energy. This corresponds to the general rule already mentioned for oriented intergrowth.

This should be especially favored when the crystallization of the two intergrowing phases occurs in a quasi-simultaneous way.

Certain single, or multiple structure relations between these phases permit very stable minima of free energy to be established in the respective interfaces which prove to be the defined orientation relations between these phases. If the intergrowing phases have very similar, or equal, properties of crystal symmetry, these defined orientation relations can be obtained in a particularly perfect way. This statement also corresponds to a statement made at the beginning of this chapter.

It is due to the nature of these combinations of substances that quasi-simultaneity of crystallization of two intergrowing phases always occurs and that these phases, in some cases, fulfil the above-mentioned favorable crystal symmetry conditions as well.

These circumstances enable true epitaxial crystallization connected with a coupled growth of the two phases.

Lamellar eutectics of this kind are characterized by perfect and defined orientation relations. Bacular eutectics, however, very frequently have an additional degree of freedom normal to the rod axis and, in such cases, do not have any clearly defined intergrowth planes (Fig. 12).

Heteroepitaxy exists when only a simple segregation of the melt takes place without any chemical reaction involved to form new phases during solidification.

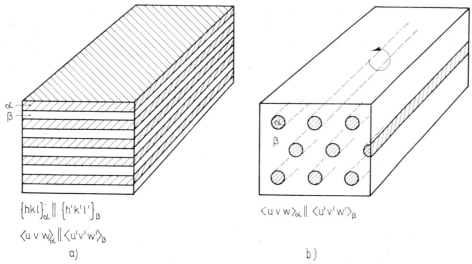

$$\{hkl\}_\alpha \parallel \{h'k'l'\}_\beta$$

$$\langle u v w \rangle_\alpha \parallel \langle u'v'w' \rangle_\beta$$

$$\langle u v w \rangle_\alpha \parallel \langle u'v'w' \rangle_\beta$$

a)   b)

Fig. 12. Schematic view of orientational relations in lamellar and bacular structures

Chemoepitaxy causes a chemical reaction to occur between the components of the substances involved during the segregation of a melt, and leads to the new formation of one or both phases (in a binary system) and one or several phases (in multi-element systems) as chemical compounds of the components present in the melt.
Typical examples of binary, heteroepitaxial eutectics are:

— Ag—Cu: (211) [110]$_{Ag}$ // (211) [110]$_{Cu}$ and

— Pb— In, Cd—Zn, Sn—Cd, Ag—Bi,

all of them having a multi-lamellar structure and revealing simple orientation relations [138], [139].
For the Au–Ge, Ag–Ge, Al–Ge, Au–Si, Ag–Si, and Al–Si systems which are summarized in detail in [140], defined intergrowth should also be possible if the specific growth conditions have been ascertained and applied.
Typical examples of chemoepitaxial eutectic systems are:

— Al$_{fcc}$—CuAl$_{2tetragonal}$: (11$\bar{1}$) [101]$_{Al}$ // (21$\bar{1}$) [120]$_{CuAl_2}$

— SnSe—SnSe$_2$ [141]:

(001) [110]$_{SnSe, \, p\text{-conducting}}$ // (001) [110]$_{SnSe_2, \, n\text{-conducting}}$
    orthorhombic   hexagonal
                          CdJ$_2$ type

Al–Al$_2$Cu and SnSe–SnSe$_2$ are multi-lamellar.

— Al$_{fcc}$–Al$_3$Ni$_{orthorhombic}$: [110]$_{Al}$ // [010]$_{Al_3Ni}$ and

— (111)$_{Al}$ // (102)$_{Al_3Ni}$ or (110)$_{Al}$ // (103)$_{Al_3Ni}$.

Al–Al$_3$Ni is of a multi-bacular structure where Al$_3$Ni is the rod-shaped phase [142]. Other bacular, chemoepitaxial eutectic systems are the following metal–oxide systems: MgO–W, Cr$_2$O$_3$–Mo, Cr$_2$O$_3$–Re, Cr$_2$O$_3$–W, and UO$_2$–W; the metal phases always constituting the rod-shaped portions [143]. Essential factors, when considering multi-lamellar eutectic heterocrystals, are the absolute lamella thickness and the lamella thickness relation.

The absolute lamella thickness π depends directly upon the conditions of solidification:

$$\lambda = \text{const.} \cdot v^{-u}$$

where v = mean growth rate and
$\quad u$ = mean empirical value of about 1/2.

The lamella thickness relation results from the volume portions of the single phases. They are given by the system itself. Due to the very unfavorable volume–surface relation caused by interface tensions, eutectic systems with very different portions of volume of the phases tend to form partially lamellar structures (island, or spheroid, formation of the phase with the smaller portion of volume), or bacular structures where the phase with the smaller portion of volume represents the rod-shaped component.

Any dopings intentionally added to the melt may decisively influence the viscosity of the melt and the resulting interface stresses, and, among other possibilities, for instance, by changing the nucleation conditions and mechanisms, thereby influence the crystallization.

The results of theoretical considerations agree well with experimental findings.

This enables scientifically founded technological process parameters to be specified for a defined crystal growth and also for substance combinations which, at present, are still open to study, but are of a technical interest. This also applies to eutectoid heterocrystals.

1.6.2. EUTECTOIDS

Eutectoid transformation products also do not have uniform structure either depending upon the substances or on the transformation conditions. Hence, it should be reasonable to subdivide the eutectoids into different groups as has been done with the eutectic systems.

The basic principles of energy and crystal symmetry, as suggested for eutectic systems, may be assumed also to apply for such intergrowth processes proceeding in the solid state and leading to defined orientations.

As a rule, eutectoid transformation products with defined orientation relations between adjacent phases are multilamellar and have the appearance of analogously structured eutectic systems (see Fig. 12a).

The lamellar growth of eutectoid transformation products proceeds step by step at the growth interface. As a rule, the ends of the lamellae grow into the matrix parallel with the lamellae planes [144]. In respect to the direction of growth, this agrees with the rules of lamellar eutectic solidification.

70

The growth rate of the nuclei is the rate determining parameter for the overall growth process [144]. Once the nuclei have reached a critical size and have spread over the lamellae faces, more new nuclei are formed.

All changes of concentration due to the transformation of the matrix substance leading to the eutectoid structure occur in the solid state just ahead of an incoherent growth interface, and are diffusion controlled. Because of the effect of atomic attraction and repulsion forces, and of lattice imperfections, the diffusion coefficient is, in the region of the growth interface, 10^3 to 10^5 times higher than in the largely undisturbed crystal under normal conditions [145]. The eutectoid transformation exhibits substantial analogies in respect to the growth mechanism of the new phases with the formation of cellular precipitations.

For this reason, eutectoid transformations leading to defined orientations can be considered to be endotaxial.

Heteroendotaxial eutectoids are formed by simple segregation of solid solutions without chemical reactions taking place. The resulting crystallites consist of the pure components or of corresponding solutions of a different constitution.

Chemoendotaxy involves a chemical reaction during the transformation of a matrix substance.

In the case of segregation of a solid solution, the disproportioning, or complete dissociation, of a chemical compound, at least one of the resulting phases may be expected to be a chemical compound with a constitution other than that of the matrix. This chemical compound results from a chemically reactive interaction of the components existing in the matrix. Eutectoids have also to be considered to be chemoendotaxial if only disproportioning or dissociation of a matrix substance existing as a chemical compound takes place without the formation of other new chemical compounds in addition to the phases resulting from transformation, and if these phases consist of pure components or solid solutions.

An example of a heteroendotaxial eutectoid structure is given by the Ti–W system where a β-Ti solid solution containing 9 At% of W decomposes below 715°C into W and Ti phases each having dissolved portions of solid Ti and W, respectively [146]. The orientation relations are of the most simple kind. An example of the chemoendotaxial process of an eutectoid transformation is the disproportioning of Ta_2C to form $TaC + Ta$, which starts somewhat below 1 400°C [147]. During a transformation of Ta_2C exposed to a temperature gradient, the resulting phases intergrow with each other according to the most simple orientation relations.

The expression $\lambda^4 \cdot v = $ const. describes the relation between the lamella thickness λ, and the mean growth rate v during the growth process. Furthermore, it can be stated:

(a) The growth rate depends upon temperature.
(b) This dependence is non-linear.
(c) The thickness of the lamella decreases as the temperature decreases. Constant lamella thicknesses over wide ranges require, therefore, strictly isothermal growth conditions.
(d) The lamella thickness relation of the phases is determined by the portions of the respective volumes, as is the case for a eutectic system.

Eutectoid systems suitable for the production of basic structures of electronic components are by far less frequent than eutectic ones and mostly exhibit a more complicated structure. Although this is an obvious disadvantage, this statement should not be considered as a final evaluation of the significance and usefulness of eutectoidal configurations in the above sense, since current knowledge is far less advanced in this field than it is for eutectic systems.

1.7.   CONCLUSIONS

Interfaces will become more important in electronics the more the properties of layer structures are utilized for component conceptions [148].

The phase boundary is of particular significance in relation to semiconductor technology. Here, an electron flow, as seen from the geometrical point of view, is controlled by means of various physical effects either at a phase boundary or in the volume of the semiconductor, in a wider sense, by a combination of phase boundary systems interacting with each other through volume zones [149].

Likewise, it is the reproducible uniform geometrical and structural formation of layers and phase boundaries which is exceptionally significant for the stability and reliability of passive thin film components, in particular, and similar thick film components. Any local irregularities in layer thickness, and structural inhomogeneities of the dielectric layer itself, will cause capacitor breakdowns, large dielectric loss angles and irregular space charge concentrations. The same conditions are able, likewise, to influence electrical parameters of metallic conductive layers and resistive films so that the specified tolerances cannot be observed. Knowledge of the respective interface structure is necessary, especially for layer systems of semiconductor components. In a great number of interesting investigations the effect of the substrate on the structure of thin epitaxial layers has been studied and important results for the semiconductor technique were obtained. The most substantial results are given in [150]–[152].

The main problem in this connection is the structural misfit between the substrate and the deposit, and the investigation of the fitting mechanism.

[152] is the first transmission electronmicroscopic investigation of epitaxial films, grown on compact substrates with surfaces of known atomically smooth structure. The observed arrays of interface dislocations and stacking faults originate from grown-in substrate dislocations. A quantitative interpretation of the structures observed is presented in [152], based on the concept of *Mathews* and *Jesser* (cited in [150]). It was concluded in the investigated cases that a spontaneous creation of misfit is not necessary to fit the two lattices.

These results which have been obtained by intentionally studying the model system Ag–Au are of exceptionally great practical value, though they are not directly applicable to other systems. Observations at the Si–SiO$_2$ interface, which are summarized in [134], confirm this intention quite clearly. Similar relations were found, or may be expected, for other semi coherent interfaces in cases of similarly low misfit such as GaAs–Ge heteroepitaxy, but also for Si–Si and Ge–Ge homoepitaxy [153], with provable different lattice dimensions due to highly differing doping concentrations, though in adamantine crystals glide is inhibited by the *Peierls* stress [154], but may not be expected for incoherent interfaces in cases of too large a misfit, *e.g.* hetero-epitaxy of Si on spinel or sapphire, and of Al, Ag, Au with Si and Ge.

Such relations also enable specific statements on the electronic properties of those interfaces and adjoining volume areas to be made.

Due to the incorporation of both P and B atoms into the silicon lattice (evidence is given for a linear lattice contraction coefficient of $2.5 \cdot 10^{-24}$ cm^3 Si crystal versus an boron atom in [155]), heavy strains build up in the semiconductor lattice itself, which are partly compensated by dislocations. Such dislocations appear especially near the surface arranged as specific dislocation networks and, thus, also contribute to the formation of semi-coherent interfaces. In the case of epitaxy this has definite consequences for the structure of the transition zone.

In addition, the semiconductor epitaxial layers themselves exhibit typical epitaxial defects throughout their entire volume. Such structural defects are indicated in [11] for Si epitaxial layers: tripyramids, stacking faults, dislocations, which have, of course, a certain effect on the electronic properties of layers. Thus, tripyramids may cause short circuits in the circular emitter zone.

Without going into more detail we may conclude that efforts to improve semiconductor epitaxy can only be successful if the above-mentioned significance of the interfaces is taken into account, and when attempts are successful to achieve a minimum of free energy especially at the relevant interfaces and in view of the structural interconnection and functional interaction in the adjacent transition zones and within the layers. This statement emphasizes the importance of controlling oriented intergrowth, especially homoepitaxy and heteroepitaxy, in semiconductor techniques.

Another point which is often neglected is the unintended formation of chemically reactive intermediate layers between the primary intergrowth partners, *i.e.* the substrate and the film, during heteroepitaxy and homoepitaxy. This has been thoroughly discussed along with chemoendotaxial oxide layer growth on Ta, but in another context. It should be pointed out here, however, that, also in the cases of homoepitaxy and heteroepitaxy, surface impurities of the substrate, on the one hand, or impurities introduced from the atmosphere, on the other hand, may have reactive effects. Such contaminants favor the formation of intermediate layers which disturb, or even prevent, the desired direct contacts between the intergrowing phases.

Such intermediate layers, even those grown in a nonreactive manner, *i.e.* contaminant absorbing layers, are often extremely thin with mean thicknesses of a few interatomic spacings and irregular in respect to the area covered by them.

Hence, Si surfaces can always be expected to be covered by thin silicon oxide films. In respect to epitaxy, this always results in a change of the energy balance for the intergrowth process.

If pure metal–semiconductor or semiconductor–semiconductor junctions are required, care has to be taken that such perturbing factors can be excluded.

On the other hand, such an influence may also promote epitaxy if the initiation of layer growth becomes easier because of initial reactive intermediate layer formation or the existence of preferred nucleation positions on adsorbed contaminants [156]. However, such intermediate layer formations are not very reproducible and are, consequently, not desired for component layer structures.

Another mode of forming intermediate layers, however, is remarkable insofar as a structural transition zone is built up which only enables heteroepitaxy to take place between certain substances.

An example of this is the epitaxial growth of niobium layers on MgO on top of a thin NbO intermediate layer [156]. Another example is the epitaxial growth of NiO on sapphire on top of a thin $NiAl_2O_4$ intermediate layer [157]. In the latter case, the oxygen coordination schemes of the sapphire and the NiO are decisive. Such coordination relations involve a $NiAl_2O_4$ spinel formation as a structural junction between the sapphire (α-Al_2O_3) and the NiO for the sapphire–NiO heteroepitaxy in accordance with the discussion on energy [1] already presented.

The results discussed here should be essential for efforts to achieve heteroepitaxially deposited semiconductor substances, above all Si, on insulator crystal substrates such as sapphire and MgAl-spinel.

With these final remarks it was intended to point out some important problems linked with epitaxial and endotaxial intergrowth in producing basic structures for electronic components. It is evident that oriented eutectic solidification and eutectoid transformation will become more and more important.

1.8. REFERENCES

[1] *Schneider, H. G.*: Habilitationsschrift. Technische Hochschule, Ilmenau (1967).

[2] *Lewicki, A.*: Einführung in die Mikroelektronik. R. Oldenbourg Verlag, München – Wien (1966).

[3] *Schneider, H. G.* (Ed.): Zuverlässigkeit Elektronischer Bauelemente. VEB Deutscher Verlag für Grundstoffindustrie, Leipzig (1974).

[4] *Čistyakov, Yu. D.*: Dissertatsiya. Institut Stali i Splavov, Moskva (1967).

[5] *Čistyakov, Yu. D., Schneider, H. G., Weinhold, C.*: in "Epitaxie – Endotaxie" (Ed.: H. G. Schneider), VEB Deutscher Verlag für Grundstoffindustrie, Leipzig (1969) p. 15.

[6] *Schneider, H. G.*: in "Advances in Epitaxy and Endotaxy" (Eds.: H. G. Schneider and V. Ruth), VEB Deutscher Verlag für Grundstoffindustrie, Leipzig (1971) p. 13.

[7] *Khambata, A. J.*: Einführung in die Mikroelektronik (Translation from English). VEB Verlag Technik, Berlin (1967).

[8] *Auth, J.*: in "Probleme der Festkörperelektronik" Vol. 1. VEB Verlag Technik, Berlin (1969) p. 11.

[9] *Riedl, W. J.*: in "Grundlagen Aktiver Elektronischer Bauelemente" (Ed.: H. G. Schneider). VEB Deutscher Verlag für Grundstoffindustrie, Leipzig (1972) p. 297.

[10] *Mittenentzwei, H. W.*: in "Grundlagen Aktiver Elektronischer Bauelemente" (Ed.: H. G. Schneider). VEB Deutscher Verlag für Grundstoffindustrie, Leipzig. (1972) p. 330.

[11] *Schulz, M.*: in "Grundlagen Aktiver Elektronischer Bauelemente" (Ed.: H. G. Schneider), VEB Deutscher Verlag für Grundstoffindustrie, Leipzig (1972) p. 343.

[12] *Hadamovsky, H. F.* (Ed.): Halbleiterwerkstoffe, IInd Edition. VEB Deutscher Verlag für Grundstoffindustrie, Leipzig (1972).

[13] *Münch, W. v.*: Technologie der GaAs-Bauelemente. Springer Verlag, Berlin – Goettingen – Heidelberg (1969).

[14] *Thiessen, K.*: in "Grundlagen Aktiver Elektronischer Bauelemente" (Ed.: H. G. Schneider)‘ VEB Deutscher Verlag für Grundstoffindustrie, Leipzig (1972) p. 75.

[15] *Minden, H. T.*: Solid State Technology 4 (1968) 25.

[16] *Barry, B., Hicks, H. G. B.*: Elektrisches Nachrichtenwesen (ITT – SEL), 47 (2) (1972) 109.

[17] *Čistyakov, Yu. D. Weinhold, C., Georgieff, I. Ž.*: in "Epitaxie – Endotaxie" (Ed.: H. G. Schneider), VEB Deutscher Verlag für Grundstoffindustrie, Leipzig (1969) p. 212.

[18] *Cohen-Solal, G., Marfaing, Y., Bailly, F.:* in "Épitaxie—Endotaxie" (Ed.: H. G. Schneider), VEB Deutscher Verlag für Grundstoffindustrie, Leipzig. (1969) p. 187.

[19] *Cohen-Solal, G., Riant, Y.:* Appl. Phys. Letters 19 (10) (1971) 436.

[20] *Wang, C. C.:* J. Appl. Physics 40. (1969) 3433.

[21] *Krause, H.* in "Grundlagen Aktiver Elektronischer Bauelemente" (Ed.: H. G. Schneider), VEB Deutscher Verlag für Grundstoffindustrie, Leipzig (1972) p. 314.

[22] *Krause, H.:* Phys. stat. solidi (a) 1, K53 (1970).

[23] *Mercier, J.:* J. Electrochem. Soc. 117 (6) (1970) 812.

[24] *Mercier, J.:* Revue de Physique Appliquée 4 (3) (1969) 345.

[25] *Distler, G. I., Kobzareva, S. A., Gerasimov, Y. M.:* Proceedings of the 2nd Colloquium on Thin Films, Budapest (1967) p. 81.

[26] *Distler, G. I.:* Proceedings of the 2nd Colloquium on Thin Films, Budapest (1967) p. 38.

[27] *Distler, G. I., Gerasimov, Y. M., Obronov, V. G.:* Thin Solid Films 10 (1972) 195.

[28] *Distler, G. I., Obronov, V. G.:* Die Naturwissenschaften 57 (1970) 495.

[29] *Distler, G. I.:* J. Cryst. Growth 3 (1968) 175.

[30] *Distler, G. I., Shenyavskaya, L. A.:* Fisika Tverdogo Tela 11 (1969) 488.

[31] *Halla, F.:* Kristallchemie und Kristallphysik Metallischer Werkstoffe. J. A. Barth Verlag, Leipzig (1957).

[32] *Burke, J.:* The Kinetics of Phase Transformations in Metals. Pergamon Press, London (1965).

[33] *Deute, V., Jacobi, W., Hornischer, R:* Z. Phys. Chemie N. F. 81 (1972) 12.

[34] *Schöne, E.:* XII. Internationales Wissenschaftliches Kolloquium, Technische Hochschule Ilmenau, Part 10 (Elektronik) (1967) p. 59.

[35] *Schneider, H. G.:* Freiberger Forschungsheft No. B 71 (1964).

[36] *Dobson, P. S., Filby, J. D.:* J. Cryst. Growth 3 (4), (1968) 209.

[37] *Dietze, W., Miller, M., Amberger, E.:* in "Boron", Vol. 3, (Ed.: T. Niemyski), PWN, Warszawa (1970) p. 72.

[38] *Giese, R. F.* Jr.: in "Boron" Vol. 3. (Ed.: T. Niemyski), PWN, Warszawa (1970) p. 151.

[39] *Economy, J., Matkovich, V. I.:* in "Boron", Vol. 3, (Ed.: T. Niemyski), PWN, Warszawa (1970) p. 159.

[40] *Matkovich, V. I., Economy J.:* in "Boron", Vol. 3, (Ed.: T. Niemyski), PWN, Warszawa (1970) p. 168.

[41] *Potworowski, J. A., Lonc, W. P.:* in "Boron", Vol. 3, (Ed.: T. Niemyski), PWN, Warszawa (1970) p. 193.

[42] *Werheit, H.:* in "Festkörperprobleme X — Advances in Solid-State Physics". Akademie-Verlag, Berlin (1970) p. 189.

[43] *Joshi, M. L., Dash, S.:* IBM J. Research and Development 11 (3) (1967) 271.

[44] *Jungbluth, E. B.:* J. Appl. Phys. 38 (1) (1967) 133.

[45] *Duffy, M. C., Barson, F., Fairfield, J. M., Schwuttke, G. H.:* J. Electrochem. Soc. 115 (1) (1968) 84.

[46] *Duffy, M. C., Barson, F., Fairfield, J. M., Schwuttke, G. H., Parker, T. J.:* J. Electrochem. Soc. 115 (12) (1968) 1290, discussion to [45].

[47] *Joshi, M. L., Wilhelm, F.:* J. Electrochem. Soc. 112 (2) (1965) 185.

[48] *Joshi, M. L., Dash, S.:* J. Electrochem. Soc. 113 (1) (1966) 45.

[49] *McDonald, R. A., Ehlenberger, G. G., Huffmann, T. R.:* Solid-State Electronics 9 (8) (1966) 807.

[50] *Schmidt, P. F., Stickler, R.:* J. Electrochem. Soc. 111 (11) (1964) 1188.

[51] *O'Keeffe, T. W., Schmidt, P. F.:* J. Elektrochem. Soc. 112 (8) (1965) 878.

[52] *Beck, C. G., Stickler, R.:* J. Appl. Phys. 37 (12) (1966) 4683.

[53] *Schmidt, P. F., von Gelder, W., Drobek, J.:* J. Electrochem. Soc. 115 (1) (1968) 79.

[54] *Kovi, E.:* J. Electrochem. Soc. 111 (12) (1964) 1383.

[55] *Fairfield, J. M., Schwuttke, G. H.:* J. Electrochem. Soc. 115 (4) (1968) 414.

[56] *Jaccodine, R. J.:* J. Appl. Physics 39 (7) (1968) 3105.

[57] *Wolley, E. D., Stickler, R., Chu, T. L.:* J. Electrochem. Soc. 115 (4) (1968) 409.

[58] *Yukimoto, Y.* Jap. J. Appl. Phys. 8 (5) (1969) 568.

[59] *Nes, E., Washburn, J.:* J. Appl. Physics 42 (9) (1971) 3562.

75

[60] *Nes, E., Washburn, J.:* Correction to [59].

[61] *Trumbore, F. A.:* Bell Syst. Techn. Journal 1 (1960) 205.

[62] *Maissel, L. I., Glang, R.* (Ed.): Handbook of Thin Film Technology. McGraw-Hill Book Co., New York et al. (1970).

[63] *Chopra, K. L.:* Thin Film Phenomena. McGraw-Hill Book Co., New York et al. (1969).

[64] *Brenner, A.:* Electrodeposition of Alloys I/II Academic Press Inc., New York (1963).

[65] *Hockaday, C. H.:* Electronic Engng. (5) (1969) 57.

[66] *Holmes, P. J., Corkhill, J. R.:* Electronic Components (10) (1969) 1171.

[67] *Teuschler, H.−J.:* in "Grundlagen Passiver Elektronischer Bauelemente" (Ed.: H. G. Schneider), VEB Deutscher Verlag für Grundstoffindustrie, Leipzig (1973) p. 231.

[68] *Schneider, H. G.:* Elektronika (Warszawa) XIII (12) (1972) 506.

[69] *Čistyakov, Yu. D., Mendelejevič, A. Yu.:* Isv. vusov, Čvetn. metallurgiya No. 3 (1965) 127.

[70] *Čistyakov, Yu. D.:* Rost Kristallov (Moskva) VIII (1968) 258.

[71] *Schneider, H. G.:* Rost Kristallov (Moskva) VIII (1968) 255.

[72] *Schneider, H. G., Langer, H.−D.:* Wiss. Z. TH Ilmenau 14 (2) (1968) 165.

[73] *Langer, H.−D., Schneider, H. G.:* Kristall und Technik 7 (5) (1972) 561.

[74] *Langer, H.−D., Meyer, S., Schneider, H. G.:* in "Kristallisation" (Ed.: H. Ringpfeil), VEB Deutscher Verlag für Grundstoffindustrie, Leipzig (1969) p. 72.

[75] *Shewchun, J., Hardy, W. R., Kuenzig, D., Tam, C.:* Thin Solid Films 8 (1971) 81.

[76] *Smialek, L., Mitchell, T. E.:* Phil. Mag. 22 (180) (1970) 1105.

[77] *Kofstadt, P.:* J. Less-Common Metals 7 (1964) 241.

[78] *Gebhardt, E., Seghezzi, H. D.:* Plansee Proc. Reutte (Tirol) (1958) 280.

[79] *Vaughan, D. A., Stewart, O. M., Schwartz, C. M.:* Trans. Met. Soc. AIME 221 (1961) 937.

[80] *Haul, R.:* Physisorption an Festkörperoberflächen im UHV-Bereich. Paper of the Leopoldina-Congress "Physik und Chemie der Kristalloberfläche", Halle 8. − 11. 10. 1966.

[81] *Ang, C. Y.:* Acta Met. 1 (1953) 123.

[82] *Vermilyea, D. A.:* Acta Met. 6 (1958) 166.

[83] *Gulbransen, E. A., Andrew, F. K.:* Trans. Met. Soc. AIME 188 (1950) 586.

[84] *Powers, R. W., Doyle, M. V.:* Trans. Met. Soc. AIME 215 (1959) 655.

[85] *Albrecht, W. M., Klopp, W. D., Koehl, B. G., Jaffee, R. J.:* Trans. Met. Soc. AIME 221 (1961) 110.

[86] *Cowgill, M. G.:* Thesis, University of Liverpool (1962).

[87] *Gebhardt, E., Seghezzi, H. D.:* Z. Metallkunde 48 (1957) 503.

[88] *Samsonov, G. V.:* Fisiko-Chim. Svoistva Elementov, Naukova Domka, Kiev (1965).

[89] *Langer, H.−D.:* in "Diffusion in metallischen Werkstoffen" (Ed.: H. Ringpfeil), VEB Deutscher Verlag für Grundstoffindustrie, Leipzig (1970) p. 279.

[90] *Gebhardt, E., Seghezzi, H. D.:* Z. Metallkunde 50 (1959) 248.

[91] *Kubaschewski, O., Hopkins, B.:* Oxidation of Metals and Alloys, Butterworths, London (1962).

[92] *Kofstadt, P.:* J. Electrochem. Soc. 110 (1963) 491.

[93] *Davies, J. A., Pringle, J. P. S., Graham, R. L., Brown, F.:* J. Electrochem. Soc. 109 (1962) 999.

[94] *Kubaschewski, O., Hopkins, B.:* J. Less-Common Metals 2 (1960) 172.

[95] *Waber, J. T., Sturdy, G. E., Wise, E. M., Tripton, C. R.:* J. Electrochem. Soc. 99 (1952) 121.

[96] *Waber, J. T.:* J. Chem. Phys. 20 (1952) 734.

[97] *Cathcart, J. V., Bakish, R., Norton, O. R.:* J. Electrochem. Soc. 107 (1960) 668.

[98] *Petersson, R. C., Fassell, W. M., Wadsworth, M. E.:* Trans. Met. Soc. AIME 200 (1954) 1038.

[99] *Cowgill, M. G., Stringer, J.:* J. Less-Common Metals 2 (1960) 233.

[100] *Uiks, K. E., Blok, F. E.:* Termodinamičeskie Svoistva 65, Elementov ikh Okislov, Galogenidov, Karbidov i Nitridov, Metallurgiya, Moskva (1965).

[101] *Brauer, G., Müller, G., Kühner, G.:* J. Less-Common Metals 2 (1960) 172.

[102] *Norman, N.:* J. Less-Common Metals 4 (1962) 52.

[103] *Schönberg, N.:* Acta Chem. Scand. 8 (1954) 240.

[104] *Steeb, S., Renner, J.:* J. Less-Common Metals 9 (1965) 181.

[105] *Wasilewski, R. J.:* J. Amer. Chem. Soc. 75 (1953) 1001.

76

[106] *Steeb, S., Renner, J.:* J. Less-Common Metals 10 (1966) 246.

[107] *Adelsberg, L. M., Pierre, G. R. St., Speiser, R.:* Trans. Met. Soc. AIME 236 (1966) 1363.

[108] *Brauer, G., Müller, H., Kühner, G.:* J. Less-Common Metals (1962) 533.

[109] *Brauer, G., Müller, H.:* Plansee Proc. Reutte (Tirol), (1958) 257.

[110] *Müller, H.:* Thesis, University of Freiburg (1958).

[111] *Ingram, M. G., Chupka, W. A., Berkowitz, J.:* J. Chem. Phys. 27 (1959) 569.

[112] *Wykoff, R.:* Crystal Structures. Interscience Publ., New York 1948).

[113] *Lehovec, K.:* J. Less-Common Metals 7 (1964) 397.

[114] *Kelly, K. K.:* J. Amer. Chem. Soc. 62 (1940) 818.

[115] *Simanov, Yu. P., Lapičkii, A. V., Artamanova, E. P.:* Vestnost. Mosk. Univers. 9 (9), Ser. Fis. Mat. i Estestven. Nauk 9 (1954) 109.

[116] *Harvey, J., Wilman, H.:* Acta Cryst. 19 (1961) 1278.

[117] *Lapičkii, A. V., Simanov, Yu. P., Semenenko, K. N., Jarembas, E. J.:* Vestnost. Mosk. Univers. 9 (3), Ser. Fis. Mat. i Estestven. Nauk 2 (1959) 85.

[118] *Saslavskiij, A. J., Svinčuk, R. A., Tutov, A. G.:* Dokl. AN SSSR 104 (1955) 409.

[119] *Pawel, E., Cathcart, J. V., Campbell, J. J.:* J. Elektrochem. Soc. 107 (1960) 965.

[120] *Orr, R. L.:* J. Americ. Ceram. Soc. 75 (1954) 2808.

[121] *Pearson, W. B.:* Handbook of Lattice Spacings. Pergamon Press, Oxford (1958).

[122] *King, B. W., Schultz, J., Durbin, E. R., Duckworth, W. H.,* Battelle Mem. Inst. Report 1106, Columbus, Ohio, 9 (1965).

[123] *Hahn, R. B.:* J. Amer. Chem. Soc. 73 (1951) 5091.

[124] *Calvert, L. D., Droper, P. H. G.:* Can J. Chem. 40 (1962) 1943.

[125] *Pawel, R. E., Cathcart, J. V., Campbell, J. J.:* Acta Met. 10 (1962) 149.

[126] *Lagergren, S., Magneli, A.:* Acta Chem. Scand 6 (1952) 4444.

[127] *Reismann, A., Holtzberg, F. Berkenblit, M., Berry, M.:* J. Amer. Chem. Soc. 78 (1956) 4514.

[128] *Gebhardt, E., Seghezzi, H. D.:* Z. Metallkunde 50 (1959) 521.

[129] *Zinsmeister, G.:* Bull. Schweiz. Elektrotechn. Vereins 54 (1963) 699.

[130] *Norman, N., Kofstadt, P., Krudtaa, O. J.:* J. Less-Common Metals 4 (1962) 124.

[131] *Vermilyea, D. A.:* J. Electrochem. Soc. 102 (1955) 207.

[132] *Bakish, R. J.:* J. Elektrochem. Soc. 105 (1958) 574.

[133] *Fromm, E.:* Z. Metallkunde 57 (1966) 540.

[134] *Obernik, H.:* in "Probleme der Festkörperelektronik" Vol. 2, VEB Verlag Technik, Berlin (1970) p. 67.

[135] *Schneider, H. G., Langer, H.-D.:* Wiss. Z. TH. Karl-Marx-Stadt 15, (1) (special edition), (1973) pp. 119—133.

[136] *Langer, H.-D., Treiber, H. G., Schneider, H. G.:* Research Report, Technische Hochschule, Karl-Marx-Stadt (1973).

[137] *Kerr, H. W.:* in "Epitaxie—Endotaxie" (Ed.: H. G. Schneider) VEB Deutscher Verlag für Grundstoffindustrie, Leipzig (1969) p. 116.

[138] *Ellwood, E. C., Bagley, K. Q.:* J. Metals 76 (1949) 631.

[139] *Cline, H. E., Stein, D. F.:* Trans. Met. Soc. AIME 245 (1969) 841.

[140] *Hellawell, A.:* Progress in Materials Science. Vol. 15, No. 1 (1970).

[141] *Albers, W., Verberkt, J.:* J. Mat. Science 5 (1970) 24.

[142] *Kerr, H. W., Lewis, M. H.:* in "Advances in Epitaxy and Endotaxy" (Eds.: H. G. Schneider, V. Ruth), VEB Deutscher Verlag für Grundstoffindustrie, Leipzig (1971) p. 147.

[143] *Hart, P. E.:* Ceramic Age 88 (3) (1972) 29.

[144] *Carpay, F. M. A.:* Acta Met. 20 (7) (1972) 929.

[145] *Hornbogen, E., Warlimont, H.:* Metallkunde. Springer-Verlag Berlin—Heidelberg—New York (1967).

[146] *Maykuth, D. J., Ogden, H. R., Jaffee, R. J.:* Trans. AIME 197 J. Metals (1953) 231.

[147] *Tardif, A., Piquard, G., Wach, J.:* Rev. Int. Hautes Tempér. et Refract. 8 (1971) 143.

[148] *Flietner, H.:* in "Grundlagen Aktiver Elektronischer Bauelemente" (Ed.: H. G. Schneider) VEB Deutscher Verlag für Grundstoffindustrie, Leipzig (1972) p. 27.

[149] *Paul, R.:* in "Grundlagen Aktiver Elektronischer Bauelemente" (Ed.: H. G. Schneider) VEB Deutscher Verlag für Grundstoffindustrie, Leipzig (1972) p. 51.

[150] *Matthews, J. W.:* in "Epitaxie—Endotaxie" (Ed.: H. G. Schneider) VEB Deutscher Verlag für Grundstoffindustrie, Leipzig (1969) p. 135.

[151] *van der Merwe, J. H.:* in "Advances in Epitaxy and Endotaxy" (Eds.: H. G. Schneider,V. Ruth), VEB. Deutscher Verlag für Grundstoffindustrie, Leipzig (1971) p. 129.

[152] *Woltersdorf, J.:* Thesis, University of Halle (1973); and a paper to be published in phys. stat. sol. (a) (1975).

[153] *Vaulin, Yu. D., Migal, N. N., Migal, V. P., Stenin, S. I.:* Physica status solidi (a) 15 (2) (1973) 697.

[154] *Matthews, J. W., Mader, S., Light, T. B.:* J. Appl. Phys. 41 (9) (1970) 3800.

[155] *Brümmer, O., Höche, H. R.:* Paper of the colloquium "Physikalische und chemische Meß- und Analysentechnik", Technische Hochschule Karl-Marx-Stadt, 7. 5. — 10. 5. 1973.

[156] *Mayer, H.:* in "Advances in Epitaxy and Endotaxy" (Eds.: H. G. Schneider,V. Ruth), VEB Deutscher Verlag für Grundstoffindustrie, Leipzig (1971) p. 63.

[157] *Neuhaus, A.:* Fortschritte der Mineralogie 29/30, (1950/51) 136.

2. THE FORMATION OF EPITAXIAL SEMICONDUCTOR FILMS BY CHEMICAL VAPOR DEPOSITION (CVD)

2.1. HETEROGENEOUS NUCLEATION IN CHEMICAL VAPOR DEPOSITION SYSTEMS

C. M. Jackson

Nonferrous Metallurgy Division, Batelle, Columbus Laboratories, Columbus, Ohio, USA

J. P. Hirth

Department of Metallurgical Engineering, The Ohio State University, Columbus, Ohio, USA

2.1.1. INTRODUCTION

The processes that utilize vapor–solid transformations for the formation of a deposit on a substrate can be divided into two major classes, *Physical Vapor Deposition* (PVD) and *Chemical Vapor Deposition* (CVD). The primary difference between the two processes involves the relationship between the material that arrives at the substrate from the source and the material of which the deposit is composed; in physical vapor deposition the two materials are the same, while in chemical vapor deposition they are not. In chemical vapor deposition the source material undergoes a chemical reaction (often, thermal decomposition or disproportionation) at the substrate. In the simplest case, one product of the reaction is deposited and the other is liberated as a gas. CVD is usually performed at near atmospheric pressure in an open or closed chamber in which matter is transported via a temperature, pressure, or concentration gradient [1]. Excluding reactions within the vapor phase, the overall sequence of possible steps is: transport to the boundary layer; diffusion through the boundary layer at the substrate; adsorption of reactants, products, and carrier gas on the substrate; and the deposition reaction which can include surface diffusion, nucleation, diffusional growth and agglomeration. At various stages of the reaction, local equilibrium[1] may or may not obtain.

In both CVD and PVD, the deposit is usually formed by a mechanism that involves nucleation and growth. The term *nucleation* refers to the process whereby critical-sized clusters (called critical nuclei) of atoms, adsorbed on the substrate, gain an atom and thereby become supercritical. The importance of the critical nucleus is that the growth of clusters of this size or larger is energetically favorable, whereas the growth of clusters of smaller sizes is not. As supercritical clusters grow, they unite with others to form a macroscopic deposit. So it is that nucleation presents an energy barrier (the energy required to form a nucleus) to the formation of deposits, and also plays a role in the growth of successive overlayers.

Nucleation is particularly important in cases where the degree of supersaturation, *i.e.*, the chemical driving force, is relatively small. In this type of situation nucleation

---

[1] The concept of local equilibrium is that chemical reactions are so rapid that although the system as a whole is not at equilibrium, a small (but still macroscopic) volume element of the system is at equilibrium.

may become rate controlling, so that growth of the deposit can occur at less than the maximum realizable rate as determined by vapor-phase mass transfer considerations. Furthermore, under low supersaturations the nucleation and growth of particular orientations can predominate, as in epitaxy. On the other hand, under conditions of high supersaturation the nucleation rate is so high that growth rather than nucleation is the rate-controlling step. Under these conditions, growth tends to produce a more or less randomly-oriented, fine-grained metal aggregate [2], and growth rates tend to approach the maximum realizable growth rate.

2.1.2.   KINETICS OF HETEROGENEOUS NUCLEATION IN PVD

Heterogeneous nucleation is defined as nucleation that occurs on a substrate of some type. The substrate need not be an intentional one; dust particles have been found to serve as effective substrates. In fact, it has been suggested [3] that ions or charged clusters may have served as unintentional substrates in research previously considered to represent homogeneous nucleation. Although the exact shape of nuclei is not known, some assumption regarding their shape must be made if nucleation kinetics are to be developed. Both spherical cap-shaped and circular disk-shaped nuclei have been assumed and are discussed. Atomistic variants and other geometric shapes can be readily treated in a similar manner by the introduction of the appropriate geometric factors.

2.1.2.1.   **Spherical cap-shaped nuclei**

Assuming the existence of monomers, subcritical clusters and spherical cap-shaped critical nuclei on a surface, both absolute rate theory [4] and classical nucleation theory [5], [6] express the nucleation rate, J, as the product of the concentration of critical nuclei, n_i^*, and the frequency, ω, with which monomers join nuclei, *i.e.*,

$$J = \omega n_i^*. \tag{1}$$

The concentration of critical nuclei is given by

$$n_i^* = q \, \exp \left(\frac{-\Delta G^*}{kT} \right), \tag{2}$$

where q is the number of substrate sites per unit area, ΔG^* is the free energy of formation of a critical nucleus, and k and T have their usual meanings. ΔG^* depends on Θ, the contact angle between the nucleus and the substrate; σ, the specific interfacial free energy of the nucleus–vapor interface; and ΔG_v, the bulk *Gibbs free energy* change per unit volume of condensed phase, and has the form

$$\Delta G^* = \frac{16\pi \, \sigma^3 f(\Theta)}{3\Delta G_v^2} \tag{3}$$

80

where

$$f(\Theta) = \frac{(2 + \cos \Theta)(1 - \cos \Theta)^2}{4} \qquad (4)$$

ΔG_v is given by

$$\Delta G_v = \frac{-kT}{\Omega} \ln \left(\frac{p}{p_e}\right), \qquad (5)$$

in which Ω is the atomic volume of the vapor species at the temperature of the nucleus, p is the pressure of the vapor species,[1] and p_e is the equilibrium pressure of the vapor species at the temperature of the nucleus. The ratio $\frac{p}{p_e}$ is known as the supersaturation ratio, and is designated as S,

$$S = \frac{p}{p_e}. \qquad (6)$$

If the monomers join critical nuclei by surface diffusion, the frequency with which they join is

$$\omega = na \, 2\pi r^*(\sin \Theta)v \, \exp \left(\frac{-\Delta G_d^*}{kT}\right) \qquad (7)$$

where n is the concentration of adsorbed monomers, a is the jump distance, r^* is the radius of a critical nucleus, v is the vibration frequency of adsorbed monomers, and ΔG_d^* is the activational *Gibbs free energy* for a surface jump. The *Gibbs–Thomson* relationship yields

$$r^* = \frac{-2\sigma}{\Delta G_v} \qquad (8)$$

n, the concentration of adsorbed monomers, is given by

$$n = \frac{p\alpha_c}{(2\pi \, m \, kT)^{1/2}} \exp \frac{\Delta G_{des}^*}{kT}. \qquad (9)$$

ΔG_{des}^* is the activational *Gibbs free energy* for desorption of an adsorbed monomer, α_c is the condensation coefficient of the monomer, and m is the mass of a monomer. Thus, the rate of heterogeneous nucleation is given by

$$J = \frac{-\Omega \alpha (\sin \Theta) \, qpa \, \Delta G_v}{2kT \, [2\pi \, m\sigma f(\Theta)]^{1/2}} \exp \left[\frac{\Delta G_{des}^* - \Delta G^* - \Delta G_d^*}{kT}\right] \qquad (10)$$

[1] This pressure may be the pressure of a homogeneous vapor surrounding the substrate, or the effective pressure imposed by means of a vapor beam flux.

Equation (10) contains the *Zeldovich factor* [3], which provides for the tendency for the equilibrium concentration of nuclei to become depleted because of both the finite rate of nucleation and the loss of atoms or molecules, as a multiplier. A more detailed derivation of the nucleation rate equation is given elsewhere [3], [7], [8].

In analyzing the data from heterogeneous nucleation experiments using eq. (10), the nucleation rate is taken as a sensible value, usually 1 nucleus cm^{-2} s^{-1}. For such a rate, eq. (10) becomes

$$\left(\frac{1}{\Delta G_{v\text{crit}}}\right)^2 = \frac{3}{16\pi\,\sigma^3 f(\Theta)} \left\{ kT\left[\ln C + \ln\left(\frac{-p\Delta G_{v\text{crit}}}{T}\right)\right] + \Delta G^*_{\text{des}} - \Delta G^*_{\text{d}} \right\} \tag{11}$$

where $C = \dfrac{\Omega g \alpha_c(\sin\Theta)qa}{2k[2\pi\,m\sigma f(\Theta)]^{1/2}}$. $\Delta G_{v\text{crit}}$ is the bulk free energy change per unit volume for a nucleation rate of 1 nucleus cm^{-2} s^{-1} and is given by

$$\Delta G_{v\text{crit}} = \frac{-kT}{\Omega}\ln S_{\text{crit}}, \tag{12}$$

where S_{crit} is termed the critical supersaturation ratio. Inasmuch as nucleation rate is sensitive to small changes in supersaturation [6] (at least in some systems), defining the critical supersaturation ratio as that which corresponds to the particular nucleation rate of 1 nucleus cm^{-2} s^{-1} is permissible. The critical supersaturation varies with temperature and depends on the particular condensate-substrate set being considered. The test of eq. (11) is to plot $\left(\dfrac{1}{\Delta G_{v\text{crit}}}\right)^2$ versus $T\cdot\left[\ln C + \ln\left(\dfrac{-p\Delta G_{v\text{crit}}}{T}\right)\right]$.

If eq. (10) reflects the true state of affairs on the substrate, the plot should be a straight line with a slope of $\dfrac{3k}{16\pi\,\sigma^3 f(\Theta)}$ and an intercept at $T\left[\ln C + \ln\left(\dfrac{-p\Delta G_{v\text{crit}}}{T}\right)\right] = 0$ of $\dfrac{3(\Delta G^*_{\text{des}} - \Delta G^*_{\text{d}})}{16\pi\,\sigma^3 f(\Theta)}$.

This nucleation rate expression (eq. (10)) has been tested using data from experiments on the heterogeneous nucleation of a variety of materials (Cd on Cu, [7], Na on CsCl, [9], Zn on glass, [10], and H_2O on Ag, [11]). In all of these cases the aforementioned plot was a straight line. Moreover, the values of Θ and ΔG_{des} that were deduced from the plot appeared reasonable for experimental systems that were not perfectly clean. Thus, eq. (10) appears to offer a valid interpretation of heterogeneous nucleation for these condensate-substrate sets.

2.1.2.2. Circular disk-shaped nuclei

Although the preceding formalism, which considers the critical nucleus to have the shape of a spherical cap, may apply in the cases just mentioned, the agreement with experimental data is by no means universal. For instance, when experimental data for certain other condensate-substrate sets (Zn on Pyrex glass, [12], [13] Cr_2J_4 on quartz, [14], [15] and Cr_2J_4 on single-crystal sapphire [14], [16] were plotted in the manner that should have given a straight line if the cap-shaped-nucleus formalism holds, curved plots were instead obtained. On the other hand, a good fit with theory (*i.e.*, a straight line) is obtained in these cases when the critical nuclei are assumed to have the shape of a circular disk, one monomer high [12] to [16]. For the disk-shaped nucleus, eq. (11) is replaced by the expression

$$\left[\frac{-1}{\Delta G_{v_{crit}} + \frac{\Sigma \sigma}{h}} \right] = \frac{hk}{\pi \varepsilon^2} \left[T(\ln C' + \ln p) + \frac{\Delta G_{des}^* - \Delta G_d^*}{k} \right], \tag{13}$$

where h is the height of the disk, ε is the nucleus-vapor edge energy, $\Sigma \sigma$ is the sum of the specific interfacial free energies of the condensate-vapor and condensate-substrate interfaces, minus that of the substrate-vapor interface. C' is an abbreviation

$$C' = Zr^*gaq \, \alpha_c \left(\frac{2}{mkT} \right)^{1/2} \tag{14}$$

where Z is the *Zeldovich factor* for disk-shaped nuclei.
The derivation of eq. (13) is given in [12] and [14].

2.1.3. ALTERNATE ATOMISTIC ASSUMPTIONS AND THEIR CORRESPONDING NUCLEATION-RATE EQUATIONS

The preceding equations, both for a cap-shaped nucleus and for a disk-shaped nucleus are predicated upon the following assumptions:

a) equivalence of impingent and desorption fluxes,
b) metastable equilibrium between adsorbed monomers and critical nuclei, and
c) a surface-diffusion mechanism for the addition of monomers to critical nuclei.

In any given situation, it is not immediately apparent which, if any, of these considerations hold.
Accordingly, nucleation-rate equations have been developed for ten different atomistic mechanisms; none of these mechanisms includes all three of these assumptions [14], [16]. Moreover, several of these mechanisms involve the activation of translational and/or rotational degrees of freedom in the nuclei, rotational degrees of freedom in the adsorbed monomers, and/or the addition of single molecules to nuclei by means of diffusion in a two-dimensional gas. These phenomena are particularly likely when the substrate temperature is relatively high, as in CVD systems.
The major assumptions made in the ten mechanisms and the nucleation-rate equations appropriate for each are as follows:

(1) Impingent and desorption fluxes are equal. Adsorbed monomers and critical nuclei are in metastable equilibrium; nuclei neither rotate nor translate.

6*

Monomers are added to critical nuclei by direct impingement from the vapor phase.

$$J = \frac{\Omega \sigma^{1/2}(1 - \cos \Theta)qp}{[2\pi\, m\, \mathrm{f}(\Theta)]^{1/2}\, kT} \exp \frac{-\Delta G^*}{kT}.$$ (15)

(2) Impingent and desorption fluxes are equal. Adsorbed monomers and critical nuclei are not in metastable equilibrium (surface diffusivity ≈ 0).
Monomers are added to critical nuclei by direct impingement from the vapor phase.

$$J = \frac{\Omega \sigma^{1/2}(1 - \cos \Theta)pq}{[2\pi\, m\, \mathrm{f}(\Theta)]^{1/2}\, kT} \left[\frac{p}{(2\pi\, m\, kT)^{1/2}\, vq} \exp \frac{\Delta G^*_{\mathrm{des}}}{kT} \right]^{-\frac{32\pi\sigma^3 \mathrm{f}(\Theta)}{3\Omega(\Delta G_{\mathrm{v}})^3}}.$$ (16)

(3) Desorption flux $= 0$; adsorbed monomer concentration is time dependent. Adsorbed monomers and critical nuclei are in metastable equilibrium; nuclei neither rotate nor translate.
Monomers are added to critical nuclei by surface diffusion.

$$J = \frac{-\Omega(\sin \Theta)av\, qt\, F_{\mathrm{imp}}(\Delta G_{\mathrm{v}})}{2[kT\, \sigma\, \mathrm{f}(\Theta)]^{1/2}} \exp \frac{-(\Delta G^*_{\mathrm{d}} + \Delta G^*)}{kT}$$ (17)

where

$$\Delta G_{\mathrm{v}} = \frac{-kT}{\Omega} \ln S_{\mathrm{t.d.}} = \frac{-kT}{\Omega} \ln \left(S\, tv\, \exp \frac{-\Delta G^*_{\mathrm{des}}}{kT} \right)$$ (18)

with
$S_{\mathrm{t.d.}}$ = supersaturation ratio corresponding to time-dependent adsorption,
S = supersaturation ratio as defined by eq. (6),
t = time duration of experiment,
F_{imp} = impingent flux of species being nucleated.

(4) Desorption flux $= 0$; adsorbed monomer concentration is time-dependent. Adsorbed monomers and critical nuclei are in metastable equilibrium; nuclei neither rotate nor translate.
Monomers are added to critical nuclei by direct impingement from the vapor phase.

$$J = \frac{\Omega \sigma^{1/2}(1 - \cos \Theta)F_{\mathrm{imp}}\, q}{[kT\, \mathrm{f}(\Theta)]^{1/2}} \exp \frac{-\Delta G^*}{kT},$$ (19)

where ΔG_{v} and F_{imp} are as defined with eq. (18).
(5) Desorption flux $= 0$; adsorbed monomer concentration is time dependent. Adsorbed monomers and critical nuclei are not in metastable equilibrium (surface diffusivity ≈ 0).

84

Monomers are added to critical nuclei by direct impingement from the vapor phase.

$$J = \frac{\Omega(1 - \cos\Theta)^{1/2}qF_{imp}}{[kT\,f(\Theta)]^{1/2}} \left[\frac{p}{(2\pi\,m\,kT)^{1/2}vq} \exp\frac{\Delta G^*_{des}}{kT}\right]^{-\frac{32\pi\sigma^3 f(\Theta)}{3\Omega(\Delta G_v)^3}}, \tag{20}$$

where ΔG_v and F_{imp} are as defined with eq. (18).

(6) Impingent and desorption fluxes are equal. Adsorbed monomers and critical nuclei are in metastable equilibrium; nuclei rotate. Monomers are added to critical nuclei by surface diffusion.

$$J = -8\left(\frac{2\Omega}{15m}\right)^{1/2} \frac{\pi a(\sin\Theta)\sigma^2 p^2}{hv\,kT(\Delta G_v)^{3/2}} \exp\frac{2\Delta G^*_{des} - \Delta G^*_d - \Delta G^*}{kT}, \tag{21}$$

where h is *Planck's constant*.

(7) Impingent and desorption fluxes are equal. Adsorbed monomers and critical nuclei are in metastable equilibrium; nuclei translate. Monomers are added to critical nuclei by surface diffusion.

$$J = \frac{16[2m\,f(\Theta)]^{1/2}a(\sin\Theta)\pi^{3/2}\sigma^{5/2}p}{3(h\Delta G_v)^2} \exp\frac{\Delta G^*_{des} - \Delta G^* - \Delta G^*_d}{kT}. \tag{22}$$

(8) Impingent and desorption fluxes are equal. Adsorbed monomers and critical nuclei are in metastable equilibrium; nuclei rotate and translate. Monomers are added to critical nuclei by surface diffusion.

$$J = \frac{-1024}{3}\left(\frac{kT}{15\Omega}\right)^{1/2} \frac{a(\sin\Theta)\,f(\Theta)m\pi^{7/2}\sigma^5 p}{h^3(\Delta G_v)^{9/2}} \exp\frac{\Delta G^*_{des} - \Delta G^* - \Delta G^*_d}{kT}. \tag{23}$$

(9) Impingent and desorption fluxes are equal. Adsorbed monomers and critical nuclei are in metastable equilibrium; nuclei rotate and translate. Monomers are added to critical nuclei by diffusion in a two-dimensional gas.

$$J = \frac{1024}{9}\left(\frac{m}{30\Omega}\right)^{1/2} \frac{(\sin\Theta)\,kT\,f(\Theta)\pi^3\sigma^5 p}{vh^3(\Delta G_v)^{9/2}} \exp\frac{\Delta G^*_{des} - \Delta G^*}{kT}. \tag{24}$$

(10) Impingent and desorption fluxes are equal. Adsorbed monomers and critical nuclei are in metastable equilibrium; nuclei neither rotate nor translate. Monomers are added to critical nuclei by diffusion in a two-dimensional gas; monomers rotate.

$$J = \frac{-\alpha'\Omega q(\sin\Theta)(\Delta G_v)p}{4\pi\,m\,v[\sigma f(\Theta)kT]^{1/2}} \exp\frac{\Delta G^*_{des} - \Delta G^*}{kT} \tag{25}$$

where α' is the condensation coefficient for a monomer joining a critical nucleus.

Equations (15) through (25) provide ten additional formalisms against which heterogeneous nucleation data can be tested. The derivations of these equations are given elsewhere [16].

Even these equations do not exhaust the list. For small critical-sized nuclei, a bulk description of ΔG^* is inapplicable and an alternate description in terms of atomistic models and statistical mechanical partition functions is more appropriate ([17]–[20]). Aside from minor geometric corrections to the nucleus circumference and area, the changes all reside in ΔG^*. Thus with the atomistic ΔG^*, the various other modifications of this section still apply.

Following the work of *Lewis* and *Campbell* [21], although work along similar lines preceded it ([22]–[24]), there has been a burst of activity related to nucleation processes and sometimes referred to as nucleation theory. Recent work is reviewed in [25]–[28]. In actuality this work focuses on supercritical cluster growth and the kinetics include nucleation, growth and agglomeration of islands on a substrate. Insofar as nucleation kinetics enter these processes, the above descriptions hold. Indeed, when the nucleation kinetics are separated out of the overall kinetics, agreement with theory is fair but not complete [29]; for other cases analyzed in terms of nucleation theory, it is clear that growth processes have intruded into the analysis so that the more general kinetic analysis is required ([21]–[29]).

2.1.4. KINETICS OF HETEROGENEOUS NUCLEATION IN CVD SYSTEMS

The primary modification that must be made in extending the kinetics of heterogeneous nucleation to the CVD case lies in the method of calculating the supersaturation ratio, S, in the system. This calculation is not as straightforward as in the case of PVD systems because the pressure of the material being nucleated is not readily measured. This pressure is the result of a chemical compound undergoing a reaction at the substrate, and liberating a material of high vapor pressure as well as the material being nucleated. The pressure of the material being nucleated is related to the equilibrium constant for the chemical reaction (usually, thermal decomposition or reduction) that is taking place. Consider, as an example, the dissociation reaction

$$MX(g) \rightleftarrows M(g) + X(g). \tag{26}$$

The equilibrium constant, K_e, for this reaction, assuming that the gases behave ideally, is

$$K_e = \frac{(p_M)(p_X)}{p_{MX}}. \tag{27}$$

At equilibrium,

$$p_M = \frac{(K_e)(p_{MX})}{p_X} = \frac{p_{MX}}{p_X} \exp \frac{-\Delta G^0}{kT}, \tag{28}$$

86

where ΔG^0 is the *standard free energy* of the dissociation reaction (eq. (26)). Therefore, the supersaturation is

$$S = \frac{p_M}{p_{M_e}} = \frac{p_{MX}}{p_{M_e} p_X} \exp \frac{-\Delta G^0}{kT}, \qquad (29)$$

where p_{M_e} is the equilibrium vapor pressure of the M species at the temperature of the nucleus. The supersaturation ratio for a reduction reaction can be deduced in a similar manner.

Thus, if the assumption is made that the impingent fluxes of MX and X from the vapor source equilibrate at the substrate, the effective pressures of these two constituents (calculated from the impingent fluxes) can be inserted into eq. (29) and the supersaturation ratio can be calculated.[1] The procedure that is used when this assumption is invalid is discussed in connection with the experiments on the CVD of chromium from chromous iodide.

In dealing with heterogenous nucleation in CVD systems, consideration should be given to the possibility of the reactant and/or the gaseous product of the chemical reaction functioning as impurities on the substrate, and affecting nucleation kinetics. For instance, the adsorption of these materials on the substrate can decrease the activation energy for desorption, decrease the condensation coefficient, and decrease the nucleus-vapor interfacial free energy; that is, the effects of these materials would be the same as expected for the adsorption of an exogenous impurity. The result, in most cases, would be an increase in the pressure, p, and the supersaturation ratio, S, required for a given nucleation rate.

Other factors in the nucleation rate equations also change, but the appropriate substitutions are obvious: *e.g.*, ΔG_d^* for one of the reactants replacing that of the monomeric deposit species if dissociation of reactant occurs at the stage of growth of a nucleus. Specific examples are given in [2], [14], [30–33].

2.1.5.   HETEROGENEOUS NUCLEATION OF EPITAXIAL DEPOSITS
          IN CVD SYSTEMS

The formation of epitaxial films is important in CVD systems as it is in PVD systems. Epitaxy can occur as a result of the way in which nuclei, which have no epitaxial relationship to the substrate, agglomerate and grow to form macroscopic deposits; or it can occur as a result of *epitaxial nucleation*. The latter subject is considered here, using the theory previously outlined for the heterogeneous nucleation of cap-shaped nuclei at an average surface concentration of much less than a monolayer.

In order for epitaxial nucleation to occur, the nucleation rate for the epitaxial orientation must be greater than that for other orientations. Denoting the nucleation rate for an epitaxial orientation, Θ_1, as J_1, and that for a nonepitaxial orientation Θ_2, as J_2, from eq. (10) the ratio, β, of the nucleation rates for the two orientations at the

[1] In customary CVD systems, the impingent flux of the nucleating species, M, is small enough o be ignored.

critical supersaturation is

$$\beta = \frac{J_1}{J_2} = \frac{\sin \Theta_1}{\sin \Theta_2} \left[\frac{f(\Theta_2)}{f(\Theta_1)} \right]^{1/2} \exp \{B[f(\Theta_2) - f(\Theta_1)]\} \,, \tag{30}$$

where

$$B = \frac{16\pi\sigma^3}{3 \Delta G^2_{vcrit} kT} \,. \tag{31}$$

The specific interfacial free energy of the nucleus-substrate interface is presumably smaller in epitaxial nucleation than in nonepitaxial nucleation; hence, $\Theta_2 > \Theta_1$ and $f(\Theta_2) > f(\Theta_1)$. Since ΔG_{vcrit} decreases rapidly with increasing temperature, B and β increase rapidly with increasing temperature. Thus, epitaxial nucleation is favored at higher temperatures. Inasmuch as CVD generally requires high substrate temperatures, CVD systems are inherently favorable for epitaxial nucleation.

When the contact angle is very small, e.g., less than perhaps 5 or 10 degrees, the model on which eq. (30) is based is no longer applicable [34]. Also, at small angles for cap-shaped nuclei line tension terms become important [35]. Moreover, nucleus models other than the cap-shaped one become increasingly important at small contact angles. The afore-mentioned possible effect of the reactant or the gaseous product of the chemical reaction functioning as indigenous impurities is also important in epitaxy. Impurities, whether indigenous or exogenous, can make epitaxial nucleation less favorable by lowering σ and, therefore, B and β. Impurity adsorption can also affect Θ. Based on the above development, the primary requisites for epitaxial nucleation are [30]:

a) A low index substrate plane, a substrate with close lattice match with the condensate, and a substrate that strongly interacts with the condensate, all factors which favor a low Θ for epitaxial nucleation relative to Θ_2 and, hence, a large $\Delta f(\Theta)$ in eq. (30).
b) A low supersaturation (high substrate temperature), which tends to prevent nucleation of other orientation that have higher Θ values, and which prevents incoherent crystal growth following nucleation, a type of growth that would destroy epitaxy.
c) Use of a molecular-beam system, if possible, rather than a flow system, because in a beam system the reaction product X is more efficiently removed. If not removed, adsorbed X can act as an indigenous impurity, and promote both nucleation of other orientations with higher Θ values and incoherent growth once nucleation has occurred.

We note that there are alternate views of the role of nucleation in epitaxy [36], [37]. In these cases, the orientation of small three- or four-atom critical nuclei establish the final epitaxial relation. As discussed extensively for PVD cases, [3], [28], [34], the former model is favored because epitaxy is usually observed at high substrate temperatures and low supersaturations where the critical sized nucleus is so large, up to 10^4 atoms, that the capillarity model on which eq. (30) is based should definitely apply. The *Walton* models [36], [37] should apply at low temperatures and high supersaturations where the critical size is as small as three or four atoms, but there, as indicated by eq. (30) or its atomistic equivalent [20] the nucleation of non-epitaxially oriented nuclei becomes highly favored. Finally, it is well established that growth and agglomeration processes play a role in epitaxy [3], [21–[29], [38]; the part of nucleation in overall epitaxy is, however, well established [28], [34], [38]. While most of this discussion centers on PVD, all of the conclusions with regard to the range of supersaturation and substrate temperature for epitaxial nucleation apply as well to CVD systems.

2.1.6.   **APPLICATION OF HETEROGENEOUS NUCLEATION THEORY TO THE CVD OF CHROMIUM FROM CHROMOUS IODIDE**

The application of heterogeneous nucleation theory to a specific system, the CVD of chromium from chromous iodide, is now considered. The experimental method consisted of impinging a molecular beam from a *Knudsen cell* containing chromous iodide onto a heated single-crystal sapphire substrate. By scanning the effusant with a time-of-flight mass spectrometer, it was found that chromous iodide effuses as the dimer, Cr_2J_4, rather than as the monomer, CrJ_2 [39].

In order to minimize the possibility that impurity adsorption might have affected the nucleation, the experiments were carried out in an ion-pumped ultrahigh vacuum system capable of attaining pressures of 10^{-9} torr. Ion pumping was chosen in order to avoid the possibility of contamination by oil or mercury derived from pumps employing such fluids. The apparatus consisted of a pumping section and a deposition chamber, separated by a large trap cooled with liquid nitrogen to collect evolved iodine. Details of the experimental procedure, and a photograph of the apparatus utilized, are given elsewhere [15], [24].

The values deduced for the critical supersaturation ratio in CVD depend strongly on the assumptions made regarding the condensation and desorption behavior of the species involved. In the present research, the condensation coefficients of chromous iodide and iodine are pertinent, as are the desorption behaviors of chromium, chromous iodide, and iodine.

From data obtained during the calibration of the impingent flux during this research, it was concluded that the condensation coefficient of chromous iodide on sapphire is essentially unity at the temperatures studied, 116 to 164°C [14], [15]. Others have found the condensation coefficient of iodine (on Pyrex glass) to be about 0.01 at 0°C and to decrease with increasing temperature [40]. Accordingly, it appears that all of the impingent chromous iodide, but essentially none of the impingent iodine, adsorbed on the substrate in the present experiments. Furthermore, at the high temperatures involved in this research it is likely that the mean stay time of iodine adatoms and admolecules would be extremely short. Thus, it appears that iodine from the *Knudsen cell* and iodine from the decomposition of chromous iodide at the substrate would have little, if any, effect on the extent of the decomposition reaction on the substrate. Furthermore, the impingent flux of chromium from the cell is many orders of magnitude smaller than that of chromous iodide and, therefore, can be ignored.

Accordingly, instead of determining the effective pressures of both chromous iodide and iodine at the substrate and substituting them for p_{MX} and p_X, respectively, in eq. (29) in order to calculate the supersaturation ratio, the effective pressure of chromous iodide alone was utilized. Since calculations indicate that the dissociation of chromous iodide should be greater than 99.9% at the high temperatures and very low pressures involved, the decomposition of chromous iodide was assumed to be complete and the impingent flux of chromium was taken as twice the impingent flux of chromous iodide. This flux was then converted into an effective pressure of chromium at the substrate, the supersaturation ratio being this pressure divided by the equilibrium vapor pressure of chromium at the substrate temperature.[1]

Fig. 1. Electron micrograph of a replica of a CVD deposit of chromium formed on a sapphire substrate at 648°C

The experiments involved exposing a single-crystal sapphire substrate to the vapor beam for a heated *Knudsen cell*[2] containing chromous iodide, for a period of 1 hour. If no deposit appeared in that time, the temperature of the cell was increased to the extent necessary to increase the flux from the cell by a factor of five. This process was repeated at as many flux levels as necessary to produce a deposit.

Chromium was nucleated at substrate temperatures from 648 to 939°C. Experiments at lower temperatures, from 105 to 540°C, resulted either in the nucleation of chromous iodide (*i.e.*, PVD), or the formation of no deposit at all; these experiments are described elsewhere [14], [16].

An electronmicrograph of a replica of a representative deposit of chromium formed in this research is given in Fig. 1. It may be seen that the deposit, although observable with the unaided eye during deposition, is not continuous but consists of a large number of supercritical nuclei. Furthermore, the fact that very fine scratches on the polished sapphire substrate are more potent catalysts for heterogeneous nucleation is obvious from this figure. This increased potency arises because less energy is re-

[1] To be exact, the equilibrium vapor pressure of chromium should be evaluated at the temperature of the nucleus. This temperature, although not known, is assumed to be reasonably close to the temperature of the substrate as suggested by the work of *Feder et al.* [41].
[2] The *Knudsen cell* was fabricated from 99.7 + percent pure recrystallized alumina.

90

quired to form a nucleus at such surface irregularities than is required on a smooth surface.

It has already been noted that one would expect the tendency towards epitaxial nucleation to increase with increasing substrate temperature in most CVD systems. This is exactly what was observed in the present research. Specifically, at substrate temperatures of 648 and 739°C the deposit consisted entirely of randomly oriented fine crystals of chromium. At 842 and 939°C, however, in addition to randomly oriented fine crystals of chromium there were also a number of relatively large chromium crystals that exhibited a preferred orientation with respect to the substrate. The data were analyzed in terms of nucleation theory by converting at each temperature, the smallest impingent flux that resulted in an observable deposit to a pressure of chromium, and defining the associated supersaturation ratio as the critical supersaturation ratio for nucleation at that temperature.

Given in Fig. 2 is the relationship between the natural logarithm of the critical supersaturation ratio for the nucleation of chromium from chromous iodide on single-crystal sapphire, as a function of the reciprocal of the substrate temperature. A curve of this form would follow from eqs. (10) or (15) to (25) if the value of ΔG^* for nucleation were nominally constant. Then from eq. (3), ΔG would be constant and, from eqs. (5) and (6), $\ln S$ should vary linearly with $\left(\dfrac{1}{T}\right)$ as observed. Thus, Fig. 2 is in rough accord with expectation from nucleation theory.

However, when the experimental data are put into the form of eq. (11), corresponding to the mechanism of surface diffusion to a cap-shaped nucleus, and plotted (Fig. 3) it is found that there is no linearity upon which one might base an extrapolation and arrive at a value of $(\Delta G^*_{des} - \Delta G^*_d)$. Alternatively, one can fit the data to the disk model (eq. (13)). Here a fit would be obtained, but only for the absurd value $\Sigma\sigma$ of $-16\,000$ ergs cm^{-2}. There is a precedent for the above in that a number of plots of PVD nucleation data have curvatures of the form of Fig. 3 [34]. As discussed recently [28], PVD data which fit eq. (11) tend to be for cases where the substrate is isotropic, either because of adsorption of a tightly bound monolayer of the depositing substance or because of adsorption of gaseous impurities. Cases of deposition onto clean substrates in ultrahigh vacuum often fail to fit eq. (11). The formalism of eq. (11) can be forced onto the data in a number of ways, including a decrease in contact angle or of σ with increasing temperature, increasing lack of thermal accommodation with increasing temperature, and incomplete dissociation of chromous iodide at the lower substrate temperatures. In view of other observations in PVD systems, however, the most plausible explanation appears to be noncritical nucleation associated with a spectrum of sites on the substrate with varying degrees of catalytic potency for nucleation. The configuration of clusters in Fig. 1 suggests preferential nucleation on steps on the substrate, perhaps at kink sites. Nuclei could have formed at supersaturations lower than the apparent one at low substrate temperatures, but, because of decreased agglomeration tendencies at low temperatures, not appeared to have nucleated with the technique used in the experiments.

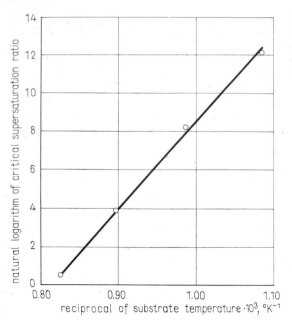

Fig. 2. Natural logarithm of critical supersaturation ratio versus reciprocal of substrate temperature for the nucleation of chromium from chromous iodide on single-crystal sapphire

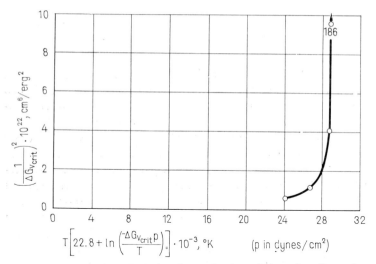

Fig. 3. CAP nucleus analysis for the nucleation of chromium from chromous iodide on single-crystal sapphire

2.1.7.   OTHER CVD NUCLEATION EXPERIMENTS

There have been a number of observations of the overall nucleation, growth, and agglomeration processes which indirectly apply to nucleation. In particular, results on epitaxial CVD deposition are pertinent, [32], and [42]–[44]. There has been much work on the epitaxial deposition of silicon on silicon, chiefly by the reduction of

92

silane, as reviewed by *Chapelle et al.* [42]. Notable among these researches are the work of *Widmer* [45], who showed the marked effect of ultrahigh vacuum in decreasing the temperature range for epitaxy, analogous to the PVD work of *Matthews* [24], and of *Bhola* and *Mayer* [46], who found epitaxial nucleation above 1 050°C but nonepitaxial nucleation below 1 000°C. *Widmer's* findings tend to indicate the possible role of noncritical defect nucleation in CVD analogous to the PVD cases cited above [47]–[49]. The latter work [46] tends to support the capillarity-based view of the role of nucleation in epitaxy, agreeing with criterion b) above. A more rigorous treatment of the epitaxial deposition of silicon is given by *Riedl* in this book [44].

A very extensive work is that of *Joyce et al.* [43] (a series of earlier papers by *Joyce* and coworkers is cited in the reference) on the epitaxial deposition of silicon on silicon from a molecular beam of silane. For both {111} and {100} deposition, an induction period was observed which was associated with the reduction of an SiO_2 surface film [43]. Following the induction period, the nucleation rate was extremely rapid for most cases. Consequently, most of the kinetic analysis dealt with the saturation density of nuclei, which involves the extended nucleation-growth-agglomeration theories rather than nucleation alone [21]–[29]. With regard to the purely nucleation theories, the results are interesting in indicating the extreme sensitivity of nucleation rate to surface condition.

2.1.8.   **SUMMARY**

a) Experimental data for heterogeneous nucleation in certain condensate-substrate sets in PVD can be interpreted successfully using formalisms based on cap-shaped nuclei, while data for other condensate-substrate sets in PVD can be interpreted best if the nucleus is assumed to have the shape of a circular disk.

b) Alternate nucleation formalisms have been derived that are not predicated upon the three assumptions (equivalence of impingent and desorption fluxes, metastable equilibrium between adsorbed monomers and critical nuclei, and a surface-diffusion mechanism for the addition of monomers to critical nuclei) that are usually made in heterogeneous nucleation theory. Moreover, several of these mechanisms involve the activation of translational and/or rotational degrees of freedom in the nuclei, rotational degrees of freedom in the adsorbed monomers, and/or the addition of single molecules to nuclei by means of diffusion in a two-dimensional gas. These phenomena are particularly likely when the substrate temperature is relatively high, as in CVD systems.

c) The primary modification that must be made in extending the kinetics of heterogeneous nucleation to the CVD case lies in the method of calculating the supersaturation ratio, S, in the system.

d) On the basis of the capillarity-based theories of nucleation, CVD systems are concluded to be inherently favorable for epitaxial nucleation, because of the high substrate temperatures involved.

e) An epitaxial relationship has been found during the nucleation of chromium crystals from chromous iodide onto single-crystal sapphire substrates in CVD at temperatures of 842 and 939°C. The results agree qualitatively with expectation based upon the capillarity theory of nucleation.

f) Data on the CVD of chromium from chromous iodide onto single-crystal sapphire substrates at temperatures between 648 and 939°C do not fit the customary heterogenous nucleation formalisms based on cap-shaped and disk-shaped nuclei.

g) Most, if not all, nucleation experiments in CVD systems involve growth and agglomeration also in the kinetic observations and analysis. A clear need exists for data on the purely nucleation aspects of the CVD process.

2.1.9.   **REFERENCES**

[1] *Schäfer, H.:* Chemical Transport Reactions. Academic Press, New York (1964).

[2] *Gretz, R. D.:* in: Vapor Deposition (Eds. C. F. Powell, J. H. Oxley and J. M. Blocher) J. Wiley, New York (1966) p. 149.

[3] *Hirth, J. P., Pound, G. M.:* Condensation and Evaporation. Pergamon Press, Oxford (1963).

[4] *Glasstone, S., Eyring, H., Laidler, K. J.:* Theory of Rate Processes. McGraw-Hill Book Co. New York (1941).

[5] *Volmer, M., Weber, A.:* Z. Physik. Chem. 119. (1926) 277.

[6] *Volmer, M.:* Kinetik der Phasenbildung. Steinkopff, Dresden (1939).

[7] *Pound, G. M., Simnad, M. T., Yang, L.:* J. Chem. Phys. 22 (1954) 1215.

[8] *Lothe, J., Pound, G. M.:* J. Chem. phys. 36 (1962) 2080.

[9] *Yang, L., Birchenall, C. E., Pound, G. M., Simnad, M. T.:* Acta Met. 2 (1954) 462.

[10] *Hudson, J. B.:* J. Chem. Phys. 36 (1962) 887.

[11] *Hudson, J. B., Koop, S. A.* (to be published).

[12] *Hruska, S. J.:* Acta Met. 12 (1964) 1211.

[13] *Ruth, V., Moazed, K. L., Hirth, J. P.:* J. Chem. Phys. 44 (1966) 2093.

[14] *Jackson, C. M.:* Doctoral Dissertation, The Ohio State University, 1966. (Available from University Microfilms, P. O. Box 1764, Ann. Arbor, Michigan 48106).

[15] *Gretz, R. D., Jackson, C. M., Hirth, J. P.:* Surface Sci. 6 (1967) 171.

[16] *Jackson, C. M., Hirth, J. P.* (to be published).

[17] *Frenkel, J.:* Z. Phys. 26 (1924) 117.

[18] *Walton, D.:* J. Chem. Phys. 37 (1962) 2182.

[19] *Hirth, J. P.:* Ann. N. Y. Acad. Sci. 101 (1963) 805.

[20] *Halpern, V.:* Brit. J. Appl. Phys. 18 (1967) 163.

[21] *Lewis, B., Campbell, D. S.:* J. Vac. Sci. Tech. 4 (1967) 209.

[22] *Pashley, D. W., Stowell, M. J., Jacobs, M. H., Law, T. J.:* Phil. Mag. 10 (1964) 127.

[23] *Chakraverty, B. K.:* J. Phys. Chem. Sol. 28 (1967) 2413.

[24] *Matthews, J. W.:* Phil. Mag. 12 (1965) 1143.

[25] *Stowell, M. J., Hutchinson, T. E.:* Thin Solid Films 8 (1971) 41.

[26] *Sigsbee, R. A.:* J. Appl. Phys. 42 (1971) 3904.

[27] *Lewis, B.:* Surface Sci. 21 (1970) 273.

[28] *Hirth, J. P.:* Proc. 2nd Int. Conf. Epitaxy, Jerusalem, May, 1972.

[29] *Poppa, H.:* J. Appl. Phys. 38 (1967) 3883.

[30] *Gretz, R. D., Hirth, J. P.:* in: Proceedings of the Conference on Chemical Vapor Deposition of Refractory Metals, Alloys, and Compounds. American Nuclear Society (1967) p. 73.

[31] *Sigsbee, R. A.:* in: Nucleation (Ed.: A. C. Zettlemoyer) Dekker, New York (1969) p. 151.

[32] *Chopra, K. L.:* Thin Film Phenomena. New York Mc-Graw-Hill, (1969).

[33] *Hirth, J. P.:* in: Vapor Deposition (Eds.: C. F. Powell, J. H. Oxley, J. M. Blocher, Jr.) Wiley, New York (1966) p. 126.

[34] *Hirth, J. P., Moazed, K. L.:* in: Physics of Thin Films, Vol. 4. (Eds.: G. Haas and R. E. Thun) Academic Press, New York and London (1967) p. 97.

[35] *Gretz, R. D.:* Surface Sci. 5 (1966) 239.

[36] *Walton, D.:* Phil. Mag. 7 (1962) 1671.

94

[37] *Rhodin, T. N., Walton, D.:* in: Metal Surfaces ASM, Cleveland (1963) p. 259.

[38] *Matthews, J. W.:* Phys. Thin Films 4 (1967) 137.

[39] *Alexander, C. A.* (unpublished research).

[40] *Stern, J. H., Gregory, N. W.:* J. Phys. Chem. 61 (1957) 1226.

[41] *Feder, J., Russell, K. C., Lothe, J., Pound, G. M.:* Advan. Phys. 15 (1966) 111.

[42] *La Chapelle, T. J., Miller, A., Morritz, F. L.:* Prog. Solid State Chem. 3 (1967) 1.

[43] *Joyce, B. A., Bradley, R. R., Watts, B. E.:* Phil. Mag. 19 (1969) 403.

[44] *Riedl, W. J.:* in this book

[45] *Widmer, H.:* Appl. Phys. Lett. 5 (1964) 108

[46] *Bhola, S. R., Mayer, A.,* R. C. A. Review 24 (1963) 511.

[47] *Kenty, J. L., Hirth, J. P.:* Trans. AIME 245 (1969) 2373.

[48] *Hirth, J. P., Moazed, K. L., Ruth, V.,* in: Epitaxie und Endotaxie (Ed: H. G. Schneider) VEB Deutscher Verlag für Grundstoffindustrie, Leipzig (1969) p. 25.

[49] *Kenty, J. L.:* in: Advances in Epitaxy and Endotaxy (Eds.: H. G. Schneider and V. Ruth) VEB Deutscher Verlag für Grundstoffindustrie, Leipzig (1972).

THERMODYNAMIC APPROACH TO DIFFUSION CONTROLLED CHEMICAL VAPOR DEPOSITION IN MULTICOMPONENT FLOW SYSTEMS AND ITS APPLICATION TO THE EPITAXIAL SILICON DEPOSITION FROM (SiCl₄ + H₂) MIXTURES

W. J. Riedl

UNITRA Research and Development Center for Single Crystals, Warsaw, Poland

2.2.1.   MAIN STEPS IN CHEMICAL VAPOR DEPOSITION OF CRYSTAL FILMS

Chemical Vapor Deposition (*CVD*) on substrates or on already present crystal films usually occurs near atmospheric pressure. Excluding reactions in the gas phase, the deposition consists of the following steps:

(a) Transport of reactants through the gas phase to the deposition surface.
(b) Surface processes such as adsorption (both physical and activated) of reactants, chemical reactions, nucleation, crystal growth, and desorption of products.
(c) Transport of final products away from the deposition surface.

The second step is usually very complex. The relative importance of the kinetic and diffusional factors has been discussed by *Spacil* and *Wulff* [1]. Depositions, the rates of which are controlled by steps (a) and (c) or by step (b) only, are referred to as *diffusion controlled* or *kinetically controlled*, respectively.

The terms *diffusional* or *kinetic control* refer not to a deposition rate *per se*, but to the relative resistance offered to deposition by diffusion or by chemically activated steps. It is obvious that under steady-state conditions diffusion and kinetic rates must be equal. Depending on conditions (concerning temperature, pressure, flow dynamics, and geometry) the same CVD process may at one time be controlled by diffusion, but at another time by kinetic steps. For instance, at relatively low temperatures when chemical reaction rates are small, deposition is kinetically controlled. An increase in temperature results in an increase in the rate of surface processes, which finally become so fast that they do not control the total deposition rate and the deposition becomes then diffusion controlled. Similarly, the control mechanism is influenced by a change of pressure; chemical reactions are usually accelerated by an increase in pressure. Hence, many deposition processes are kinetically controlled at low pressure, but diffusion controlled at increased pressure.

If for a definite flow *Reynold's number* is larger than its critical value, the flow becomes turbulent. According to the classical theory, at the gas–solid interface remains a laminar boundary gas layer through which gas components are transported by diffusion and not by convection. The concept of a boundary layer was first suggested by *Noyes* and *Whitney* [2] and later extended by *Nernst* [3]. Although the thickness of such a layer is relatively small, its resistance to transport is much larger than the resistance in the main gas stream; the latter can therefore be neglected.

For the mathematical formalism it is usually assumed that partial pressures or concentrations outside the boundary layer are constant and their gradients exist within the boundary layer only. Hence, the flux of a component i, J_i, is given by

$$J_i = k_{ci}(C_{oi} - C_i) \tag{1}$$

or

$$J_i = k_{pi}(p_{oi} - p_i) \tag{2}$$

where

C_{oi}, p_{oi} are the concentration and the partial pressure of a component i in the main gas stream, respectively,

C_i, p_i are the concentration and the partial pressure of a component i at the gas–solid interface, respectively,

k_{ci}, k_{pi} are proportionality factors or the mass-transfer coefficients.

If pure diffusion occurs across a length δ, at moderate pressure, at which the mean free path is relatively small, the mass-transfer coefficients can be expressed by

$$k_{ci} = \frac{D}{\delta} \cdot \tag{3}$$

and

$$k_{pi} = \frac{DP}{\delta\,RT} \tag{4}$$

where

D is the diffusion constant,
P is the pressure,
R is the gas constant, and
T is the absolute temperature.

Equation (4) has been used by *Schäfer* to correlate the vapor transport of metals by their halides in non-flow closed systems [4], [5]. This equation is usually applicable for pressures higher than 1 torr.

In engineering the effective thickness δ of the boundary layer, related to the mass-transfer coefficient, is used instead of the true thickness. The concept of δ has been explained in Fig. 1. The effective gradient equals then $(C_0 - C)/\delta$. An effective boundary layer theory has been developed in the course of many years and is now widely used in engineering and in technological designing.

Analysis of the gas phase diffusion should provide a method of controlling the morphology and perfection of the layers being deposited, by controlling the supersaturation at the gas–solid interface. Such an analysis requires the knowledge of the ratios of mass-transfer coefficients which are considered in the following.

98

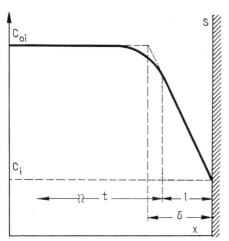

Fig. 1. Concentration distribution (C_{oi}, C_i) in the vicinity of a solid surface for turbulent flow. S represents the solid surface, x is the distance from the surface, l is the width of the laminar zone, t is the region of turbulence, δ is the effective thickness of the boundary layer

2.2.2.   MASS-TRANSFER COEFFICIENTS

Since eqs. (3) and (4) involve the quantity δ, their application in practice is inconvenient, but the mass-transfer coefficients can be found either by direct measurements or by analogy with heat transfer. Methods for their estimation are given below.

2.2.2.1.   Laminar flow

The mass transfer at a laminar flow parallel to the plane of a flat plate obeys the equation [6]:

$$Sh = a\ Re^{1/2}\ Sc^{1/3} \tag{5}$$

where

$$Sh = \frac{k_{ci}\,d}{D} \tag{6}$$

$$Re = \frac{d\bar{w}\,\gamma}{\eta} \tag{7}$$

$$Sc - \frac{\eta}{\gamma D} \tag{8}$$

Sh is the dimensionless *Sherwood number*
Re is the dimensionless *Reynolds number*
Sc is the dimensionless *Schmidt number*
d   is the diameter of the tubular reactor
$\bar{w}$ is the mean lincar velocity of the gas flow

γ is the density of the gas
η is the viscosity of the gas
D is the diffusion constant of the gas
a   is a proportionality factor (approximately a $\approx$ 2/3).

From eqs. (5)–(8) follows:

$$k_{ci} = a \left(\frac{\overline{w}}{d}\right)^{1/2} \cdot \left(\frac{\gamma}{\eta}\right)^{1/6} \cdot D^{2/3} \tag{9}$$

$$k_{pi} = k_{ci} \frac{P}{RT} . \tag{10}$$

The mass-transfer coefficient for the laminar flow is proportional to the square root of the mean linear velocity of gases. This relation has been confirmed experimentally even for very complex systems (*e.g.* in the deposition system resulting from ($SiCl_4$ + + H_2) mixtures [7]).
Chemical vapor deposition of epitaxial films involves gas mixtures in which one component is present in large excess and the others are very diluted. The diluted components scarcely affect each others' diffusion rates. The ratio of the mass-transfer coefficients of two components (see eqs. (9) and (10)) very diluted by a third one is given by

$$\frac{k_{c_1}}{k_{c_2}} = \frac{k_{p_1}}{k_{p_2}} = \left(\frac{D_1}{D_2}\right)^{2/3} . \tag{11}$$

2.2.2.2. Turbulent flow

If for a definite flow *Reynolds number* is larger than its critical value, the flow becomes turbulent. In most cases the existence of a laminar boundary layer at the phase boundary one has been assumed.
Higbie [8] assumed that the mass transfer of gases is associated with the movement of whirlpools formed between the phase boundary and the main stream of flowing gases interrupted by short periods in which non-stationary diffusion occurs. According to this model, the mean velocity of the mas-stransfer depends on the residence time of a whirlpool at the phase boundary, and on the total mass of the diffusing component. Simplifying the problem, *Higbie* assumed the residence times of all whirlpools reaching the phase boundary to be equal to each other. From the penetration theory of *Higbie* it follows that

$$k_{ci} = 2 \sqrt{\frac{D}{\pi \tau_c}} . \tag{12}$$

Usually, the mean residence time τ_c of whirlpools is not known, but if the penetration theory is correct, the mass-transfer coefficients for a turbulent flow should be proportional to the square root of the diffusion constant. In most cases the correlation

100

for the mass transfer may be described by

$$Sh = a\, Re^b\, Sc^c \tag{13}$$

or by the equivalent relation

$$\frac{k_{ci}\, d}{D_i} = a\left(\frac{d\bar{w}\,\gamma}{\eta}\right)^b \left(\frac{\eta}{\gamma D_i}\right)^c \tag{14}$$

hence

$$k_{ci} = a\, d^{b-1}\, \bar{w}^b \left(\frac{\gamma}{\eta}\right)^{b-c} D_i^{1-c}. \tag{15}$$

It has been shown experimentally that k_{ci} is proportional to the square root of the diffusion constant, hence $c = 0.5$ (in contradiction to the theory of the laminar boundary layer, where $c = 0$; eq. (3). It has been shown by numerous experiments that c is closer to 0.5 than to 0. Hence, the interpretation of mass transfer by the penetration theory is more exact than by the theory of the laminar boundary layer. The further development of the penetration theory has been carried out by *Danckwerts* [9], who found also, as *Higbie* did earlier, that the mass-transfer coefficient was proportional to the square root of the diffusion constant. The validity of the penetration theory for the mass transfer in regions near solid surface has been confirmed by *Hanratty* [10].

The mass-transfer coefficients are proportional to $D^{2/3}$ for a laminar flow and to $D^{1/2}$ for a turbulent one. The change of exponent occurs almost exclusively within the region of transition flow and therefore using the exponent 2/3 for a laminar flow and 1/2 for a turbulent one is a fairly good approximation. If more exact results are necessary, the coefficients a, b, c in eqs. (13) and (15) may be determined by experiments. The experimental results, for this purpose, should be plotted in two coordination systems: log Sh versus log Re and log Sh versus log Sc.

2.2.3. DIFFUSION CONSTANTS

2.2.3.1. The effect of temperature

From the elementary kinetic theory of gases follows the proportionality of the diffusion constant to $T^{3/2}$. A more rigorous treatment yields an exponent larger than $\frac{3}{2}$.

Sutherland [11] recommended an equation which yields even more exact results:

$$\frac{D(T_2)}{D(T_1)} = \left(\frac{T_2}{T_1}\right)^{2.5} \frac{T_1 + C_{AB}}{T_2 + C_{AB}} \tag{16}$$

where

C_{AB} is *Sutherland's constant* for the given gas-pair A, B.

For technological purpose it has been adopted by *Arnold* [12], who has also given a convenient way of calculation of *Sutherland's constant*:

$$D_{1.2} = \frac{8.37 \cdot 10^{-3} \, T^{2,5}}{P(V_1^{1/3} + V_2^{1/3})(T + C_{1,2})} \left(\frac{M_1 + M_2}{M_1 \cdot M_2} \right)^{1/2} \tag{17}$$

where

$$C_{1,2} = F\sqrt{C_1 \cdot C_2}, \tag{18}$$

$$C_1 = 1.47 \, T_{s_1}, \tag{19}$$

$$C_2 = 1.47 \, T_{s_2}, \tag{20}$$

$D_{1,2}$ is the diffusion constant for the gas-pair 1, 2;
M_1, M_2 are the molecular weights of the gaseous components 1 and 2, respectively
V_1, V_2 are the molvolumes at normal boiling temperature of the condensed gases 1 and 2, respectively.

T_{s_1}, T_{s_2} are the normal boiling temperatures of components 1 and 2, respectively
F is a factor interpolated from Table 1.

Table 1. Values of *F*-factors according to reference [12]

V_1/V_2	1	2	3	4	6	8	10
F	1.00	0.98	0.953	0.920	0.875	0.838	0.805

From Table 1 follows the interpolation formula:

$$F = 1.0233 - 2.33 \cdot 10^{-2} \frac{V_1}{V_2}. \tag{21}$$

Combining eqs. (18) to (21), one obtains

$$C_{1,2} = \left(1.504 - 3.43 \cdot 10^{-2} \frac{V_1}{V_2} \right) \sqrt{T_{s_1} \cdot T_{s_2}}. \tag{22}$$

Although *Arnold* published his equation earlier than *Gilliland* [13], it is more exact, particularly in describing temperature effects. From *Arnold's equation* (17) we obtain for the same gas mixture for different temperatures

$$\frac{D(T_2)}{D(T_1)} = \left(\frac{T_2}{T_1} \right)^{2.5} \frac{T_1 + C_{1,2}}{T_2 + C_{1,2}}. \tag{16a}$$

Many semiconducting epitaxial films are deposited from mixtures of chlorides and hydrogen, the latter being present in large excess. One of the gaseous products is

102

then HCl for which experimental data are available [14]:

Table 2. HCl diffusion constants according to reference [14]

T [°K]	294	327	372	473	523
D_{HCl, H_2} [cm²s⁻¹]	0.795	0.954	1.187	1.798	2.10

For HCl: $T_s = 168.13$ K, $V_s = 30.64$ [cm³/mole]

H₂: $T_s = 20.39$ K, $V_s = 14.3$ [cm²/mole]

From the above data using eq. (22) one obtains

$$C_{HCl, H_2} = 84.03 . \tag{23}$$

According to Table 2 is $D_{HCl, H_2} (294) = 0.795$. From eqs. (16a) and (23) it follows that

$$D_{HCl, H_2} (523) = 2.09 \ [cm^2 \ s^{-1}] . \tag{24}$$

This value differs from the experimental one by less than 0.5%. Calculating the exponent m from the system of equations:

$$\frac{D(523)}{D(294)} = \left(\frac{523}{294} \right)^m \tag{25}$$

$$\frac{D(523)}{D(294)} = \left(\frac{523}{294} \right)^{2.5} \frac{294 + 84}{523 + 84} \tag{16b}$$

we obtain $m = 1.678$, and from the experimental data [14] $m = 1.670$. The agreement is sufficient to use the simpler formula (25) with the m-value calculated from *Arnold's equation*. There are many systems in which the m-values are so close together that by assuming the values to be equal no serious errors are made. This statement is of practical value in considering multicomponent systems and systems with a temperature gradient in the boundary diffusion layer.

2.2.3.2.   Corrections for diffusion in gaseous mixtures

In practice, particular components are diffusing in a dilute gaseous mixture. Therefore, the assumption of diffusion in a pure one-component gas is erroneous to some extent. If a sufficient number of data is available, it is possible to calculate corrections using *Wilke's* [15] approximation formula

$$D'_A = \frac{1 - y_A}{\dfrac{y_B}{D_{AB}} + \dfrac{y_C}{D_{AC}} + \dfrac{y_D}{D_{AD}} + \ldots} \tag{26}$$

D'_A is the diffusion constant of the component A into the mixture B, C, D, . . .
y_A, y_B, y_C are the mole fractions of the components A, B, C, respectively.
D_{AB}, D_{AC}, D_{AD} are the interdiffusion constants of the binary mixtures AB, AC, AD
respectively.

Fairbanks and *Wilke* [16] have shown that, employing eq. (26), D'_A can be calculated
with a relatively high accuracy.

2.2.3.3.   Corrections for the decomposition of gaseous products

If several components are decomposing within the boundary layer, this phenomenon
changes the mass-transfer coefficients. By decomposition of a gas product, its concentration gradient within the boundary layer is increased and, in consequence, this
also increases its mass-transfer coefficient. In many cases, if some experimental data
are available, and using some simplifications on the basis of material balance in fluxes,
it is possible to calculate adequate corrections which depend on the particular features
of the considered system.

2.2.4.   LIMITATION OF A MATHEMATICAL DESCRIPTION TO DEFINITE CON DITIONS

For a full mathematical description of CVD it would be necessary to know at least
the thermodynamic data of reactants and products, the thermodynamic and kinetic
data of all participating chemical reactions, surface free energy data of the phases
being deposited, activation energies of surface diffusion, chemical reactions, and
adsorption of reactants and the appropriate description of products.
Since usually not all of these data are available, we try to limit a problem to certain
definite conditions. For example, at sufficiently high temperatures, the description
of deposition at steady state is possible by considering diffusion in the gas phase;
the surface reaction rates are then so high that they do not affect the rate of the deposition.

2.2.5.   METASTABLE EQUILIBRIA AT CVD

2.2.5.1.   The concept of activity at metastable equilibria

A consideration of diffusion-controlled CVD usually includes thermodynamic equilibria at the gas–solid interface. Such an assumption contradicts the basic condition
of crystal layer growth: supersaturation at the interface. This contradiction may be
avoided by the assumption of a metastable rather than a stable (*i.e.* thermodynamic)
equilibrium at the phase boundary. At steady state the *Gibbs free energy* of a component being deposited is in the gas phase and in the solid phase larger than at thermodynamic equilibrium.

104

For a thermodynamic description of a system in which a condensed phase is a mixture, the definition of activity of a component i is usually used:

$$a_i = \exp \frac{G_i - G_{io}}{RT} \tag{27}$$

where

G_i is the *partial molar Gibbs free energy* of a component in the mixture and G_{io} is the *molar Gibbs free energy* of a component i at a *standard state*.

In the case of non-electrolytes the value G_{io} represents most frequently the *molar Gibbs free energy* of a pure component i. If a vapor is an ideal gas, then

$$a_i = \frac{p_i}{p_{io}} \tag{28}$$

where

p_i is the partial pressure of a component i over the appropriate solution and p_{io} is the pressure of saturated vapor of the component i.

Because of the assumption of the above-mentioned standard state (pure component i) and of the corresponding value G_{io} it is impossible to distinguish between small crystallites or nuclei and large perfect crystals, since according to the definition of a standard state — pure component i — the value of G_{io} was in all these cases the same. Therefore, in this chapter the standard state is defined in a different manner. It has been assumed that the standard state corresponds to a large crystal in which no thermodynamically unstable defects (dislocations, grain boundaries etc.) exist. The last limitation does not apply to the point defects in thermodynamically stable concentration. According to this definition G_{io} corresponds to one mole of such a standard crystal; consequently, its activity is equal to unity. The activity of all other pure crystals is larger than unity.

Since polycrystals or nuclei are thermodynamically unstable, their equilibrium with an adjacent gas phase must also be unstable (metastable).

Hirth, Moazed and *Ruth* [17] considered an example of a simple chemical reaction

$$KX_{(g)} \rightleftharpoons K_{(s)} + X_{(g)}.$$

The gas KX dissociates on a hot substrate into a solid deposit $K_{(s)}$ and a gaseous product $X_{(g)}$. The driving force for nucleation is, as usual, the difference of the *Gibbs free energy* per unit volume

$$\Delta G_v = -\frac{kT}{V} \ln \frac{n_K}{n_{Ke}} \tag{29}$$

where

k is *Boltzmann's constant,*
T is the absolute temperature,
V is the molar volume,
n_K is the actual concentration of adsorbed K-atoms per unit area, and
n_{Ke} is the concentration of adsorbed K-atoms per unit area at equilibrium.

The corresponding concentration of K-atoms at steady state is given by

$$n_K = \frac{\alpha_K \, p_K}{\sqrt{2\pi \, mkT}} \, \frac{1}{v} \, \exp \frac{\Delta G_{des(K)}}{kT} \tag{30}$$

where

v is the vibration frequency of the adsorbed atom,
α_K is the condensation coefficient,
p_K is the partial pressure of K,
m_K is the mass of a single K-atom, and
ΔG_{des} is the *Gibbs free energy* for activation of desorption.

At equilibrium, in eq. (30) n_K and p_K become n_{Ke} and p_{Ke}, respectively. Since the atomic flux J_K is defined by the *Hertz–Knudsen* equation

$$J_K = \frac{\alpha_K \, p_K}{\sqrt{2\pi \, m_K \, kT}} \tag{31}$$

it follows from the definition of activity and from eqs. (39) and (31):

$$\frac{n_K}{n_{Ke}} = \frac{p_K}{p_{Ke}} = \frac{J_K}{J_{Ke}} = a_K . \tag{32}$$

From eq. (29) one obtains

$$\Delta G_v = -\frac{kT}{V} \, \ln \frac{p_K}{p_{Ke}} = -\frac{kT}{V} \, \ln \frac{J}{J_{Ke}} = -\frac{kT}{V} \, \ln a . \tag{33}$$

Similarly, one obtains for the *molar Gibbs free energy*

$$\Delta G_K = -RT \ln \frac{p_K}{p_{Ke}} = -RT \ln \frac{J_K}{J_{Ke}} \tag{34}$$

where

R is the gas constant
or

$$\Delta G_K = -RT \ln a_K \tag{35}$$

106

ΔG_K corresponds to a negative change of the *Gibbs free energy* at the changes:
a) n_K to n_{Ke} b) to p_K to p_{Ke} c) $a > 1$ to $a = 1$. According to the definitions utilized in this chapter for a steady state, ΔG_K and a_K are measures for deviation from equilibrium at the gas–solid interface. Since the change described by conditions (a)–(c) can occur spontaneously, $\Delta G_K < 0$ and

$$- \Delta G_K = G_K - G_{K(o)} \tag{36}$$

where

G_{Ko} is the *Gibbs free energy* at standard state. $- \Delta G_K$ may also be considered as a measure of supersaturation. Denoting the value of the *Gibbs free energy* of K in the gas phase as G_K and the activity of the solid K-phase at metastable equilibrium with this gas phase as a_K, the stable (thermodynamic) equilibrium requires the condition $\Delta G_K = 0$, corresponding to $a_K = 1$.

2.2.5.2. Multicomponent chemically reacting systems

If the deposition is a result of j simultaneous chemical reactions, the differential of the *Gibbs free energy* of each reaction j, in which from r reactants k products are formed, is given by

$$dg_j = \sum_{i=1}^{k} v'_{ji}(V'_{ji}\, dP - S'_{ji}\, dT + RT d \ln a'_{ji}) - \sum_{i=1}^{r} v_{ji}(V_{ji}\, dP - S_{ji}\, dT + RT d \ln a_{ji}) \tag{37}$$

where

V is the molar volume,
S is the molar entropy,
P is the total pressure,
R is the gas constant,
T is the absolute temperature,
a is the activity, and
v is the stoichiometric coefficient.
The index i refers to the i-th component,
the index j to the j-th reaction, and
the index $'$ to the products.

Taking into account

$$S_{ji} = \frac{H_{ji}}{T} \tag{38}$$

where

H is the molar enthalpy one obtains

$$\frac{dg_j}{RT} = \frac{\sum_{i=1}^{kj} v'_{ji} V'_{ji} - \sum_{i=1}^{rj} v_{ji} V_{ji}}{RT}\, dP - \frac{\sum_{i=1}^{kj} v'_{ji} H'_{ji} - \sum_{i=1}^{rj} v_{ji} H_{ji}}{RT^2}\, dT + d \ln \frac{\prod_{i=1}^{kj} a'_{ji}{}^{v'_{ji}}}{\prod_{i=1}^{rj} a_{ji}{}^{v_{ji}}}. \tag{39}$$

For ideal gases one may write:

$$V'_{ji} = V_{ji} = \frac{RT}{P}$$

(40)

and

$$a_{ji} = x_{ji} = \frac{p_i}{P}$$

(41)

where

x is the mole fraction and
p_i the partial pressure.

Defining

$$\sum_{i=1}^{kj} v'_{ji} - \sum_{i=1}^{rj} v_{ji} = \Delta v$$

(42)

and

$$\sum_{i=1}^{kj} v'_{ji} H'_{ji} - \sum_{i=1}^{rj} v_{ji} H_{ji} = \Delta h_j$$

(43)

one obtains

$$\frac{dg_j}{RT} = - \frac{\Delta h_j}{RT^2} \, dT + d \ln \frac{\prod\limits_{i=1}^{kj} p'^{\,v'_{ji}}_{ji}}{\prod\limits_{i=1}^{rj} p^{\,v_{ji}}_{ji}} + \Delta v \, \frac{dP}{P} \; .$$

(44)

Integrating eq. (44) (and assuming $\Delta h_j = $ const.) between the appropriate limits, the lower one corresponding to the stable (thermodynamic) equilibrium (index e) and the upper one to the metastable equilibrium (index ae),

$$\int_{(e)}^{(ae)} \frac{dg_j}{RT} = - \int_{T_0}^{T} \frac{\Delta h_j}{RT^2} \, dT + \int_{(e)}^{(ae)} d \ln \frac{\prod\limits_{i=1}^{kj} p'^{\,v'_{ji}}_{ji}}{\prod\limits_{i=1}^{rj} p^{\,v_{ji}}_{ji}}$$

(45)

one obtains

$$\frac{\Delta g_j - 0}{RT} = \frac{\Delta h_j}{R} \left(\frac{1}{T} - \frac{1}{T_0} \right) + \ln \left[\frac{\prod\limits_{i=1}^{kj} p'^{\,v'_{ji}}_{ji}}{\prod\limits_{i=1}^{rj} p^{\,v_{ji}}_{ji}} \right]_{(ae)} - \ln K(T_0)$$

(46)

where

$K(T_0)$ is the equilibrium constant at T_0. Metastable equilibrium between a pure solid

108

phase with activity a_{js} and gaseous reactants and products, can hence be described for $\Delta g = 0$ by eq. (47):

$$\frac{\Delta h_j}{R}\left(\frac{1}{T} - \frac{1}{T_0}\right) + \ln a_{js} + \ln \left[\frac{\prod_{i=1}^{kj-1} p'_{ji}{}^{v'_{ji}}}{\prod_{i=1}^{rj} p_{ji}{}^{v_{ji}}}\right]_{(ae)} - \ln K(T_0) = 0. \tag{47}$$

Hence

$$a_{js} = K(T_0)\exp\left[-\frac{\Delta h_j}{R}\left(\frac{1}{T} - \frac{1}{T_0}\right)\right] \cdot \left[\frac{\prod_{i=1}^{rj} p_{ji}{}^{v_{ji}}}{\prod_{i=1}^{kj-1} p'_{ji}{}^{v'_{ji}}}\right]_{(ae)} \tag{48}$$

or

$$a_{js} = \exp\left(-\frac{\Delta g_j}{RT}\right). \tag{49}$$

In a system in which, as a result of a chemical reaction j, only one pure component is being deposited, a_{js} is the activity of this solid component in the metastable equilibrium with a gas phase. From eq. (48) one can conclude that an increase in the partial pressures of reactants and a decrease in the partial pressures of products result in an increase in activity of a solid phase in metastable equilibrium with a given gas phase. Taking into account eq. (48) or eq. (49), the necessary condition of stability of a crystal nucleus may be obtained:

$$a < a_{js}. \tag{50}$$

If the activity of some kind of nuclei on the solid surface is larger than a_{js}, such nuclei are unstable.

2.2.6.   GROWTH OF CRYSTAL FILMS

2.2.6.1.   Conclusions from the theory of Burton, Cabrera and Frank, and from flux balances

According to the theory of *Burton, Cabrera* and *Frank* [18], in a one-component system the growth rate of a pure crystal being deposited from the vapor is given by

$$J = \frac{\tanh\left(\dfrac{y_0}{\sqrt{2}\,x_s}\right)}{\dfrac{y_0}{\sqrt{2}\,x_s}} \cdot \frac{1}{\sqrt{2\pi\, mkT}} \beta\,\Delta p \tag{51}$$

where

$$y_0 = \frac{4\pi\, \eta s}{kT \ln \dfrac{p}{p_e}} \approx \frac{4\pi\, \eta s\, p_e}{kT\, \Delta p} \tag{52}$$

and

$$x_s = \sqrt{\frac{2D}{v_s}} = \sqrt{2Dt_s} \tag{53}$$

y_0 is the distance between single ledges,
x_s is the root mean square surface diffusion distance,
v_s the evaporation frequency,
D the surface diffusion constant,
β a reflection coefficient,
t_s the mean time of residence of an atom on the surface,
m the mass of an atom or a molecule,
k is *Boltzmann's constant*,
T the absolute temperature,
η the *Gibbs function* of the edge in units of erg cm^{-1},
s is the mean area of an atom or a molecule at the surface,
p_e the saturated vapor pressure,
p the actual vapor pressure, and
$\Delta p = p - p_e$.

In the theory of *Burton, Cabrera*, and *Frank* two special cases are considered:

a) Ledges of a spiral dislocation are close together: supersaturation is relatively high, *i.e.*

$$y_0 \ll x_s.$$

Then $\tanh\left(\dfrac{y_0}{\sqrt{2}\, x_s}\right) \longrightarrow \dfrac{y_0}{\sqrt{2}\, x_s}.$ \hfill (54)

From eqs. (51) and (54) we obtain the well-known linear law

$$J = \frac{\beta}{\sqrt{2\pi\, mkT}}\, \Delta p = C''\, \Delta p. \tag{55}$$

b) Ledges of a spiral dislocation are far apart; supersaturation is relatively small, then

$$y_0 \gg x_s \text{ and}$$

$$\tanh \frac{y_0}{\sqrt{2}\, x_s} \longrightarrow 1. \tag{56}$$

110

Combining eqs. (51) and (56) yields

$$J = \frac{\sqrt{2}\, x_s}{y_0} \frac{\beta}{\sqrt{2\pi\, mkT}}\, \Delta p \tag{57}$$

and from eqs. (52) and (57) one obtains the well-known parabolic law

$$J = \frac{\sqrt{2}\, x_s\, kT\, \beta (\Delta p)^2}{4\pi\, \eta s\, p_e \sqrt{2\pi\, mkT}} = C''(\Delta p)^2. \tag{58}$$

Equations involving $\Delta p = p - p_e$ can be expressed in terms of the supersaturation ratio, which in a one-component system is equal to the activity

$$a = \frac{p}{p_e}. \tag{59}$$

Hence

$$\Delta p = p - p_e = p_e \left(\frac{p}{p_e} - 1 \right) = p_e(a - 1) \tag{60}$$

and

$$(\Delta p)^2 = p_e^2 (a - 1)^2. \tag{61}$$

Combining eqs. (55) and (60) we obtain

$$J = \frac{\beta}{\sqrt{2\pi\, mkT}}\, p_e(a - 1) = C_1(a - 1). \tag{62}$$

Similarly, from eqs. (58) and (61) it follows that

$$J = C_2(a - 1)^2. \tag{63}$$

In multicomponent systems in which a condensing component is transported to and away from the interface in the form of various chemical compounds, the constants C_1 and C_2 are no longer the same as in a one-component system. At dynamic equilibrium with respect to the deposit the rate of condensation is equal to the rate of evaporation (or gas etching) and the net condensation rate is then equal to zero. The fluxes onto and from the interface can be larger, by even several orders of magnitude, than in a one-component system. If the component i is being deposited at steady-state conditions, then

$$\Delta J = \vec{J}_{i(g)} - \overleftarrow{J}_{i(g)} = \vec{J}_{i(s)} - \overleftarrow{J}_{i(s)} \tag{64}$$

$J_{i(g)}$ is the flux of a component i in the form of various compounds through a gas phase onto the interface,

111

$\vec{J}_{i(g)}$ is the flux of a component i in the form of various compounds through a gas phase away from the interface,

$\vec{J}_{i(s)}$ is the flux of a component i into a crystal lattice, and

$\overleftarrow{J}_{i(s)}$ is the flux of a component i from a crystal lattice toward the interface.

Expressing

$$\vec{J}_{i(g)} = a_i J_{i(g)e} \tag{65}$$

where

$J_{i(g)e}$ is the flux of a component i through a gas phase at dynamic equilibrium with respect to a solid deposit i,

we obtain

$$\Delta J_{i(g)} = \vec{J}_{i(g)} - a_i J_{i(g)e}. \tag{66}$$

According to the theory of *Burton, Cabrera* and *Frank* [18] for large supersaturations holds

$$\Delta J_{i(s)} = C(a_i - 1). \tag{67}$$

From eqs. (66) and (67) one can deduce that an increase in activity results in a decrease of net flux in the gas phase, $\Delta J_{i(g)}$, and an increase of a net flux into the crystal lattice, $\Delta J_{i(s)}$. For steady state

$$\Delta J_{i(g)} = \Delta J_{i(s)} \tag{68}$$

must be fulfilled, and hence

$$\vec{J}_{i(g)} - a_i J_{i(g)e} = C(a_i - 1) \tag{69}$$

after rearranging

$$\vec{J}_{i(g)} = a_i(J_{i(g)e} + C) - C \tag{70}$$

when $a_i = 1$, then

$$\vec{J}_{i(g)} = J_{i(g)e}. \tag{71}$$

According to eq. (70), for steady state, the activity at the gas–solid interface is given by

$$a_i = \frac{\vec{J}_{i(g)} + C}{J_{i(g)e} + C} = \frac{1 + \dfrac{\vec{J}_{i(g)}}{C}}{1 + \dfrac{J_{i(g)e}}{C}} \tag{72}$$

112

If $C \to \infty$, then $a \to 1$, and if $C \to 0$, then $a \to \dfrac{\vec{J}_{i(g)}}{\vec{J}_{i(g)e}}$; hence, at the phase boundary applies $1 < a_i < \dfrac{\vec{J}_{i(g)}}{\vec{J}_{i(g)e}}$.

The increase in C-values results in an increase of fluxes in the gas phase toward and from the interface corresponding to the same activity value. Since C depends on the nature of the chemical processes of the chemical compounds transporting, and since C depends also on the mass-transfer coefficients the i deposit, by appropritate choice of the process to achieve a large C-value it is possible to reach a relatively high deposition rate at a small activity. Such a situation is advantageous for perfection of crystal deposits.

In processes meeting the condition of a high C-value at dynamic equilibrium with respect to i, i.e. at $a_i = 1$ and $\Delta J_{i(s)} = 0$, the fluxes $\vec{J}_{i(g)e}$, $\overleftarrow{J}_{i(g)e}$; $\vec{J}_{i(s)e}$, $\overleftarrow{J}_{i(s)e}$ are also relatively large.

The transport of a component being deposited towards and from the interface occurs in the form of different chemical compounds; at the phase boundary chemical reactions occur, although the condition $\Delta J_{i(s)} = \Delta J_{i(g)} = 0$ is fulfilled. The increase of a C-value in a CVD system can be achieved if the system involves a gaseous etching component, which results in an increase of the i-flux in the form of its gaseous compound. In the case of deposition of epitaxial silicon from $SiCl_4 + H_2$ this etching component is for instance HCl formed at the phase boundary, which is increasing the silicon flux away from the interface in the form of $SiCl_2$. The result is a relatively small silicon activity at the phase boundary, smaller than at UHV condensation, and therefore in the first case the perfection of deposits is usually better than in the last one.

In a one-component system an increase of C-values can be achieved by increasing the temperature at the interface [18].

2.2.6.2.　　**The effect of activity at the gas–solid boundary on the crystal growth mechanism**

Rearranging an equation given by *Pound* [19] one obtains:

$$f(\Theta) \le \frac{3\,\Delta G_v^2}{16\pi\,\sigma^3}\left[kT\left(\ln \frac{C_1}{J_c} + \ln P\right) + \Delta G_{des} - \Delta G_D \right] \tag{73}$$

where

$$f(\Theta) = 0.5 - 0.75\cos\Theta + 0.25\cos^3\Theta \tag{74}$$

Θ　　is the contact angle of the deposit at the substrate,
J_c　　the smallest observable nucleation rate,
P　　the actual vapor pressure of a component being deposited,
T　　the absolute temperature,

k *Boltzmann's constant,*

ΔG_{des} the *Gibbs activation free energy* for desorption,

ΔG_{D} the *Gibbs activation free energy* for surface diffusion,

σ the interfacial tension of the gas–deposit system,

ΔG_{v} the smallest supersaturation in free energy terms corresponding to the observable limit of nucleation, and

C_{1} is a constant [17].

Since

$$\Delta G_{v} = -\frac{RT}{V}\ln a_{i} \tag{35a}$$

a decreasing activity at the interface results in a decreasing contact angle Θ at which the smallest observable nucleation rate occurs. Since on the surface of a real crystal adsorption sites with various Θ values are distributed, nucleation on sites with high Θ values can be practically eliminated by decreasing the activity at the gas–solid interface; e.g. by appropriately decreasing a_{i} it is possible to decrease the nucleation rate on flat crystal surfaces below practical values, and to keep it above practical values along ledges. Such a condition improves film perfection. *Chakraverty* and *Pound* [20] derived equations determining the ratio of nucleation rates at macroscopic steps I_{L}, and on a flat crystal surface, I_{F}:

$$\ln\frac{I_{L}}{I_{F}} = \ln g + \frac{16\pi\sigma^{3}}{3\,\Delta G_{v}^{2}\,kT}\,[f(\Theta) - \Phi(\Theta)] \tag{75}$$

where

g is the fraction of crystal surface covered by macroscopic steps,

$f(\Theta)$ a function defined by eq. (74), and

$\Phi(\Theta)$ a function defined by eq. (76).

$$\Phi(\Theta) = \frac{1}{4}\left\{(\sin\Theta - \cos\Theta) + \frac{2}{\pi}\cdot\cos^{2}\Theta\right.$$
$$(\sin^{2}\Theta - \cos^{2}\Theta)^{1/2} + \frac{2}{\pi}(\cos\Theta\,\sin^{2}\Theta\,\arcsin(ctg\,\Theta) - \cos\Theta\,\sin^{2}\Theta) -$$
$$-\frac{2}{\pi r_{c}}\int_{r_{c}\cos\Theta}^{r_{c}\sin\Theta}\arcsin\frac{r_{c}\cos\Theta}{(r_{c}^{2} - y^{2})^{1/2}}\,dy\left.\right\} \tag{76}$$

where

r_{c} is the radius of a critical nucleus and

y an integration variable.

A decreasing activity results in decreasing $-\Delta G_{v}$ (eq. (35a)), and increasing the ratio $\dfrac{I_{L}}{I_{F}}$. Thus, controlling the activity at the gas–solid interface should provide a method of controlling the morphology and perfection of the crystal films.

It is worthwhile to consider one more approach. Taking into account the concept of contact angle [21], *Hirth, Moazed,* and *Ruth* [17] discussed the effect of Θ on the rate of critical nuclei formation, J_c, and on epitaxial growth:

$$n\, J_c = A + \ln \sin \Theta - B\, f(\Theta) \tag{77}$$

where

$$B = \frac{16\pi\, \sigma^3}{3\, kT\, \Delta G_v^2} \tag{78}$$

A is a constant independent of Θ and
$f(\Theta)$ is the function defined by eq. (74).

If J_1 is the rate of the formation of critical nuclei with a favorable orientation for epitaxy, corresponding to the contact angle Θ_1, and J_2 is the rate of the formation of critical nuclei with a non-favorable orientation for epitaxy, corresponding to the contact angle Θ_2 then — according to the authors [17] — β is the measure of the favored epitaxial orientation:

$$\beta = \frac{J_1}{J_2} > 1 \tag{79}$$

Combining eqs. (77) and (79) we obtain

$$\beta = \frac{\sin \Theta_1}{\sin \Theta_2} \exp B[f(\Theta_2) - f(\Theta_1)] \tag{80}$$

where β is always positive (relation (79)) and contact angles larger than $180°$ have no physical meaning. Hence $f(\Theta_2) > f(\Theta_1)$ and consequently $\Theta_2 > \Theta_1$. An increase of temperature results in an increasing B-value, as can be shown as follows:
According to definition (35a):

$$-\Delta G_v = \frac{RT}{V} (\ln p - \ln p_e(T)) \tag{81}$$

where

$p_e(T)$ is the saturated vapor pressure at the temperature T.

Applying the *Clausius–Clapeyron equation*

$$\ln p_e(T) = \ln p_e(T_0) - \frac{L}{R}\left(\frac{1}{T} - \frac{1}{T_0}\right) \tag{82}$$

where

T_0 is some reference temperature and
L is the molar heat of sublimation.

8*

115

Combining eqs. (81) and (82), we obtain

$$- \Delta G_v = \frac{L}{V} - \frac{T}{V} \left(\frac{L}{T_0} - R \ln \frac{p}{p_e(T_0)} \right) > 0 . \tag{83}$$

At epitaxial deposition, $\dfrac{p}{p_e(T_0)}$ is slightly larger than 1; the expression in brackets in eq. (83) has a positive value, hence the increase of T at $p = $ const. results in a rapid decrease of $-\Delta G_v$ and $T \Delta G_v^2$, as well as an increase of B (eq. (78)) and β (eq. (80)). Hence, at high temperature epitaxial nucleation is more favored than at lower temperature. From eqs. (35a), (77), and (78) one can deduce that an increase of activity results in a decrease of B and β. Hence, the epitaxial growth is more favored at low a values at the gas–solid interface.

Although the theory of *Burton*, *Cabrera* and *Frank* [18], *Pound*'s equation (73), *Chakraverty*'s and *Pound*'s equation (75) as well as eqs. (77)–(80) are primarily developed for one-component systems, the tendencies of changes caused by activity changes at the gas–solid interface are the same in one- and multicomponent systems and the conclusions derived for the former case are also qualitatively valid for the latter one. These conclusions may be considered as a guide for choosing technological processes and conditions to obtain epitaxial films with a desired perfection. In CVD flow systems there is namely a possibility of choosing an appropriate process, involving desired chemical reactions to achieve relatively large deposition rates at small a values. The activity defined in this chapter can be considered as a measure of a system's departure from equilibrium.

2.2.7.   **THE DEPOSITION OF PURE ELEMENTAL CRYSTAL FILMS IN CVD FLOW SYSTEMS**

In this section the deposition of pure elemental crystal films in CVD flow systems is described by means of equations of metastable equilibria, of transport, and of material balances.

2.2.7.1.   **Assumptions**

a) The deposition of a pure elemental crystal phase occurs in a steady-state process.
b) The deposition is controlled by diffusion in the gas phase.
c) The gaseous reactants do not react in the gas phase, but exclusively on or at the solid interface.
d) The gaseous components obey ideal gas laws.
e) The conversion with respect to the mass of gases is negligible.

The last condition should be satisfied in practice, otherwise the deposited films in the flow direction have a non-uniform thickness.

2.2.7.2. Equations of metastable equilibria

For each reaction occurring at the gas–solid interface, one equation of a metastable equilibrium is necessary (see Section 2.2.5.2) and for the entire system one further equation:

$$\sum_i p_i = P. \tag{84}$$

(The sum of all partial pressures equals the total pressure.)

2.2.7.3. Mass-transport equations

Equations of transport have the form given by eq. (2). It is advisable to introduce relative parameters:

$$i_i = \frac{J_i}{k_1} = \frac{k_i}{k_1}(p_{io} - p_i) \qquad \cdots\cdots \tag{85}$$

$$\frac{k_i}{k_1} = b_i \tag{86}$$

where

j_i is a relative flux with the dimension of a pressure,
b_i is a dimensionless quantity, and
k_1 is the mass-transfer coefficient of the component 1 diffusing from the main stream of gases across the boundary layer towards the gas–solid interface, and transporting the material being deposited.

For the component 1 we have then

$$j_1 = p_{10} - p_1 \tag{87}$$

where

p_{10} is the partial pressure of the component 1 in the main gas stream and
p_1 is the partial pressure of the component 1 at the gas–solid interface.

According to assumption (e) for products formed at the interface, p_{io} is very small, hence

$$j_i = b_i(p_{io} - p_i) \approx -b_i\,p_i. \tag{88}$$

2.2.7.4. Mass-balance equations

The systems considered contain gaseous compounds whose partial pressures are inter-related by equations of metastable equilibria. Chemical formulae of these compounds have the general form:

$$
\begin{array}{llll}
A_{\alpha_1} & B_{\beta_1} & C_{\gamma_1} & \cdots\cdots \\
\vdots & \vdots & \vdots & \\
A_{\alpha_2} & B_{\beta_2} & C_{\gamma_2} & \cdots\cdots \\
\vdots & \vdots & \vdots & \\
A_{\alpha_i} & B_{\beta_i} & C_{\gamma_i} & \cdots\cdots \\
\vdots & \vdots & \vdots & \\
A_{\alpha_n} & B_{\beta_n} & C_{\gamma_n} & \cdots\cdots
\end{array}
\tag{89}
$$

where

A, B, C ... are elements involved in the system and
α_i, β_i, γ_i are stoichiometric numbers.

If the pure element A is being deposited, its mass-balance equation has the form:

$$
\sum_{i=1}^{n} \alpha_i \cdot j_{A_{\alpha_i} B_{\beta_i} C_{\gamma_i} \cdots} = j_0 = \frac{J_0}{k_1}
\tag{90}
$$

where

J_0 is the deposition flux of the element A and
k_1 is the mass-transfer coefficient of the component 1 in the form of which the element A is transported towards the interface.

The balance equations of other elements have similar forms:

$$
\sum_{i=1}^{n} \beta_i \cdot j_{A_{\alpha_i} B_{\beta_i} C_{\gamma} \cdots} = 0 \quad \text{(B–balance)}
\tag{91}
$$

$$
\sum_{i=1}^{n} \gamma_i \cdot j_{A_{\alpha_i} B_{\beta_i} C_{\gamma_i} \cdots} = 0 \quad \text{(C–balance)}.
\tag{92}
$$

Solving the system of equations (of metastable equilibria, of transport, and of balances) for a steady state at given $K_j(T)$, P, and b_i yields partial pressures p_i at the interface, relative fluxes j_i, and the net condensation flux of the element A as functions of the activity a_A and of initial partial pressure of the component 1 in the form of which the element A is transported to the interface.

The increase of activity a_A results in an increase of evaporation of the element A in the form of its gaseous compounds and therefore the net flux j_0 in the gas phase decreases. For crystal growth some supersaturation is necessary, hence a_A has to be larger than unity. An increase of a_A yields an increase of the net flux j_0 into the crystal lattice. If $a_A = 1$, j_0 in the gas phase reaches its maximum value, but j_0 into the lattice equals zero in this case. Such a situation contradicts the condition of steady state for which:

$$j_{0(g)}(a) = j_{0(s)}(a).$$
(93)

Therefore, a has to be larger than unity in order to satisfy eq. (93)

2.2.7.5. **Determination of the activity (a) and the deviation from equilibrium $(-\varDelta g)$ at the phase boundary[1]**

If, for a given k_{p_1}-value, experimental $\varDelta J_{(s)}$-values for corresponding p_0-values are known, then

$$j_{0\,exp} = \frac{\varDelta J_{(s)\,exp}}{k_{p_1}}.$$
(85a)

Solving the system of eqs. (48), (84)–(92) by means of a computer at $T = $ const. and $P = $ const. yields numerical values of the quantities shown in Tables 3.1 to 3.n:

Table 3.1 for $a = a_1$

$p_0, j_0, p_1, \ldots \ldots p_i,$ $j_1, \ldots \ldots \ldots j_i \ldots$

Table 3.2 for $a = a_2$

$p_0, j_0, p_1, \ldots \ldots p_i,$ $j_1, \ldots \ldots \ldots j_i \ldots$

Table 3.n for $a = a_n$

$p_0, j_0, p_1, \ldots \ldots p_i \ldots, j_1, \ldots \ldots \ldots j_i, \ldots$

[1] $-\varDelta g$ and a are correlated by eq. (49) $(-\varDelta g = RT \ln a)$

119

For each a-value one pair of p_0 and j_0-values equal to the experimental pair of $j_{0\,exp}$ and $p_{0\,exp}$ should be chosen from an appropriate table. Such pairs should be inserted into the table for the steady state arranged as follows

$a_0, j_0, a, \log j_0, \log(a - 1), p_1 \dots p_i \dots, j_1 \dots j_i \dots$

$\begin{matrix} \cdot & \cdot & \cdot & \cdot & \cdot & \cdot & \cdot & \cdot & \cdot \\ \cdot & \cdot & \cdot & \cdot & \cdot & \cdot & \cdot & \cdot & \cdot \\ \cdot & \cdot & \cdot & \cdot & \cdot & \cdot & \cdot & \cdot & \cdot \\ \cdot & \cdot & \cdot & \cdot & \cdot & \cdot & \cdot & \cdot & \cdot \end{matrix}$

Equation (94)

$$\Delta J_{(s)} = C(a - 1)^n \quad (1 \leq n \leq 2) \tag{94}$$

holds for the linear ($n = 1$) and for the parabolic ($n = 2$) law of crystal growth as well, covering all supersaturation ranges. The curve $\log j_0$ versus $\log (a - 1)$ exhibits only a slight curvature and in not too large intervals the exponent n is approximately constant. Thus, the relation

$$n = \frac{d \log j_0}{d \log(a - 1)} \tag{95}$$

is justified and in this way n for various a-values is available by graphical or numerical methods. Finally, we obtain

$$C = \frac{j_0}{(a - 1)^n}. \tag{96}$$

The results can be arranged in a table:

$p_0, j_0, a, n, C, \vec{J}_{(g)}, \overleftarrow{J}_{(g)}, \vec{J}_{(s)}, \overleftarrow{J}_{(s)}, p_1, \dots p_i, j_1 \dots j_i \dots$

$\begin{matrix} \cdot & \cdot & \cdot & \cdot & \cdot & \cdot & \cdot & \cdot & \cdot & \cdot \\ \cdot & \cdot & \cdot & \cdot & \cdot & \cdot & \cdot & \cdot & \cdot & \cdot \\ \cdot & \cdot & \cdot & \cdot & \cdot & \cdot & \cdot & \cdot & \cdot & \cdot \\ \cdot & \cdot & \cdot & \cdot & \cdot & \cdot & \cdot & \cdot & \cdot & \cdot \end{matrix}$

2.2.8. DEPOSITION OF EPITAXIAL SILICON IN CVD SYSTEMS

2.2.8.1. Review of methods

The following sources for deposition of epitaxial silicon in CVD systems are used: $SiCl_4$, $SiHCl_3$, SiH_4, $SiBr_4$, and SiI_4. The method most frequently applied and most frequently described in the literature is the reduction of silicon halides by hydrogen (I), particularly $SiCl_4$ [22]–[45], $SiHCl_3$ [7], [46], [47] and less frequently $SiBr_4$ [48], [49] or SiI_4 [50]. Very often the methods of thermal decomposition (II) of SiH_4, diluted by H_2 [51]–[57], or of thermal decomposition of a molecular beam (III) of SiH_4 [58]–[62] have been utilized. For depositing epitaxial films in closed systems the following techniques have been developed: Decomposition of iodides [38], [63]–[69],

120

chemical transport methods (IV) [38], [64], [65], [68], [69] or the sandwich method (V), the latter differing from the former by a very close distance (several hundreds Å) between the source and the substrate [60], [66], and [67]; decomposition of chlorides [70]–[72] is less frequently used.

For the techniques I and II a pressure of about 1 atmosphere is usually employed. The compounds serving as the sources of silicon are considerably diluted with hydrogen up to the partial pressure in the range 10^{-3} to $2 \cdot 10^{-2}$ atm. The main element of equipment for realizing the techniques I or II is a vertical or horizontal reactor in which a silicon substrate is placed upon a support made of graphite or of high resistant silicon, heated in a r. f. field, sometimes together with the whole reactor [32]. Through the reactor is flowing hydrogen with some amount of silicon halide or SiH_4. The range of temperature used for the reduction of $SiCl_4$ is 900 to 1 350°C, usually 1 200 to 1 300°C, for $SiBr_4$ and SiI_4 1 000 to 1 200°C [50], for $SiHCl_3$ 1 110 to 1 280°C [46], for the pyrolysis of SiH_4 920 to 1 260°C [52], 1 000 to 1 400°C [51] and 1 180°C [56], [57]. Partial or advanced investigations concerning the kinetics of reduction are mentioned in numerous papers: $SiCl_4$ [22], [27]–[31], [33]–[36], 40], [42] and [43], $SiBr_4$ [49], SiI_4 [50], $SiHCl_3$ [7], [47] and concerning the kinetics of pyrolysis of SiH_4 [52]–[62].

2.2.8.2. Steps controlling the deposition rate

In agreement with the considerations in Section 2.2.2, depending on the deposition conditions (*i.e.* temperature, pressure, composition, flow dynamics, and geometry) the deposition rate can be controlled either by surface processes or by gas-phase transport. Thus, the higher the temperature, the stronger is the tendency for the system to be diffusion controlled. The increase of flow intensity shifts the kinetically controlled temperature range towards higher tempratures [7], [22], [27]. The attempt to correlate a controlling mechanism with one particular parameter such as temperature, pressure, or flow intensity is responsible for great discrepancies between the results of various authors; this is especially due to different flow intensities or reactor geometries used. Since the quality of silicon epitaxial films is better at higher deposition temperatures, the temperature range 1 200 to 1 270°C is frequently used; at normally applied deposition, *i.e.* if the flow intensities are not too high, the deposition is diffusion controlled [7], [27], [35], [40], [44]. According to several authors, the deposition was kinetically controlled [22], [33], [46] under the conditions applied. The deposition rate is considerably affected by the $SiCl_4$ concentration; the rate curves exhibit a maximum [7], [22], [23], [27], [31]. In the pure $SiCl_4$ or H_2 silicon is etched according to the reactions:

$$Si_{(s)} + SiCl_4 \longrightarrow 2 SiCl_{2(g)}$$

$$Si_{(s)} + 2 H_2 \longrightarrow SiH_{4(g)}$$

Hence, there must be a definite concentration of $SiCl_4$ in a $SiCl_4/H_2$ mixture at which the maximum deposition rate occurs. Different values have been reported by various authors, ranging from 5 to 11 mole per cent of $SiCl_4$. The $SiCl_4$ mole fraction at which neither deposition nor etching occurs ($j_0 = 0$), is reported to be in the range 8 to 30 mole percent of $SiCl_4$. Since both — the maximum and the point of $j_0 = 0$ at constant pressure — depend on temperature, flow intensity, and geometry, the discrepancies between the results of various authors are sometimes very large. The second point of $j_0 = 0$ for a low $SiCl_4$-concentration has not been reported, but from thermodynamic estimations its magnitude is in the order of 10^{-5} mole per cent of $SiCl_4$.

2.2.8.3. **Thermodynamics of epitaxial silicon deposition**

The first attempt to calculate the deposition rate of epitaxial silicon by means of thermodynamic, particularly of equilibrium equations had been realized by *Steinmaier* [31] in 1963, who took into account only two chemical reactions:

$$SiCl_4 + 2 H_2 \rightleftharpoons Si_{(s)} + 4 HCl$$

$$SiCl_4 + Si_{(s)} \rightleftharpoons 2 SiCl_2.$$

He introduced the concept of a *hypothetical silicon pressure* p_{Si}, the latter being calculated from the utmost possible yield in the system of the two above-mentioned reactions; *i.e.* this utmost possible yield corresponds to a change from a given state — defined by temperature and initial partial pressure of $SiCl_4$ (at $p = 1$ atm) — to the equilibrium state, upon the assumption that the silicon remained in the gas phase under the hypothetical pressure p_{Si}. According to *Steinmaier*, the experimental deposition rate was proportional to this hypothetical pressure p_{Si}, in the temperature range over 1 500 K, *i.e.* in the range of diffusion control. *Steinmaier's model* is sometimes called the *quasi-equilibrium model*. It was applied later by *Bloem* and *Scholte* [70] in the analysis of epitaxial silicon deposition in a closed system. *Lever* [73] developed a method of computing equilibria in the system with 7 chemical reactions. involving the following gaseous components: $SiCl_4$, $SiHCl_3$, $Si_2H_2Cl_2$, SiH_3Cl, SiH_4, $SiCl_2$, $SiCl$, HCl, H_2. Unfortunately, the calculation was based on older thermodynamic data, criticized by *Schäfer* and *Heine* [74], *Seiter* and *Sirtl* [75].
Sedgwick [35] restricted *Lever's* method to the four components H_2, HCl, $SiCl_2$, $SiHCl_3$ and calculated the silicon deposition rate on the basis of the *quasi-equilibrium model*. A similar model was applied by *Arizumi* [41] in his continuous deposition method. *Shepherd* [40] developed a method of computing the mass-transfer coefficients and the deposition rate, based on the model of only one chemical reaction: $Si + 2 HCl \rightleftharpoons SiCl_2 + H_2$. *Bradshaw* [7] used a model considering the mass transfer across the boundary layer and three reactions:

$$2 SiHCl_3 \rightleftharpoons SiCl_4 + SiCl_2 + H_2$$

$$SiCl_4 + H_2 \rightleftharpoons SiCl_2 + 2 HCl$$

$$Si + SiCl_4 \rightleftharpoons 2 SiCl_2.$$

Summarizing, papers on thermodynamics of silicon deposition may be divided into two main groups:

122

a) One group comprises papers based on a quasi-equilibrium model, *e.g.* by *Stein-maier* [31], *Bloem* and *Scholte* [70], *Lever* [75], and *Sedgwick* [35].

b) The other group comprises papers based on mass transfer and on boundary layers, *e.g.* by *Shepherd* [40] and *Bradshaw* [7].

The common shortcoming of all papers mentioned is the neglect of departure from equilibria and of the supersaturation occurring at the gas–solid interface being the necessary condition for silicon film growth, *i.e.* the direct or indirect assumption of thermodynamic equilibria at the interface.

2.2.8.4. Thermodynamic data of the Si–H–Cl system

Unfortunately, most authors suggesting various methods of mathematical analysis applied non-reliable data and thus the value of results was diminished. Reliable calculations were often not performed because of lack of knowledge of the important gaseous species, their thermodynamic data and unsatisfactory equilibrium conditions in the reaction zone. Up to 1964 the formation enthalpy ΔH^0_{298} of $SiCl_4$, $SiBr_4$, and of SiJ_4 had been derived from its value for quartz. This value has been revised within the last 34 years by as much as 26 kcal. Meanwhile, *Schäfer* and *Heine* [74] developed a new method for the determiation of ΔH^0_{298} and published a list of most reliable data, extended later to subhalides [76], [77]. In the case of $SiCl_2$, exact measurements have repeatedly been performed [77] to [79] and the reliable value $- 38.0$ kcal $\cdot$ $\cdot$ mole^{-1} has been obtained; its accuracy has been proven for the system Si–H–Cl [81]. A critical review of thermodynamic data [82] to [86] for the Si–H–Cl system has been given by *Seiter* and *Sirtl* [75]. The equilibrium constants used in this chapter have been calculated from eq. (97) or eq. (98):

$$\ln K(T) = - \frac{\Delta g_0}{RT} \tag{97}$$

$$\log K(T) = - \frac{\Delta g_0}{4.575\,T} = - \frac{1}{4.575\,T} \left[\sum_i v'_i\, G'^0_i(T) - \sum_j v_i\, G^0_i(T) \right] \tag{98}$$

where

v_i is the stoichiometric coefficient,
G°_i is the *standard Gibbs free energy* of a component i at absolute temperature T,
$K(T)$ is the equilibrium constant at absolute temperature T,
R   is the gas constant and the prime ' corresponds to the reaction products.

Defining an equilibrium factor of a component i at the temperature T:

$$\log k_i(T) = - \frac{G^0_i(T)}{4.575\,T} \tag{99}$$

we obtain

$$\log\ K(T) = \sum_i v'_i \log k'_i(T) - \sum_i v_i \log_i k_i(T). \tag{100}$$

Since

$$G_i(T) = H_i(T) - TS_i(T) \tag{101}$$

$$H_i(T) = H_i(298) + \int_{298}^{T} C_{pi}(T)\, dT \tag{102}$$

$$S_i(T) = S_i(298) + \int_{298}^{T} \frac{C_{pi}(T)}{T}\, dT \tag{103}$$

where

$H_i(T)$ is the molar enthalpy of a component i at temperature T,
$S_i(T)$ is the molar entropy of a component i at temperature T, and
$C_{pi}(T)$ is the molar heat capacity of a component i at temperature T.

Combining eqs. (99) and (101)–(103) we obtain

$$\log k_i(T) = -\frac{1}{4.575}\left[\frac{H_i(298)}{T} - S_i(298) + \frac{1}{T}\int_{298}^{T} C_{pi}(T)\, dT - \int_{298}^{T} \frac{C_{pi}(T)}{T}\, dT\right]. \tag{104}$$

Table 4. Logarithms of equilibrium factors

$\log_{10} k_i$

Species	1 450 K	1 500 K	1 523 K	1 550 K	1 600 K
Si(s)	1.9533	1.9888	2.0049	2.0236	2.0576
$SiCl_{4(g)}$	45.0131	44.3720	44.0933	43.7782	43.2271
$SiHCl_{3(g)}$	37.5233	37.0771	36.8837	36.6656	36.2855
$SiH_2Cl_{2(g)}$	29.5706	29.3204	29.2127	29.0920	28.8832
$SiH_3Cl_{(g)}$	21.1587	21.1054	21.0840	21.0611	21.0247
$SiH_{4(g)}$	12.1374	12.2819	12.3470	12.4223	12.5587
$SiCl_{2(g)}$	22.8649	22.7535	22.7059	22.6528	22.5617
$HCl_{(g)}$	14.2978	14.2309	14.2022	14.1701	14.1150
$H_{2(g)}$	8.0318	8.0746	8.0940	8.1165	8.1574

The last equation contains only data given by *Seiter* and *Sirtl* in their paper [75]. The computed equilibrium factors are given in Table 4. Inserting them into eq. (100) the logarithms of the respective equilibrium constants have been obtained (Table 5). The use of equilibrium factors $\log k_i(T)$ is very convenient, because by simple addition and subtraction they immediately give logarithms of equilibrium constants for any combination of species involved in the system.

The detailed mathematical descriptions of the system Si–H–Cl have been given by *Lever* [73], who unfortunately applied older thermodynamic data, *Teichmann* and *Wolf* [79], *Niederkorn* and *Wohl* [87], *Harper* and *Lewis* [88], and recently by *Sirtl* [89], and by *Hunt* and *Sirtl* [90].

According to *Sirtl* [89], the initial $SiCl_4$ partial pressure at which silicon is neither deposited nor etched is a sensitive indicator to the reliability of thermodynamic

124

Table 5. Logarithms of equilibrium constants

$\log_{10} K_i$

Reaction	1 450 K	1 500 K	1 523 K	1 550 K	1 600 K
$Si + 4\,HCl = SiCl_4 + 2\,H_2$	1.932	1.609	1.468	1.307	1.025
$Si + 3\,HCl = SiHCl_3 + H_2$	0.708	0.470	0.366	0.248	0.040
$Si + 2\,HCl = SiH_2Cl_2$	-0.978	-1.130	-1.197	-1.272	-1.405
$Si + HCl + H_2 = SiH_3Cl$	-3.124	-3.189	-3.217	-3.249	-3.305
$Si + 2\,H_2 = SiH_4$	-5.880	-5.856	-5.846	-5.834	-5.814
$Si + 2\,HCl = SiCl_2 + H_2$	0.348	0.378	0.391	0.406	0.432

calculations. This statement is valid if the composition throughout the whole gas phase is uniform. Otherwise, the threshold point between deposition and etching is shifted, depending on the mass-transfer coefficients which are affected by geometry and dynamics of flow.

2.2.8.5. Description of the nine-component system with six chemical reactions

Taking into account the conclusions deduced by *Lever* [73], *Sirtl* [89], and by *Hunt* and *Sirtl* [90], as to the kind of important reactions occurring in the Si–H–Cl system, the analysis given below is based on a 9-component system with the following reactions:

$$Si + 4\,HCl \rightleftharpoons SiCl_4 + 2\,H_2 \tag{105}$$

$$Si + 3\,HCl \rightleftharpoons SiHCl_3 + H_2 \tag{106}$$

$$Si + 2\,HCl \rightleftharpoons SiH_2Cl_2 \tag{107}$$

$$Si + HCl + H_2 \rightleftharpoons SiH_3Cl \tag{108}$$

$$Si + 2\,H_2 \rightleftharpoons SiH_4 \tag{109}$$

$$Si + 2\,HCl \rightleftharpoons SiCl_2 + H_2. \tag{110}$$

The components are denoted by three-cipher indices; the first cipher indicates the number of silicon atoms in the molecule, the second hydrogen, the third chlorine. Hence, $SiCl_4$ corresponds to the index 104, $SiHCl_3$ to 113, SiH_2Cl_2 to 122, SiH_3Cl to 131, SiH_4 to 140, $SiCl_2$ to 102, HCl to 011, and H_2 to 020.
According to the considerations given in Section 2.2.5.2, the equations of metastable equilibria of reactions (105)–(110) have the form:

$$a\,K_{104} = P_{104}\,p_{020}^2\,p_{011}^{-4} \tag{111}$$

$$a\,K_{113} = p_{113}\,p_{020}\,p_{011}^{-3} \tag{112}$$

$$a\,K_{122} = p_{122}\,p_{011}^{-2} \tag{113}$$

125

$$a\,K_{131} = p_{131}\,p_{020}^{-1}\,p_{011}^{-1} \tag{114}$$

$$a\,K_{140} = p_{140}\,p_{020}^{-2} \tag{115}$$

$$a\,K_{102} = p_{102}\,p_{020}\,p_{011}^{-2} \tag{116}$$

where

a is the activity of silicon at metastable equilibrium,

K_j is the equilibrium constant of the corresponding reaction, and

p_i is the partial pressure of the component i at the gas–silicon interface, at metastable equilibrium.

The sum of partial pressures is equal to the total pressure P:

$$p_{020} + p_{011} + p_{104} + p_{113} + p_{122} + p_{131} + p_{140} + p_{102} = P \tag{117}$$

In agreement with assumption (c) Section 2.2.7.1, all components except $SiCl_4$ and H_2 are formed only at the gas–silicon interface and diffuse away from it across the boundary layer. From the conditions defined by assumptions (c) and (e) it follows that the partial pressures of these components are negligible in the main gas stream.

From this and from definitions (85) and (86), we obtain

$$j_{104} = p_0 - p_{104} \tag{118}$$

$$j_{113} = \quad - b_{113}\,p_{113} \tag{119}$$

$$j_{122} = \quad - b_{122}\,p_{122} \tag{120}$$

$$j_{131} = \quad - b_{131}\,p_{131} \tag{121}$$

$$j_{140} = \quad - b_{144}\,p_{140} \tag{122}$$

$$j_{102} = \quad - b_{102}\,p_{102} \tag{123}$$

$$j_{011} = \quad - b_{011}\,p_{011} \tag{124}$$

where

p_0 is the $SiCl_4$ partial pressure in the main gas stream, approximately equal to the initial partial pressure.

The chlorine balance yields:

$$4\,j_{104} + 3\,j_{113} + 2\,j_{122} + j_{131} + 2\,j_{102} + j_{011} = 0. \tag{125}$$

The silicon balance yields:

$$j_0 = j_{104} + j_{113} + j_{122} + j_{131} + j_{140} + j_{102}. \tag{126}$$

Since the system comprises non-linear equations, it is difficult to obtain a straight-

126

forward solution and therefore it is advisable to introduce the parameter x [73] defined as:

$$x = \frac{p_{011}}{p_{020}}. \tag{127}$$

By inserting x into eqs. (111)–(116), and (118)–(124), the number of independent variables is reduced to p_{011}, a, and x:

$$p_{104} = p_{011}^2 \, a \, K_{104} \, x^2 \tag{128}$$

$$p_{113} = p_{011}^2 \, a \, K_{113} \, x \tag{129}$$

$$p_{122} = p_{011}^2 \, a \, K_{122} \tag{130}$$

$$p_{131} = p_{011}^2 \, a \, K_{131} \, x^{-1} \tag{131}$$

$$p_{140} = p_{011}^2 \, a \, K_{140} \, x^{-2} \tag{132}$$

$$p_{102} = p_{011} \, a \, K_{102} \, x \tag{133}$$

$$p_{020} = p_{011} \, x^{-1} \tag{134}$$

$$j_{104} = p_0 - p_{011}^2 \, a \, K_{104} \, x^2 \tag{135}$$

$$j_{113} = -p_{011}^2 \, b_{113} \, a \, K \, K_{113} \, x \tag{136}$$

$$j_{122} = -p_{011}^2 \, b_{122} \, a \, K_{122} \tag{137}$$

$$j_{131} = -p_{011}^2 \, b_{131} \, a \, K_{131} \, x^{-1} \tag{138}$$

$$j_{140} = -p_{011}^2 \, b_{140} \, a \, K_{140} \, x^{-2} \tag{139}$$

$$j_{102} = -p_{011} \, b_{102} \, a \, K_{102} \, x \tag{140}$$

$$j_{011} = -p_{011} \, b_{011} \, . \tag{141}$$

Inserting the expressions (128)–(141) into eqs. (117), (125), (126) and denoting:

$$K_{104} \, x^2 + K_{113} \, x + K_{122} + K_{131} \, x^{-1} + K_{140} \, x^{-2} = F_1(x) \tag{142}$$

$$4 \, K_{104} \, x^2 + 3 \, b_{113} \, K_{113} \, x + 2 \, b_{122} \, K_{122} + b_{131} \, K_{131} \, x^{-1} = F_2(x) \tag{143}$$

$$K_{104} \, x^2 + b_{113} \, K_{113} \, x + b_{122} \, K_{122} + b_{131} \, K_{131} \, x^{-1} + b_{140} \, K_{140} \, x^{-2} = F_3(x), \tag{144}$$

we obtain a system of equations with three unknown quantities x, p_{011}, j_0:

$$p_{011}(a \, K_{102} \, x + x^{-1} + 1) + p_{011}^2 \, a \, F_1(x) = P \tag{145}$$

$$p_{011}^2 \, a \, F_2(x) + 2 \, p_{011} \, a \, b_{102} \, K_{102} \, x + p_{011} \, b_{011} = 4 \, p_0 \tag{146}$$

$$p_0 - p_{011}^2 \, a \, F_3(x) - p_{011} \, a \, b_{102} \, K_{102} \, x = j_0 \, . \tag{147}$$

For convenience, it is advisable to assume p_0 as an unknown quantity and to insert various values of x. In this way using computer techniques one gets numerical

127

tables in which for a given activity a each value of x corresponds to appropriate values of p^0, p_{011}, and f_0. The p_0 value is usually known from the experiment. Hence, for given a and p_0 values the corresponding x, p_{011} and j_0 values are taken from the above-mentioned numerical tables. Inserting them into eqs. (128)–(141) we obtain all partial pressures p_i, and the corresponding relative fluxes j_i.

2.2.8.6. Ratios of mass-transfer coefficients

2.2.8.6.1. *Effect of diffusion in hydrogen*

In agreement with considerations of Section 2.2.2, the basic quantities by which ratios of mass-transfer coefficients are affected are the ratios of the diffusion constants of both components in hydrogen and the exponent m depending on flow dynamics. Diffusion constants can either be determined experimentally or calculated using approximate methods. *Shepherd* [40] cited the curve of diffusion constants in hydrogen as a function of molecular weights of various gaseous compounds [91] and used it in his calculations of silicon deposition rates.

2.2.8.6.2. *Effect of temperature*

The estimation of temperature effects is possible by means of *Arnold*'s *equation* (17). In the system Si–H–Cl the highest *Sutherland constant* exhibits $SiCl_4$ ($C = 101$), the smallest one HCl ($C = 84$). The exponents n defined by the equation:

$$\frac{D(T_2)}{D(T_1)} = \left(\frac{T_2}{T_1}\right)^n \tag{148}$$

are calculated by means of *Arnold*'s *equation* to be $n = 1.623$ and $n = 1.641$ for HCl and $SiCl_4$, respectively. This justifies the assumption that in the system Si–H–Cl the coefficients b_i, defined by eq. (86), are almost independent of temperature.

2.2.8.6.3. *Effect of other compounds*

According to the consideration of Section 2.2. 3.2, particular gaseous components do not diffuse in pure hydrogen, but they diffuse into a multicomponent gas mixture. Therefore it is possible to estimate the corrections by appropriate application of *Wilke*'s equations. The ratios of diffusion constants without and with the correction are given in Table 6.

2.2.8.6.4. *Corrections for decomposition in the boundary layer*

According to *Shepherd* [40], the products $SiCl_2$, SiH_2Cl_2, and SiH_3Cl are stable in contact with the solid silicon, but they are unstable if solid silicon is absent and therefore they decompose at some distance from the silicon surface. Our computer calculations, based on experimental data, confirmed this phenomenon. As is known from

128

Table 6. Estimated ratios of diffusion constants

D_i/D_{SiCl_4}

Species i	In pure hydrogen	In gas mixture resulting from $H_2 + SiCl_4$
$SiHCl_3$	1.158	1.172
SiH_2Cl_2	1.427	1.485
SiH_3Cl	2.170	2.378
SiH_4	3.620	4.440
$SiCl_2$	1.441	1.510
HCl	3.70	4.430

the theory, the effective thickness of the boundary layer is different for different components. The decomposition of a definite component decreases the boundary layer's effective thickness with respect to this component and this results in an increase of the mass-transfer coefficient. Since, simultaneously, some amount of HCl is formed, the effective mass-transfer coefficient of HCl decreases.

The gas mixture contains only small amounts of SiH_2Cl_2 and SiH_3Cl. Therefore one can assume, in agreement with the first approximation for corrections for decomposition, that at the phase boundary only one reaction occurs: $Si + 2\,HCl = SiCl_2 + H_2$. Solving the system of eqs. (116)–(118), and (123)–(126) for $j_0 = 0$, neglecting partial pressures of gases other than HCl, $SiCl_2$, H_2, we obtain

$$\frac{1}{p_0} = \frac{4\,K_{102}\,b_{102}}{b_{011}^2} + \frac{1}{b_{102}} + \frac{2}{b_{011}}. \tag{149}$$

For the experimental [27] point $p_0 = 0.15$, $j_0 = 0$ and for the equilibrium constant $K_{102}\,(1\,523\,K) = 2.458$, the result is

$$b_{011} = \frac{1 + \sqrt{65.55\,b_{102} - 8.832}}{6.667 - \dfrac{1}{b_{102}}}. \tag{150}$$

If p_{102} and p_{011} denote partial pressures at some distance from the interface without decomposition, the designation at decomposition being p'_{102} and p'_{011}, then by assuming a decomposition according to the equation

$SiCl_2 + H_2 = Si + 2\,HCl$

$1 - y \qquad\qquad\qquad 2y$

one obtains

$$p'_{102} = p_{102}(1 - y) \tag{151}$$

$$p'_{011} = p_{011} + 2y\,p_{102}. \tag{152}$$

9 Epitaxy

The corresponding corrections d_{102} and d_{011} have the form

$$d_{102} = \frac{p_{102}}{p_{102}(1-y)} = \frac{1}{1-y} \tag{153}$$

$$d_{011} = \frac{p_{011}}{p_{011} + 2y\,p_{102}} = \frac{1}{1 + 2y\,\dfrac{p_{102}}{p_{011}}} \cdot \tag{154}$$

Combining eqs. (118), (123)–(126), (153), and (154) we obtain

$$\frac{1}{d_{011}} = 1 + \left(1 - \frac{1}{d_{102}}\right)\frac{b_{011}}{b_{102}} \cdot \tag{155}$$

Hence, the coefficients b_i with corrections can be expressed by

$$b_{102} = \left(\frac{D_{102}}{D_{104}}\right)^{m} \cdot d_{102} \quad d_{102} > 1 \tag{156}$$

$$b_{011} = \left(\frac{D_{011}}{D_{104}}\right)^{m} \cdot d_{011} \quad d_{011} < 1 \tag{157}$$

Solving the system of eqs. (150), and (155)–(157) for $m = \dfrac{2}{3}$ *i.e.* for the laminar flow, yields $b_{102} = 2.12$, $b_{011} = 1.97$, $d_{102} = 1.60$, $d_{011} = 0.73$. Employing the simplified system of eqs. (116)–(118) and (123)–(126) the value p_0 at $j_0 = 0$ is only by 4 % smaller than the corresponding value obtained by utilizing the complete system of equations. No serious error with respect to silicon deposition is thus made if it is assumed that:

$$d_{122} = d_{131} = d_{102}. \tag{158}$$

The errors in partial pressures of SiH_2Cl_2 and SiH_3Cl are, however, not negligible.

2.2.8.7. Solving the system of equations for given conditions

Taking into account all the above mentioned corrections, the ratios b_i of mass-transfer coefficients for a laminar flow at $T = 1\,523$ K, and $p_0 = 0.15$, $j_0 = 0$, are given by:

$b_{113} = 1.172^{2/3} = 1.11$

$b_{122} = 1.485^{2/3} \cdot 1.60 = 2.09$

$b_{131} = 2.378^{2/3} \cdot 1.60 = 2.85$

$b_{140} = 4.440^{2/3} \cdot = 2.70$

$b_{102} = 1.51^{2/3} \cdot 1.60 = 2.11$

$b_{011} = 4.43^{2/3} \cdot 0.73 = 1.97$

130

Furthermore are given:

$P = 1$ atm., $T = 1\,523$ K, $K_{104} = 29.35$, $K_{113} = 2.323$, $K_{122} = 6.36 \cdot 10^{-2}$, $K_{131} = 6.07 \cdot 10^{-4}$, $K_{140} = 1.426 \cdot 10^{-6}$, $K_{102} = 2.458$.

The system of eqs. (111)–(126) has been solved for the constant activities $a = 1.001$; 1.05; 1.1; 1.2; 1.3; 1.4; 1.5; 1.75; 2; 2.5; 3; 4; 5; 13 curves have been obtained, each for one value of activity (Fig. 2 (thin line)). The choice of an appropriate curve is possible if some experimental data are available.

The function j_0 versus p_0 (at $a = $ const.) for epitaxial silicon deposition from $SiCl_4 + H_2$ mixtures, exhibits a maximum. Maximum points on curves for different a-values are localized along an almost straight line. This permits an essential simplification of the general method presented in Section 2.2.7.5.

The increase of activity results in a decrease of abscissae p_0 of the maximum points. If the experimental maximum point is available, it is possible to determine the activity value from its abscissa. For example, according to experiments of *Bylander* [27], the J curve crosses the p_0 axis at $p_0 = 0.15$ and exhibits a maximum in the range

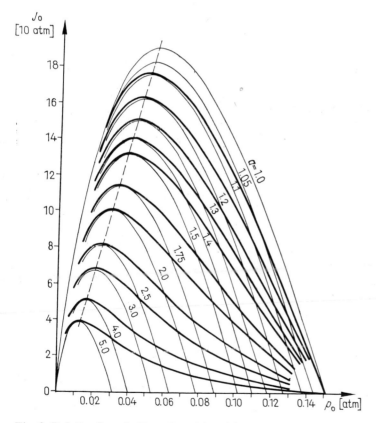

Fig. 2. Relative flux of silicon deposition (j_0) versus initial pressure (p_0) of $SiCl_4$ in ($SiCl_4 + H_2$) mixtures. a is the activity of silicon

$0.05 < p_0 < 0.06$. The number of published experimental points in the vicinity of the maximum was too small to determine $p_{0(max)}$ more accurately. The corresponding range of activity is $1 < a < 1.1$, e.g. $p_{0(max)} = 0.052$ corresponds to an activity $a = 1.05$. Taking this point as an approximation, and the relation

$$j_0 = C(a - 1) \tag{62a}$$

and calculating j_0 by solving the system of eqs. (111)–(126), we obtain $C = 0.366$. Using this value provides the possibility for the construction of the curve j_0 versus p_0 at constant C and variable a. In Fig. 1 such curves are shown displaying various activity values at the maximum points (drawn with thick lines). The ordinates of the curve for $C = 0.366$ are almost proportional to the ordinates of the experimental [27] one for J_0 versus p_0 (J_0 denoting deposition flux). When the equation

$$j_0 = C_2(a - 1)^2 \tag{63a}$$

is utilized, the agreement is worse.

From these considerations and from the experimental data [27], one can see at least that the epitaxial silicon deposition from ($SiCl_4 + H_2$) mixtures occurs at low activities, and correspondingly at low supersaturations. Otherwise, the experimental J curve, which exhibits $j_0 = 0$ at $p_0 = 0.15$ would have its maximum at lower p_0 values. From the above-mentioned experimental and theoretical data, it is possible to calculate the mass-transfer coefficient of $SiCl_4$ for one set of conditions, from the relation

$$k_{104} = \frac{J_0}{j_0} = \frac{5.68 \cdot 10^{-7} \, \text{mole} \cdot \text{cm}^{-2} \cdot \text{s}^{-1}}{1.83 \cdot 10^{-2} \, \text{atm}} = 3.1 \cdot 10^{-5} \, \text{mole} \cdot \text{cm}^{-2} \cdot \text{s}^{-1} \, \text{atm}^{-1}.$$

The value in the numerator was taken from experiments [27]; the value in the denominator was obtained by solving the system of eqs. (111)–(126). According to definition (86), the mass-transfer coefficients of all other components are given by

$$k_i = b_i \, k_{104} = 3.1 \cdot 10^{-5} \, b_i.$$

2.2.8.8.   Discussion and conclusions

2.2.8.8.1.   *Kinetically controlled CVD*

Because of the complexity of CVD of silicon and the possibility of many mechanisms accounting for this process, up to now only the kinetics of the simplest case, the pyrolysis of the molecular beam of SiH_4, has been described mathematically [58]–[62]. However, other CVD silicon processes used in semiconductor technology have not been treated in this way, and the problem of their mechanism and kinetics remains still open.

2.2.8.8.2.   *Diffusion controlled CVD*

Because of the lack of a sufficient number of necessary experimental data in the literature, the description presented here includes many simplifications and estimations. Thus the results may be considered as a first approximation only. Availability of reliable experimental data should considerably improve the accuracy of the presented model. In these data should be included: diffusion constants of all components, mass-transfer coefficients as functions of flow dynamics and geometry, kinetic data on decomposition of $SiCl_2$, SiH_2Cl_2, and SiH_3Cl in the gas phase, composition of gas at the outlet from the reactor as function of temperature, pressure, initial composition, flow dynamics, and geometry. It seems reasonable to express some experimental data on silicon deposition in terms of *Reynolds* and *Schmidt* or *Sherwood numbers*. Since the control of perfection of crystal deposits achieved by the control of parameters defining deposition conditions is essential for technology, it seems that a further development should follow the above mentioned lines.

2.2.9.   REMARKS ON DIFFUSION CONTROLLED CVD

The most important technical goal is to obtain crystal layers with a good quality and — for economic reasons — with practically significant deposition rates. The deposition depends on various measurable parameters. A correlation of them is shown in Fig. 3. The deviation from equilibrium, and hence the deposition rate and the deposit quality, vary with these parameters. According to considerations given in Section 2.2.6, the growth mechanism and deposit perfection are both affected by the deviation from equilibrium, *i.e.* also by the activity of a pure solid phase in a multicomponent steady-state CVD system, being a measure of departure from equilibrium, which corresponds to the supersaturation ratio in a one-component system. Hence, the knowledge of its value at the gas–solid interface is essential for

a) control of morphology and perfection of growing crystal films, by controlling the conditions correlated with activity, such as temperature, pressure, initial gas composition, flow dynamics, and geometry,
b) appropriate choice of a CVD system,
c) a limited application of some conclusions of nucleation and crystal growth theories, in principle developed for one-component systems.

Since the activity depends also on the C-value (eqs. (62a) and (63a)) and C, in turn, depends on the nature of the chemical processes and on the chemical compounds transporting the component being deposited, it is possible to achieve relatively high deposition rates at small activities by an appropriate choice of a chemical process meeting a large C-value. Such a condition is advantageous for a good perfection of crystal films. The increase of C values in CVD systems can be achieved if for example the system comprises a gaseous etching component (see Section 2.2.6.1).
Although theoretical considerations [17]–[20] and those in Section 2.2.6.2 concern in principle one-component PVD systems, the effect of changes of supersaturation

133

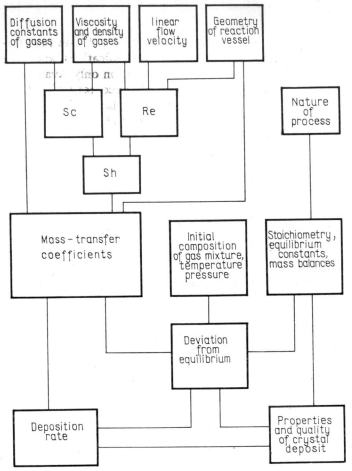

Fig. 3. Correlation of parameters for CVD flow systems (Re — *Reynolds*, Sc — *Schmidt*, Sh—*Sherwood number*)

ratio (in **PVD** systems) or of the activity (in CVD multicomponent systems) should be similar, since both supersaturation ratio and activity are related to the same excess of *Gibbs free energy*.

The determination of deviation from equilibrium is possible if — apart from stoichiometry, equilibrium constants, material balances, pressure, temperature, and the initial composition of the gas mixture at the inlet of reactor — the mass-transfer coefficients are available. They are functions of the *Reynolds*, *Schmidt*, and *Sherwood numbers*, *i.e.* functions of diffusion constants, viscosity and density of gases, linear velocity of flow, and the geometry of the reactor. This is essential for designing, especially when changing the scale (*e.g.* laboratory → industry) of a given technological process, its optimization and automation. Therefore, it is advisable to express the results of experiments in a set of terms given in Fig. 3; otherwise, the comparison of the experimental results given by various authors could be problematic.

2.2.10. REFERENCES

[1] *Spacil, H. S., Wulff, J.:* Final Report to Office of Ordnance Research under U.S. Army Contract No. DA-19-020-ORD-3760, Jan. 31, 1958.

[2] *Noyes, A. A., Whitney, W. R.:* Z. physik. Chem. 23 (1897) 689.

[3] *Nernst, W.:* Z. physik. Chem. 47 (1904) 52.

[4] *Schäfer, H., Jacob, H., Etzel, K.:* Z. anorg. u. allg. Chem. 286 (1956) 27.

[5] *Schäfer, H., Morcher, B.:* Z. anorg. u. allg. Chem. 290 (1957) 279.

[6] *Bennett, C. O., Myers, J. E.:* Momentum, mass and heat transfer. McGraw Hill, New York (1962).

[7] *Bradshaw, S. E.:* Int. J. Electronics 23 (1967) 381.

[8] *Higbie, R.:* Trans. AIChE 31 (1935) 365.

[9] *Danckwerts, P. V.:* Ind. Eng. Chem. 43 (1951) 1460.

[10] *Hanratty, T. J.:* AIChE Journal 2 (1956) 359.

[11] *Sutherland, W.:* Phil. Mag. 36 (1893) 507; 40 (1895) 421; 9 (1905) 781.

[12] *Arnold, J. H.:* Ind. Eng. Chem. 22 (1930) 1091.

[13] *Gilliland, E. R.:* Ind. Eng. Chem. 26 (1934) 681.

[14] *Trautz, M., Müller, W.:* Ann. Physik 22 (1935) 313, 329, 353; Landolt-Börnstein, Physikalisch-chemische Tabellen, 2. Teil, Bandteil A, Springer Verlag, Berlin (1960).

[15] *Wilke, C. R.:* Chem. Eng. Progr. 46 (1950) 95.

[16] *Fairbanks, D. F., Wilke, C. R.:* Ind. Eng. Chem. 42 (1950) 471.

[17] *Hirth, J. P., Moazed, K. L., Ruth, V.:* Heterogene Keimbildung und Epitaxie bei der Kondensation aus der Dampfphase, in: (Ed.: H. G. Schneider) Epitaxie — Endotaxie, VEB Deutscher Verlag für Grundtoffindustrie, Leipzig (1969) p. 250.

[18] *Burton, W. K., Cabrera, N., Frank, F. C.:* Phil. Trans. Roy. Soc. A 243 (1951) 299.

[19] *Pound, G. M., Simnad, M. T., Young, L.:* J. Chem. Phys. 22 (1954) 1215.

[20] *Chakraverty, B. K., Pound, G. M.:* Acta Met. 12 (1964) 851.

[21] *Moazed, K. L., Hirth, J. P.:* Surface Science 3 (1964) 49.

[22] *Theuerer, H. C.:* J. Electrochem. Soc. 108 (1961) 649.

[23] *Li, C. H.:* J. Electrochem. Soc. 109 (1962) 952.

[24] *Cave, E. F., Czorny, R.:* RCA Rev. 24 (1963) 523.

[25] *Theuerer, H. C.:* Proc. IRE 48 (1960) 1642.

[26] *Theuerer, H. C.:* J. Electrochem. Soc. 109 (1962) 742.

[27] *Bylander, E. G.:* J. Electrochem. Soc. 109 (1962) 1171.

[28] *Miller, K. J., Manz, R. C., Grieco, M. J.:* J. Electrochem. Soc. 109 (1962) 643.

[29] *Thomas, C. O.:* J. Electrochem. Soc. 109 (1962) 1055.

[30] *Lombos, B. A., Somogui, T. R.:* J. Electrochem. Soc. 111 (1964) 1097.

[31] *Steinmaier, W.:* Philips Res. Repts 18 (1963) 75.

[32] *Deal, B. F.:* J. Electrochem. Soc. 109 (1962) 514.

[33] *Monchamp, R. R., McAleer, W. J., Pollak, P. I.:* J. Electrochem. Soc. 111 (1964) 879.

[34] *Tung, S. K.:* The influence of process parameters on the growth of epitaxial silicon, in: Metallurgy of Semicond. Materials, Metallurgical Society Conferences 15, pp. 87—102. (Ed.: J. B. Schroder) Interscience Publishers, New York (1961).

[35] *Sedgwick, T. O.:* J. Electrochem. Soc. 111 (1964) 1381.

[36] *Wolf, E., Teichmann, R.:* Z. f. Chem. 2 (1962) 343.

[37] *Nakagawa, M.:* J. Chem. Soc. Japan 65 (1962) 466.

[38] *Theuerer, H. C.:* U.S. patent No. 3 142 596 (July 28, 1964).

[39] *Nicoll, F. M.:* J. Electrochem. Soc. 110 (1963) 1165.

[40] *Shepherd, W. H.:* J. Electrochem. Soc. 112 (1965) 988.

[41] *Arizumi, T., Nishinaga, T., Kasuga, M.:* A continuous procedure for growing epitaxial silicon films in: L. Steipe: Mikroelektronik. Oldenbourg Verlag, München–Wien (1967).

[42] *Alexander, E. G.:* Paper presented at the Meeting of the Electrochemical Society, Dallas, May 1967; J. Electrochem. Soc. 114 (1967) 650.

[43] *Grossmann, J. J.:* Paper presented at the Meeting of the Electrochemical Society, Dallas, May 1967; J. Electrochem. Soc. 114 (1967) 53C.

[44] *Shepherd, W. H.:* J. Electrochem. Soc. 115 (1968) 541.

[45] *Glang, R., Wajda, E. S.:* Trans AIME 15 (1962) 27.
[46] *Charig, J. M., Joyce, B. A.:* J. Electrochem. Soc. 109 (1962) 957.
[47] *Teichmann, R., Wolf, E.:* Z. anorg. allg. Chem. 347 (1966) 113.
[48] *Sangster, R. C., Maverich, E. F., Croutch, M. L.:* J. Electrochem. Soc. 104 (1957) 317.
[49] *Miller, K. J., Grieco, M. J.:* J. Electrochem. Soc. 110 (1963) 1252.
[50] *Seki, H., Araki, H.:* Jap. J. Appl. Phys. 4 (1965) 645.
[51] *Bhola, S. R., Mayer, A.:* RCA Rev. (1963) 511.
[52] *Joyce, B. A., Bradley, R. R.:* J. Electrochem. Soc. 110 (1963) 1235.
[53] *Meyer, S. E., Shea, D. E.:* J. Electrochem. Soc. 111 (1964) 550.
]54] *Becker, G. R., Joyce, B. A., Bradley, R. R.:* Phil. Mag. 10 (1964) 1087.
[55] *Purnell, J. H., Walsh, R.:* Proc. Roy. Soc. A293 (1966) 543.
[56] *Chu, L., Gruber, G. A.:* J. Electrochem. Soc. 113 (1966) 62C.
[57] *Chu, L., Gruber, G. A.:* J. Electrochem. Soc. 114 (1967) 522.
[58] *Joyce, B. A., Bradley, R. R.:* Phil. Mag. 14 (1966) 289.
[59] *Booker, G. R., Joyce, B. A.:* Phil. Mag. 14 (1966) 301.
[60] *Joyce, B. A., Bradley, R. R., Booker, G. R.:* Phil. Mag. 15 (1967) 1167.
[61] *Watts, B. E., Bradley, R. R., Joyce, B. A.:* Phil. Mag. 17 (1968) 163.
[62] *Joyce, B. A., Bradley, R. R., Booker, G. R.:* Phil. Mag. 19 (1969) 403.
[63] *May, J. E.:* J. Electrochem. Soc. 112 (1965) 710.
[64] *Lieth, R. M. A., Eggels, A. G. M.:* J. Appl. Phys. 35 (1964) 3015.
[65] *Wajda, E. S., Glang, R.:* Metallurgy of Elemental Compound Semiconductors, (Ed.: R. O. Grubel) Interscience Publ., New York (1961) p. 229.
[66] *Sirtl, E.:* J. Phys. Chem. Solids 24 (1963) 1285.
[67] *Nicoll, F. H.:* J. Electrochem. Soc. 110 (1963) 1165.
[68] *Glang, R.:* IBM Journal 4 (1960) 299.
[69] *Newman, R. C., Wakefield, J.:* J. Electrochem. Soc. 110 (1963) 1068.
[70] *Bloem, J., Scholte, J. W. A.:* J. Electrochem. Soc. 112 (1965) 1211.
[71] *Schäfer, H.:* Chemische Transportreaktionen. Verlag Chemie, Weinheim (1962).
[72] *Sirtl, E.:* J. Phys. Chem. Solids 24 (1963) 1285.
[73] *Lever, R. F.:* IBM Journal 8 (1964) 460.
[74] *Schäfer, H., H. Heine:* Z. anorg. allg. Chem. 332 (1964) 25.
[75] *Seiter, H., E. Sirtl:* Z. Naturforsch. 21a (1966) 1696.
[76] *Sirtl, E.:* Z. Naturforsch. 21a (1966) 2001.
[77] *Schäfer, H., Bruderreck, H., Morcher, B.:* Z. anorg. allg. Chem. 352 (1967) 122.
[78] *Schäfer, H., Nickl, J.:* Z. anorg. allg. Chem. 274 (1953) 250.
[79] *Teichmann, R., Wolf, E.:* Z. anorg. allg. Chem. 347 (1966) 113.
[80] *Sirtl, E., Reuschel, K.:* Z. anorg. allg. Chem. 332 (1964) 113.
[81] *Sirtl, E.:* in: Festkörperprobleme, vol. 6. Vieweg und Sohn, Braunschweig (1967).
[82] *Kubaschewski, O., Evans, E. L.:* Metallurgical Thermochemistry Pergamon Press, New York: (1967).
[83] *Stull, D., Sinke, G.:* Thermodynamic Properties of the Elements, Adv. Chem. Ser. 18, Am. Chem. Soc., Washington (1956).
[84] *Brimm, E. O., Humphrey, H. H.:* J. Phys. Chem. 61 (1957) 829.
[85] *Mikawa, S.:* Nippon Kagaku Zasshi 81 (1960) 1512.
[86] *Černy, C., Erdös, E.:* Coll. Czechoslov. Chem. Communs. 19 (1953) 645.
[87] *Niederkorn, J., Wohl, A.:* Rev. Roum. Chim. 11 (1966) 85.
[88] *Harper, M. J., Lewis, T. J.:* Tech. Memo. 3/M/66, Clearing-House Document AD 641 310, British Ministry of Aviation (1966).
[89] *Sirtl, E.:* Thermodynamics and Silicon Vapor Deposition, in: R. R. Haberecht and E. L. Kern: Semiconductor Silicon, Electrochemical Society, New York (1969).
[90] *Hunt, L. P., Sirtl, E.:* A Thermodynamic Analysis of the Silicon — Hydrogen — Chlorine Vapor Deposition System, presented at the Electrochemical Society Meeting in Los Angeles, May 1970.
[91] International Critical Tables 5 (1929) p. 62.

136

2.3.   **CHEMICAL VAPOR DEPOSITION OF GaP AND DEFECTS IN CVD-GROWN EPITAXIAL GaP LAYERS**

J. Noack

Central Institute of Electron Physics, Academy of Sciences of the GDR, Berlin, German Democratic Republic

2.3.1.   **GENERAL REMARKS**

The importance of GaP monocrystals and monocrystalline films is mainly due to their widespread use in optoelectronic applications. It is known that, at present, the material for light emitting diodes (LED) is prepared by solution growth from a Ga melt, deposited epitaxially on a monocrystalline GaP substrate.

Investigations have shown that GaP epitaxial layers grown from the vapor phase exhibit quantum yields which are one or two orders of magnitude less than those from solution growth. Not regarding the advantages of vapor phase reactions for the preparation of large scale films or the application of other substrates than GaP, the unsatisfactory quality of deposits grown from the vapor phase is a question of materials science, *i.e.* of the chemical and crystallographic properties, caused by the special preparation techniques and material combinations.

However, the relation between preparation technique, crystallographic perfection, and physical properties is a more general question beyond the special case considered here.

The chemical and crystallographic perfection of epitaxial layers is the most important parameter which determines their applicability, *e.g.* for optoelectronic purposes.

The above-mentioned difficulties arise as a result of the method of preparation on the one hand, and the defects caused by the substrate on the other hand. To the latter, the structure of the substrate, as well as the differences in crystallographic parameters between substrate and deposit, count.

Defects resulting from the method of deposition may be found by considering the homoepitaxy, and the variation of the deposition parameters or of the method. The influence of the substrate is revealed by comparing depositions obtained under equal conditions utilizing different substrates and analysing the defects relative to the different crystallographic parameters, as structure, thermal behavior, defects of the substrate, and − last but not least − the perfection and cleanliness of the substrate surface.

The epitaxy of III–V compounds on another substrate material is a very complex problem and the knowledge of simpler systems, like the epitaxy of elements on the same substrates should be used for comparison.

Even homostructures like silicon films with different dopants give important indications on the thermal behavior and defects resulting from it.

In the present report, we discuss the different methods of layer deposition applied by several investigators, the chemical imperfections arising from the contamination of the layers by the substrate material, like Ge in GaP depositions on germanium substrates, and questions of stoichiometry. The crystallographic defects to be examined are misorientation, twins, stacking faults, dislocations, and stress. The methods of investigation are only briefly described.

The electrical and optical properties of layers are considered in connection with defects. The purpose of this paper is not to describe semiconductor devices, or to emphasize their technical importance or application, but to give a review of the crystallographic imperfections and to demonstrate the scientific interest in heteroepitaxy.

The examples are taken from our experiments using the open tube HCl transport process and Ge substrates, and results of other cited authors.

2.3.2. PREPARATION OF FILMS

The methods for preparing epitaxial deposits of GaP to be reported are based only on vapor phase reactions. The principal means are:
a) the chemical transport reaction in a chlorine or water vapor system or
b) the use of metal organics and phosphine or triethylphosphine.

2.3.2.1. Chemical transport reactions

These reactions may be performed as follows:

(a) GaP source and HCl as the transporting agent
(b) GaP source and H_2O as the transporting agent
(c) Ga source and PCl_3 as the transporting agent and the P source
(d) Ga source and HCl as the transporting agent and PH_3 as the P source.

The thermodynamic conditions for cases (a), (b), and (d) will be described in the Ga–P–H–Cl system. The equilibrium conditions with respect to temperature and partial pressure had been measured by *Kirwan* [1] and *Seki* [2], [3]. The correlations are given in Fig. 1.

To (a): The thermodynamic parameters for the deposition process, *e.g.* temperature, temperature difference, HCl partial pressure, and the deposition rate for the transport of GaP with HCl, can be calculated.

This reaction was employed in our investigations for the deposition of GaP onto Si using a closed tube [4], and for the deposition of GaP onto Ge in an open tube system [5].

GaP–Si heterojunctions were prepared in ampoules of fused silica (200×20 mm) using the following conditions:

Table 1

source temperature	1 170—720°C
deposition temperature	1 150—700°C
temperature difference	3 to 20°C
HCl pressure $p_{HCl}^{20°C}$	150 torr
reaction time	about 100 h
film thickness	0.3 mm

138

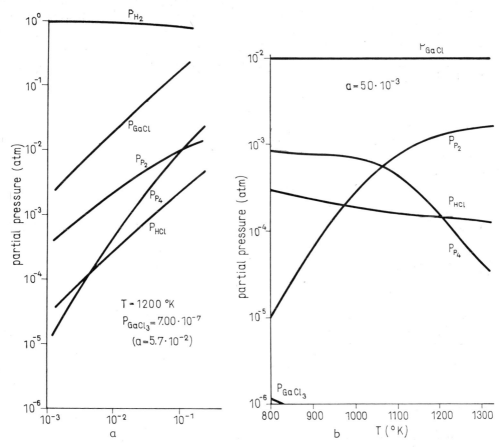

Fig. 1. Composition of the gas phase in the GaP–Cl$_2$–H$_2$ system with a total pressure of 1 atmosphere (Seki [3]) a) versus Cl/H ratio b) versus temperature and with a Cl/H ratio of $5 \cdot 10^{-3}$

For sufficiently plane epitaxial layers, deposition temperatures of 1 100°C are necessary.

For several purposes, *e.g.* testing of transport mechanism, critical deposition temperature or thermal behavior of substrate-deposit combinations, closed tube systems are sufficient. Disadvantages are the ill-defined conditions for the nucleation process and the lack of possibilities to change the deposition parameters during growth. In this connection doping problems, transition zones, etc. are important. For such problems the open tube flow methods are required. The expensive equipment for the open tube process, on the other hand, leads to an essential increase of experimental possibilities, so that its application becomes customary.

For the transport reaction of GaP with HCl several descriptions of experimental set-ups are given in [6], [7]. In our own investigations the possibility of changing the dopant during growth has been considered to prevent additional interfaces. Special attention was given to the influence of compensation. Because some dopants take part in the transport reaction, they must flow through the source room and this

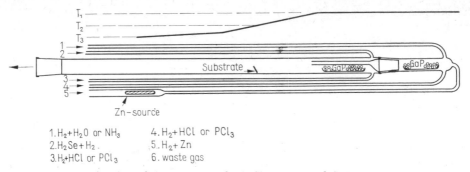

Fig. 2. Schematic drawing of the apparatus for HCl transport of GaP

can give rise to contamination. To prevent compensation, separated sources for p- and n-doping are used. Schematic drawings of the apparatus used, as well as the growth and temperature conditions, are given in Fig. 2.

For runs without doping, a simpler set-up is used. Because all fittings are only on one side of the reaction chamber, it is possible to employ a furnace with a great heat capacity to get a sufficiently constant temperature and to give the reaction mixture an effective preheating before passing the source material.

The conditions for film deposition of GaP onto Ge has been varied as follows:

Table 2

Source temperature [°C]	800–950	950	
Deposition temperature [°C]	750–920	910	
Temp. difference [°C]	35–100	40	optimum
H_2 rate [l h^{-1}]	2–5	4	values
HCl rate [ml h^{-1}]	50–250	240	
Deposition rate [μm h^{-1}]	20–150	100	

To (b): In a further transport reaction, H_2O is used in order to form a volatile Ga compound. Reaction temperatures down to 700°C are possible, but to get epitaxial layers of sufficient quality deposition temperatures of 900–1 000°C are necessary. The reaction is carried out in an open tube apparatus as described by *Frosch* [8], or in a close-spaced open tube apparatus by *Flicker* and *Goldstein* [9]. In most cases GaAs substrates were taken, but $Ga(As_{0.8}P_{0.2})$, GaP, and Ge are used, too.

Typical reaction conditions are: source temperature 1 000–1 100°C, deposition temperature 900–1 000°C with ΔT 20 to 100°C. The water content was about 40 ppm and the deposition rate 1 to 8 ml h^{-1} (25–200 μm h^{-1}). Films are grown up to a thickness of 1.6 mm.

To (c): The use of a Ga source and PCl_3 as the transporting agent and the P source is an important, and often employed, method. The experimental set-up is described in [10]–[13]. It is quite similar to that of transport with HCl or H_2O vapor. Deposition parameters for several substrates are given in Table 3.

Table 3

Parameters	Substrates	
	GaP [11]	GaP, GaAs [13]
Ga temperature [°C]	935	970
Deposition temperature [°C]	840	850–910
Deposition rate μm h^{-1}		50–80

To (d): A great advantage of the application of a separate transport of Ga by HCl and the addition of PH_3 is the possibility of varying the Ga/P ratio and to control accurately all reactant concentrations. The experimental parameters are given by *Tietjen et al.* [14], [15], and are compiled in Table 4.

Table 4

Ga source temperature [°C] 775 − 800
Deposition temperature [°C] 725 − 775

For all reactions in the Ga–H–Cl–P system the deposition temperatures for different substrates are comparable. For transport with wet hydrogen higher temperatures are necessary.

2.3.2.2.   Use of metal organics

A disadvantage of the above-cited reaction systems is the necessity of a carefully controlled two-temperature zone system and a high temperature in the entire reaction room. The application of metal organics makes it possible to form the deposit directly on the heated surface of the wafer. Thus, only the control of one temperature is required. A further advantage is that no reactive residual products are originated which could give rise to etching processes on the substrate. Furthermore, contaminations as mixed layers are prevented. For the deposition of GaP layers the substances employed are trimethylgallium (TMG) or triethylgallium (TEG) and phosphine or phosphorous triethyl (PT). The epitaxial growth apparatus used by *Wang* and *McFarlane* is shown in Fig. 3.

Typical deposition parameters are given in Table 5.

Table 5

Parameters	Substrates		
	Si	GaAs, Al_2O_3	Al_2O_3, $MgAl_2O_4$
Ga source	TEG	TEG, TMG	TMG
P source	PT	PH_3	PH_3
Deposition temperature	485°C	700°C	800°C
Deposition rate h^{-1}	1 350–2 250 Å		0.3 μm
References	[17]	[18]	[16]

141

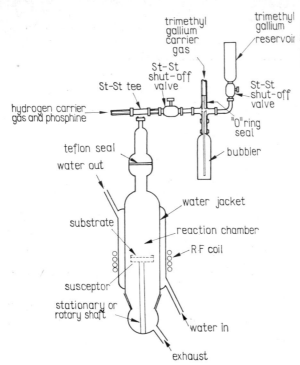

Fig. 3. Epitaxial growth apparatus for GaP deposition using trimethyl-gallium (*Wang* [16]). St–St—stainless steel

2.3.2.3.   Surface preparation

The perfection of the substrate surface is important for the formation of films of satisfactory quality. Mostly, wafers are prepared by mechanical lapping which generates a badly damaged surface layer. It is known that all these deformations and lattice damages cause defects in epitaxial layers. Therefore, usually wafers are etched and chemically polished. Some authors report that the absence of mechanical damage of the substrate surface is controlled by X-ray topography.

In the following, the etching processes are mentioned which are used in our own investigations with Ge or Si substrates and those used by other authors for various other materials.

For silicon the vapor phase etching with a H_2–HCl mixture seems to provide optimum surfaces. With varied conditions we applied this technique for Ge, too (Table 6).

Table 6

Parameters	Substrates	
	Ge	Si
H_2 rates [1 h^{-1}]	120	120
HCl rates [1 h^{-1}]	0.9	0.9
Temperature [°C]	940	1250
Time [h]	1	1

142

The etching conditions were checked by X-ray topographs. *Igarashi* employs an etching with an H_2 : HCl ratio of 50 : 1 and 1 200°C [19].

In papers dealing with sapphire or spinel such etching processes are not described. *Mottram* [13] uses GaAs substrates which were polished with diamond paste and subsequently etched with a 3 : 1 : 1 etchant of $H_2SO_4/H_2O_2/H_2O$ at 60°C and finally etched immediately prior to insertion into the apparatus with a 1 : 9 : 10 etchant of $HF/HNO_3/H_2O$.

Crafford [20] etches GaP with a mixture of HCl, HNO_3, and H_2O or a 5 : 1 : 1 $H_2SO_4/H_2O_2/H_2O$ etchant followed by vapor phase etching with HCl prior to deposition using GaP-(111)-A substrates.

2.3.3. LAYER DEFECTS

2.3.3.1. Surface defects

Usually it is desired to obtain epitaxial layers with a completely plane surface as a visible indication of perfection.

Both the GaP layers epitaxially grown onto Si and onto Ge show mirror-like surfaces if optimum deposition conditions are employed (Figs. 4a and 4b).

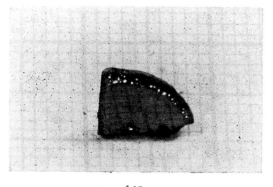

a)

1 cm

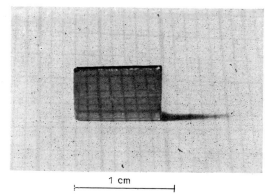

b)

1 cm

Fig. 4. Surface quality of GaP layers epitaxially grown onto a) Si, b) Ge

However, on films grown on Si, cracks occur which become clearly visible under a microscope (Fig. 5a). In Berg–Barrett topographs the cracks appear clearly broadened indicating the presence of elastic strain in the layer (Fig. 5b).

For films grown on a Si wafer with {111} orientation the cracks show a threefold symmetry. Their direction is ⟨112⟩.

As these directions are traces of cleavage planes in the {111} surface, it appears reasonable to consider the cracks as cleavage cracks of the originally monocrystalline layer created during cooling due to the difference in thermal expansion coefficients of the two materials.

Films prepared under comparable conditions by *Igarashi* [19] in the temperature range of 1 050 to 1 150°C show the same results.

Because stress increases with the difference between deposition temperature and room temperature, but decreases with decreasing areas of the epitaxial layers, there are two possibilities to prevent such serious defects.

Thomas [17] describes the preparation of GaP epitaxial layers onto Si at a deposition temperature of 485°C. This deposition process has already been mentioned in this Section. According to electron reflection measurements, the films are single-crystalline. There are no indications of cracks in the deposit.

The second possibility is the preparation of epitaxial layers with small dimensions. *Igarashi* [21] avoided cracks by the selective growth of small size (< 0.02 mm^2) areas. Silicon of different orientations covered by silicon oxide layers of a thickness

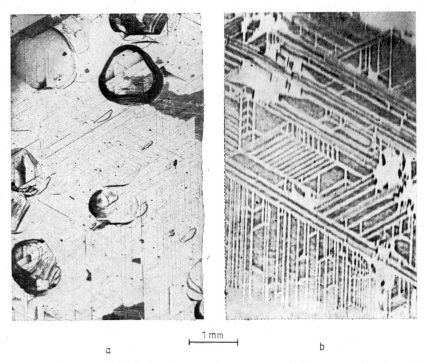

1 mm

a   b

Fig. 5. GaP layer on Si (111) showing growth structures and cleavage cracks; a) optical micrograph, b) X-ray topograph of the same area ((111) reflection, Cr-K$_\alpha$ radiation)

144

of about 10 000 Å, with windows formed by the use of photolithographic methods and etching, were taken as substrates. The deposition temperature was 1 170°C, and the deposition rate 1 μm h^{-1}, respectively.

Other surface defects are hillocks and pittings. Hillocks, as seen on Fig. 5a, are also found by *Mottram et al.* [13] in the growth of GaP in the open H$_2$(PCl$_3$)Ga system. They were attributed to dust particles or to gallium droplet formation on the surface of the substrate immediately prior to GaP nucleation. *Mottram et al.* [13] correlate the pitting effect with the final stage of the growth process and assume an excess of HCl and impurity segregation. Thus, both of these defects are not associated with the substrate-deposit system or the deposition method.

The occurrence of such serious defects as cracks are only detected for such pairs of substances characterized by a great difference in their thermal expansion coefficients, as GaP and Si. For all combinations cited it is possible to obtain GaP layers with completely plane, mirror-like, surfaces. Hillocks and pitting are growth defects which are not caused by the method applied, or the deposit–substrate combination.

2.3.3.2.   Twins and stacking faults

We were able to achieve a high perfection of the smooth epitaxial GaP layers on Ge (Fig. 4b) in the course of further investigations. Etching (HNO$_3$: HCl = 1 : 1) of these films gives rise to numerous etching lines, as shown in Fig. 7a. On a (111) plane these grooves are arranged in three crystallographically equivalent directions.

These etching grooves may be correlated to micro-twins or stacking faults in analogy with results known for silicon. In general, it was not possible to correlate the etch figures with X-ray topographic contrasts. Repeatedly, however, topographs revealed a pronounced twin structure of the {111} layers.

An example is given in Fig. 6. The dark regions of the Berg–Barrett topographs (Figs. 6a, c) correspond to the two orientations of the crystallites in the reflection position. Their orientations are sketched by pole figures (Figs. 6b, d). With the topographs for both orientations the entire surface is characterized. Further orientations may not exist. The appearance of twins is also estimated by *Thomas* [17] for GaP layers on silicon substrates. Satellite spots in electron diffraction patterns may be due to twinning or to a superlattice in the surface. There are various possibilities for the occurrence of twins. One is due to the polar character of GaP in contrast to Si or Ge. Thus, nuclei may be formed with the Ga or the P plane adjacent to the Ge or Si surface. The corresponding energies of nucleation should be comparable. Precautions have been taken to prevent one of the two nucleation processes using the conception of the influence of surface steps on nucleation mechanism and nucleation frequency for different orientations. The substrate surfaces employed were inclined by several degrees to the (111) plane. A system of crystallographic steps is employed.

The occurrence of stacking faults during epitaxial growth is described for several substances and is investigated in detail for the homoepitaxy of silicon. An improved method for preventing such kinds of defects is given by the use of substrates whose surface is tilted by some degrees against a growth face. If these conditions are fulfilled for the deposition of GaP onto Ge and special crystallographic directions are

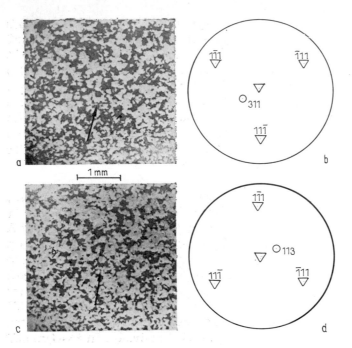

Fig. 6. Twin structure of GaP layer grown on an exactly (111) oriented Ge substrate as observed by Berg–Barrett topographs (see text)

used, it is possible to reduce the number of directions of etch grooves (Figs. 7a–c). With optimum conditions prevailing for the deposition parameters, regarding the orientation and the treatment of the substrates, it had been possible to achieve layers displaying no coarse defects after etching (Fig. 7d). A consequence of substrate tilting is the appearance of growth patterns on the surface. In order to prevent them, an angle of 3–4° was preferred.

The existence of etch grooves correlated to stacking faults in GaP epitaxial layers has already been reported by *Luther* [11].

The films are prepared on GaAs substrates in a chloride system with a deposition temperatures of 805–875°C. The etchant used is a mixture of chromic acid, water, hydrofluoric acid, and silver nitrate [22]. As well as (111)A, (111)B faces show etch grooves, which are associated with stacking faults. Contrary to films grown in the chloride system, those grown in water vapor show no etch grooves. It is concluded that wet hydrogen grown material is of better quality. This higher degree of perfection may be correlated with the higher deposition temperature, for changes in the growth rates in the chloride system have not resulted in a significant alteration of the etching behavior.

Depositing GaP onto GaAs substrates by the reaction Ga + PCl$_3$, *Mottram et al.* [13] found high stacking fault densities, whereas layers grown on GaP under similar conditions are nearly free of those defects. As the etching solution for (111)A faces,

146

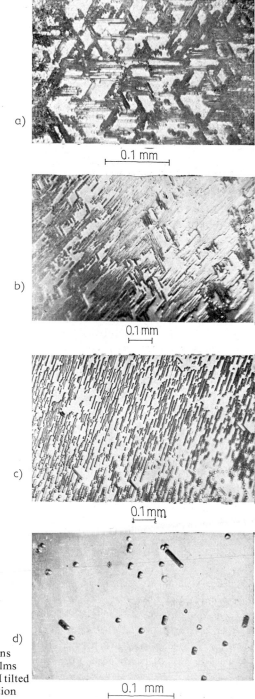

a)

0.1 mm

b)

0.1 mm

c)

0.1 mm

d)

Fig. 7. Decrease of the number of directions
and concentration of etch grooves on GaP films
on Ge with the substrate orientation {111} and tilted
some degrees of arc out of the {111}-orientation

0.1 mm

10*

a mixture of $K_3Fe(CN)_6/KOH/H_2O = 2/3/25$ at boiling temperature (60 sec) was employed [24].

Blom and *Bhargava* [23] found stacking faults on an As-grown ($\overline{1}\overline{1}1$)P surface of a GaP film grown by the water vapor reaction when the GaP substrates were unsatisfactorily prepared.

Summarizing, it can be stated that the occurrence of stacking faults and twins is neither correlated with the deposition reactions nor with the used substrate–deposit combinations. It is possible to obtain films free of these serious defects for all combinations and deposition methods mentioned.

2.3.3.3. Point defects

A serious problem in electronic studies is the contamination of epitaxial films. In the special case of the GaP–Ge combination the Ge content in the layer can reach concentrations which considerably decrease or extinguish the electroluminescence arising from Zn–O centers. The electronic and optical behavior of Ge containing GaP layers prepared in the chloride system using Ge wafers is investigated in detail by *Kasano* [25]. The essential drawbacks are the amphoteric behavior and the occurrence of further radiative, and non-radiative centers.

The Ge contamination is a problem of epitaxy, too. For crystals of high Ge content, especially near the solubility limit, a poor quality is expected. On the other hand, the high Ge concentration may be explained by the growth mechanism [12]. According to *Aoki* and *Kasano* the deposition of GaP from Ga + PCl_3 onto Ge contains a step of an alloy formation between Ge and Ga on the substrate surface and the growth is due to a VLS mechanism. The grown layers had poor crystalline quality because of their Ge content up to the solubility limit, and because of Ge inclusions.

It is possible that this mechanism will not take place if Ga is not used as a source, but the reaction of GaP with HCl, as in our case, and the films prepared in this way contain Ge with concentrations of 10^{18}, too.

On polished cross-sections of GaP grown onto Ge by the HCl technique, *Weinstein* [7] found no evidence of imperfections or diffusion at the junction. But at an etched transition zone, possibly consisting of Ge in GaP, imperfections could be observed. Probably a Ge contamination could be prevented, if metal organics and PH_3 are used for the deposition reaction. But experimental results are not known from the literature. *Aoki* and *Kasano* have coated the back surface of the wafer with polycrystalline GaP or Si films. In our investigations the Ge substrates were completely covered with an epitaxially deposited GaP layer and, thus, in a second run used as a wafer for the deposition of doped layers or *pn* junctions. As a result, the active part of this multilayer structure has Ge concentrations not affecting the electronic and optical properties. With even better reliability, this result can be achieved, when the initial epitaxial layer is separated from the Ge wafer by etching and subsequently used as a substrate.

A special problem in this connection is the origin of defects in the various interfaces.

2.3.3.4.   Stoichiometric deviations

Another group of defects are stoichiometric deviations. It is possible that differences in the CVD processes, and in the materials, are not only reflected in crystallographic perfection, but also in the stoichiometric behavior. ^{69}Ga NMR examination of GaP grown from a Ga melt and by the liquid encapsulation *Czochralski* technique (LEC) and of LEC material after heat treatment are discussed by *Peterson* [26]. The line width and peak intensities are correlated to lattice defects. GaP obtained from a Ga solution gives strong, narrow lines but LEC material tends to reveal broader and weaker peaks. They become stronger and narrower after a heat treatment at 1 000°C for 24 h. It is concluded that the defects are a deficiency of Ga. Using the transport reaction GaP + HCl, or Ga + PCl$_3$, the Ga/P ratio cannot be affected; however, it can in the reaction Ga + HCl + PH$_3$, or Ga(C$_2$H$_5$)$_3$ + PH$_3$.

There are some experimental results for GaAs$_{1-x}$P$_x$ grown on GaAs substrates by the AsH$_3$–PH$_3$–HCl–Ga system. *Stewart* [27] varied the III/V ratio in the gas phase from ~ 1.4 to ~ 7.1. The relative intensities of the band-edge and IR emission peaks of cathode-luminescence depend on the carrier concentration in the layer and the III/V ratio in the reaction gas. With increasing carrier concentration, an increase of the group III content in the gas phase is required in order to suppress the IR emission. The IR emission is suggested to be due to [V$_{Ga}$ + 3 Se] complexes. *Bhargava et al.* [28] investigated the IR emission after the annealing of GaP epitaxial layers on GaP substrates grown in water vapor, and in HCl vapor in the presence of water. A new luminescence band at 2.12 eV is ascribed to a [V$_{Ga}$ – O$_P$] complex. The existence of vacancy complexes is thought to be an essential prerequisite for radiative centers such as [Zn – O] complexes. The authors suppose that by proper control of the filling of [V$_{Ga}$ – O$_P$] complexes, *e.g.* [Zn$_{Ga}$ – O$_P$], the low radiative efficiency of vapor transport grown GaP could be improved. Group IV impurities can fill V$_{Ga}$ and produce highly non-radiative centers. The vacancy complex may be a non-radiative center, too.

Using the wet hydrogen vapor transport method, *Blom* and *Bhargava* [23] found oxygen to be incorporated and to be trapped as a nearest neighbor of a Ga vacancy. They have investigated the dissociation of the [V$_{Ga}$ – O$_P$] complex and its reaction with Zn or impurities. Considering thermodynamic equilibrium, conditions for the formation of [Zn – O] radiative centers are derived. Zn diffusion into GaP layers grown in a close-spaced water vapor transport system according to the requirements of the above-mentioned equilibrium resulted in LED's with photoluminescence yields of 0.1 %. This intensity is the highest one reported for vapor phase grown material and is comparable to that of liquid phase grown GaP. It is concluded that the oxygen content of the liquid phase and of water vapor grown material are comparable.

2.3.3.5.   Orientation defects

The relative orientation of film and substrate is usually controlled by *Laue* back-reflection patterns. The presence of non-oriented material will be indicated by the appearance of circular reflections beside the spots of oriented depositions. One source of unfavorable deposition parameters is the deposition temperature. The optimum epitaxial temperature may be accomplished by a systematic variation of the deposition temperature controlled by *Laue* patterns. The temperature necessary for epitaxial layer growth for a deposit substrate material combination depends on the deposition reactions involved. GaP depositions on Si by HCl vapor phase transport yield oriented deposits for temperatures $> 800°C$, but monocrystalline films are obtained at $485°C$ by the reaction of triethylgallium and triethylphosphine.

The epitaxial layers of GaP on Ge prepared by HCl transport [5] are monocrystalline in the limit of detection. The *Laue* spots of GaP films on sapphire show no splitting, indicating good monocrystallinity. A splitting of the patterns can occur if the used substrates are of poor crystallographic quality, for instance, flame fusion grown spinel and sapphire [29].

Epitaxial layers which are monocrystalline within the limit of detection of *Laue* patterns may contain smaller orientation defects. They can be recognized by X-ray diffractometric measurements. This method is an integral one and care must be taken in order to separate the different effects. Supposing that there are no departures in the lattice constant, the orientation distribution of the subgrains can be estimated from the reflection widths with fixed X-ray detectors. Using a diffractometer with a limit of resolution of 100 sec of arc for the Ge substrate no line broadening is observed, whereas GaP epitaxial layers on these Ge substrates show reflection widths near the resolution limit [5].

For GaP on (0001) and ($\bar{1}012$) sapphire substrates the departure of the epitaxial layer from parallel orientation was determined, by *McFarlane* and *Wang* [29], to be $0.1°$ with an error of measurement of $\pm 0.01°$.

Whereas diffractometric measurements only give the integral departure, the local and size distributions are obtained by X-ray topography. The angular aberration can be detected by the different divergence of the primary X-ray beam or by stepwise rotation of the sample. For assigning defects of GaP films on spinel and sapphire, *McFarlane* and *Wang* [29] applied the *Lang* transmission topography. Due to the high linear X-ray absorption coefficient of the GaP layer and the low linear absorption coefficient of the substrates using silver radiation, a separate imaging of the coated substrate and the epitaxial layer is possible employing the *Bragg* angle of the GaP and the substrate. Thus, it is possible to assign small, lighter lines which indicate a misorientation from the nominal deposit orientation due to scratches in the wafer surface.

Taking several topographs of the same layer but changing the crystal orientation relative to the incident X-ray beam by $0.1°$ between each topograph, the structure of the films can be detected. Substrates and GaP films deposited by the reaction of trimethylgallium and PH_3 at $800°C$ onto sapphire are investigated in [29]. Comparing such topographs, it is possible to detect the existence of small crystallites which generally follows the orientation of the grains in the substrate, but are misoriented from each other by $\pm 0.1°$.

Another possibility for the detection of the crystallite structure is described in [5] for GaP epitaxial layers grown by HCl transport on a Ge substrate.

Berg–Barrett topographs revealed a cloudy, (Fig. 8a), or granulated, (Fig. 8c), structure. Apart from the large unblackened patches in Fig. 8a which are due to some occasional growth hillocks, the layers can be regarded to be monocrystalline. The inhomogeneous blackening of the photographs, however, indicates some kind of mosaic character. Further evidence for this assumption is obtained from a comparison of Figs. 8a and b. With a divergence of the primary X-ray beam of 0.5 degrees of arc (Fig. 8a) only some unblackened spots are visible, indicating "subgrains" deviating from the reflection position. Recording the same region with a residual divergence of only 0.1°, these spots appear clearly enlarged and numerous new ones emerge. Layers showing granulated topographs are, in general, characterized by smaller orientation distributions of the subgrains. Topographs taken with a divergence of 0.1° usually showed no unblackened spots. A pronounced display of the substructure was only possible after further reduction of the primary divergence (Fig. 8d).

The size of the subgrains is about 100 μm at the surface of a 50-μm-thick layer and is about 20–50 μm at the interfaces, as was estimated from topographs taken after removal of the substrate.

The supposition underlying the determination of orientation defects by diffractometric measurements and topographs is the absence of inhomogeneities in the lattice constant near the surface of the layer. The lattice constant of GaP epitaxial layers on Ge was proved to be homogeneous within the resolution limit of about 0.05 % of the applied diffractometer and corresponds to standard values [30] near the surface of a 50 μm layer. However, a deviation of the lattice constant of about 0.2 % was observed at the interface.

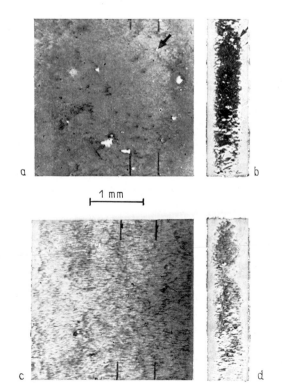

1 mm

Fig. 8. *Berg–Barrett* topographs of layers grown on substrate surfaces inclined for several degrees against the (111)-planes and observation of the mosaic character of such layers (see text)

151

Polycrystalline or partially polycrystalline layers originating from unfavorable deposition parameters like deposition temperature or deposition rate are not discussed. The orientation defects detected by X-ray topography can be assigned to different origins. Misorientations arising at scratches in the wafer surface and deviations of a part of the epitaxial layer following the orientation of a misoriented grain in the substrate can be easily understood. Both are caused by defects in the substrate.

The orientation defects described for GaP on Ge and sapphire correspond to the demonstrated departure of the subgrains in the monocrystalline layer. This deviation of the layers is detected on substrates or parts of them, which are checked by topographs or diffractometric measurements to be free of tilted grains.

This orientation defect must be correlated to the transition zone between substrate and deposit caused by the difference in crystallographic parameters.

The comparison with GaP depositions on a GaP wafer is of limited use, because GaP crystals prepared by the LEC technique show a similar structure [37].

2.3.3.6.   Dislocations

The appearance of dislocations in epitaxial layers arises from various reasons. The dislocations are introduced during the growth process caused by particular deposition conditions, concerning the substrate-deposit transition, the transition zone on the one hand and the cooling process on the other. Due to the appreciable dislocation density of the GaP bulk material used as a substrate, the association of dislocations in layers to one of the different effects is very difficult. On GaP layers deposited onto LEC substrates by a close-spaced transport with wet hydrogen, *Blom* and *Bhargava* [23] observed an EPD* in the range of 10^4–10^5 cm^{-2}, comparable with the dislocation density of the substrate crystals.

Depositing GaP onto GaP substrates at 800–850°C using the GaCl–PH$_3$ process results in a material with an EPD of $5 \cdot 10^4$–$1 \cdot 10^5$ cm^{-2} (*Crafford* [20]).

Applying an etch technique (K$_3$Fe(CN)$_6$, KOH, H$_2$O [24]) *Mottram et al.* [13] found etch pit densities in the lower 10^4 region for GaP films deposited by the Ga–PCl$_3$ technique onto GaP with low dislocation density ($\leq 10^4$). However, the EPD never exceeds $2 \cdot 10^5$ cm^{-2}, even on substrates with an EPD $> 10^6$ cm^{-2}. In GaP on GaAs wafers EPD's were always $> 2 \cdot 10^5$, even for GaAs substrates free of dislocations. GaP epitaxial layers grown by CVD and LPE techniques, as well as Czochralski material, have been examined by transmission electron microscopy (*Chase* [31]). CVD deposits from the Ga–PCl$_3$ process show the greatest concentration of defects compared with *Czochralski* and LPE grown GaP. The dislocation density is found to be between 10^7 and 10^8 cm^{-2}.

Saul [32] investigated the effect of a GaAs$_x$P$_{1-x}$ transition zone on the perfection of GaP crystals grown by deposition on GaAs substrates. The number of misfit dislocations required to compensate lattice mismatch at the growth temperature has been computed as a function of the width of the transition zone. It is predicted that

---

* EPD = etch pit density.

grading does not reduce the number of dislocations, but does reduce the density of misfit dislocations.

In GaP layers deposited onto Si by a transport reaction with HCl in a closed tube process, in general the strain is too large to allow the detection of individual dislocations. Only in very few cases small regions of reduced strain showed indications of a dislocation network with a density of about $10^5 \, cm^{-2}$ [4].

GaP deposited onto Ge by HCl transport of GaP at 910°C [5] and GaP films on sapphire deposited from the reaction of trimethylgallium and phosphine at 800°C [16], [29] generally show a dislocation density too high for a resolution of individual dislocations.

The dislocation densities cited cannot be compared without comment. Using GaP (LEC material) as a substrate, the EPD's for different deposition reactions agree with each other and are comparable with the dislocation density of the substrate. The use of GaAs substrates leads to EPD's in the range of $>2 \cdot 10^5 \, cm^{-2}$. Considering the lattice constants only, the misfit dislocation densities for the employment of GaAs or Ge as substrates should be comparable. For GaP on GaAs, *Saul* [32] predicts a dislocation density of $5 \cdot 10^9 \, cm^{-2}$. It is assumed that the misfit is completely compensated by dislocations which are almost parallel to the growth interface. Dislocations normal to the growth face are not misfit dislocations, although they may have been generated from misfit dislocations located in the transition zone. EPD's of epitaxial layers are usually determined on the grown surface. The defect concentration of 10^7–$10^8 \, cm^{-2}$ observed by *Chase* [31] using transmission electron microscopy for GaP grown onto GaAs is in good agreement with theoretical expectation. With respect to lattice constants and thermal expansion coefficients, GaAs and Ge substrates are comparable. However, epitaxial layers and their perfection cannot be compared, since X-ray topographs are missing for CVD deposited GaP on GaAs.

2.3.3.7.　Junctions

For epitaxial depositions, polished wafers are usually applied. Thus, plane boundaries between substrate and deposit could be expected. In several cases, the substrates are etched during the onset of the deposition process. Measurements carried out by *Mottram et al.* [13] for this back etching show that a surface layer of a width of 100 to 120 μm is removed. Defects arising from this etching are junctions which are not plane, and from the content of the substrate material in the layer. A typical example is shown in Fig. 9. A nearly one-degree angle-lapped GaP epitaxial layer on Ge in the vicinity of the Ge–GaP boundary shows Ge inclusions and Ge contaminations in the Ge-K_α scanning image [25]. This effect is caused by the deposition method applied using the reaction of Ga with PCl_3 as the transport reaction which gives rise to a VLS mechanism [12]. For GaP layers produced by the transport of GaP with H_2O, *Frosch* [8] found the surface adjacent to the GaAs substrate to contain cracks, defects, and a solid solution transition layer.

Layers produced under equal deposition conditions when examined by X-ray diffraction of the ($\overline{3}\overline{3}\overline{3}$) plane revealed the presence of only pure GaP and GaAs with no intermediate alloy phase [9]. Most of the heterojunctions showed plane boundaries

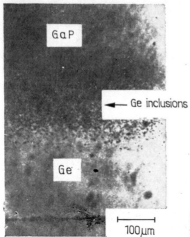

Fig. 9. Angle-lapped surface of GaP layer on Ge (scanning image, Ge-K$_\alpha$ *Kasano* [25])

between substrate and deposit (Fig. 10). These boundaries also appear using GaP as a substrate or during the preparation of multilayers with separate deposition of each layer (Fig. 10). These lines are inhomogeneities detectable in the real structure and by how they affect electronic behavior. It would be disadvantageous to produce a *pn* junction at its place. Thus, a preparation of *pn* junctions by the deposition of p- and n-material in a separate run is not expedient. It can be advantageous to deposit *pn* junctions without any interruption of the growth process. Such junctions show no boundary under an optical microscope. They are visible only by the different color of the adjacent materials. The appearance of a visible boundary on heteroepitaxial

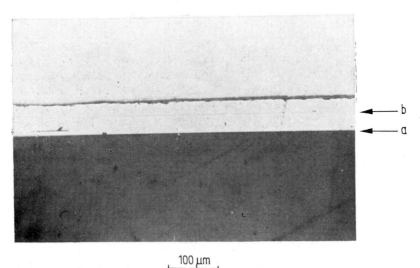

100 µm

Fig. 10. Cross-section of a GaP on Ge a) Boundary Ge/GaP, b) Boundary GaP/GaP with separated deposition of each layer

154

layers can easily be explained. The existence of a transition layer has its reason in the difference in the geometry of both lattices and the kind of binding on each side of the transition zone. The transition zone is not only a place where the geometry changes, but also where most of the physical properties change. In the transition zone, an adaption between both lattices by elastic deformation and misfit dislocations takes place. The transition zone contains the results of interactions between substrate and film, as solid solutions, etc.

A visible line occurs in a microscopic cross-section of a homostructure such as GaP grown onto GaP bulk material, or in multilayer structures, due to intermediate layers containing foreign atoms or molecules. The intermediate layers originate from adsorption layers on the wafer surface, arising *e.g.* from the reaction atmosphere. Usually this face is characterized by an increase in dislocation density. Changes in the real structure at the substrate-deposit boundary or between different layers can be investigated by X-ray topography or etching. The specimens are usually prepared for investigation by angle lapping or cleaving perpendicular to the surface. For mixed crystals, $(Ga_xIn_{1-x}P$; *Stringfellow* [33], $GaAs_{1-x}P_x$; *Stewart* [34], *Kishino* [35], or GaAs epitaxy, such investigations are known, but not for GaP. For topographs, cleavage planes are to be preferred. However, this becomes difficult for GaP deposits on Ge because of differences in the cleavage directions. A further disadvantage is the high dislocation density of these layers, and it should be difficult to get information relating to the interface or the overgrowth with A or B orientation of GaP on Ge.

2.3.3.8. Stress and bending

Stress is discussed in this context as a consequence of the crystallographic differences of the intergrown materials and as a source of defects in the substrate-deposit combination, which occurs during cooling from the reaction temperature to room temperature.

Considering the substrate-deposit combination at the end of the growth process, its state is characterized by the influences of the defects caused by the nucleation, *e.g.* misfit dislocations and orientation defects and defects caused by the process and the conditions of deposition. During the subsequent cooling process, forces, caused by differing thermal behavior, are acting upon this system, and these result in plastic deformation, stress, and bending. Thus, both substances are in the elastic range and the forces only cause elastic deformation. The observed bending can be calculated using the elastic moduli, the *Poisson* ratio, the thickness of the substrate and deposit, the temperature of deposition and of measurement. The radius of curvature can be determined by optical methods as described by *Feder* and *Light* [36] or by X-ray diffraction measurements using the angular setting of the crystal for maximum $K\alpha_1$ intensities at different points on the surface.

With this method, radii of curvature between 0.5 m and 50 m were measured on GaP and Ge [5]. For the epitaxy of GaP on GaAs the effect of a $GaAs_xP_{1-x}$ transition zone on the bending stresses is calculated by *Saul* [32]. The bending is high and depends on the width of this zone and the thickness of the growth layer and the substrate.

If deviations occur, the elastic limit is exceeded and cracks appear as in GaP layers on Si or in the range of plastic deformation. In such cases, it will be difficult to relate the defects to any particular influence.

Another possibility of detecting stress in GaP layers is the investigation of the optical properties. GaP with sphalerite structure belongs to the cubic system and ought to present an optically isotropic behavior. Microscopic cross-sections of GaP layers on Ge [5], however, under crossed Nicols, show the typical behavior of optically aniso-tropic materials (Fig. 11a). Some of the films exhibit inhomogeneous birefringence (Fig. 11b). This structure is only detectable in polarized light. It is supposed that these inhomogeneities are connected with the etching behavior.

The application of a comparator, (1st order), gives colors of equal retardation and, in some cases, enables one to recognize the birefringence as being homogeneous over the layer.

It should be mentioned that an inhomogeneous contamination of GaP films by the Ge wafer could give rise to anisotropy. But, in the environment of cracks, no birefringence is detectable, and the anisotropy has to be assigned to stress.

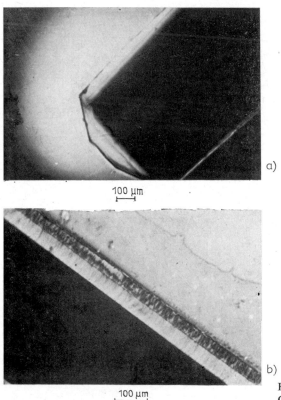

100 μm

a)

100 μm

b)

Fig. 11. a) Birefringence in GaP films on Ge, b) inhomogeneities in birefringence

2.3.4.   REFERENCES

[1] *Kirwan, D. J.:* J. Electrochem. Soc. 117 (1970) 1572.
[2] *Seki, H.:* Jap. J. Appl. Phys. 6 (1967) 1414.
[3] *Seki, H., Eguchi, H.:* Jap. J. Appl. Phys. 10 (1971) 39.
[4] *Noack, J., Möhling, W.:* phys. stat. sol. (a) 3, (1970) K229.
[5] *Kies, J. Möhling, W., Noack, J.:* Krist. u. Techn. 9 (1974) 205.
[6] *van der Does de Bye J. A. W., Peters, R. C.:* Philips Res. Reps. 24 (1969) 210—230.
[7] *Weinstein, M., Bell, R. O., Menna, A. A.:* J. Electrochem. Soc. 111 (1964) 674.
[8] *Frosch, C. J.:* J. Electrochem. Soc. 111 (1964) 180.
[9] *Flicker, H., Goldstein, B., Hoss, P. A.:* J. Appl. Phys. 35 (1964) 2959.
[10] *Kamath, G. S., Bowman, D.:* J. Electrochem. Soc. 114 (1967) 192.
[11] *Luther, L. C., Roccaseca, D. D.:* J. Electrochem. Soc. 115 (1968) 850.
[12] *Aoki, M., Kasano, H.:* Jap. J. Appl. Phys. Suppl. 39 (1970) 234.
[13] *Mottram, A., Peaker, A. R., Sudlow, P. D.:* J. Electrochem. Soc. 118 (1971) 318.
[14] *Tietjen, J. J., Amick, J. A.:* J. Electrochem. Soc. 113 (1966) 724.
[15] *Tietjen, J. J., Enstrom, R. E., Ban, V. S., Richman, D.:* Solid State Technol. 15 (1972) 42.
[16] *Wang, C. C., McFarlane III, S. H.:* J. Cryst. Growth 13/14 (1972) 262—267.
[17] *Thomas, R. W.:* J. Electrochem. Soc. 116 (1969) 1449.
[18] *Manasevit, H. M. Simpson, W. I.:* J. Electrochem. Soc. 116 (1969) 1725.
[19] *Igarashi, O.:* J. Appl. Phys. 41 (1970) 3190.
[20] *Crafford, M. G., Groves, W. O., Herzog, A. H., Hill, D. E.:* J. Appl. Phys. 42 (1971) 2751.
[21] *Igarashi, O.:* J. Electrochem. Soc. 119 (1972) 1431.
[22] *Abrahams, M. S., Buiocchi, C. J.:* J. Appl. Phys. 6 (1967) 1424.
[23] *Blom, G. M., Bhargava, R. N.:* J. Cryst. Growth 17 (1972) 38—45.
[24] *Valkovskaya, M. I., Bayanskaya, Yu. S.:* Soviet phys. Solid State (Engl. Transl.) 8 (1967) 1976.
[25] *Kasano, H.:* J. Appl. Phys. 43 (1972) 1792.
[26] *Peterson, G. E., Carneval, A., Verleur, H. W.:* Mat. Res. Bull. 6 (1971) 51—56.
[27] *Stewart, C. E. E.:* J. Cryst. Growth 8 (1971) 259—268.
[28] *Bhargava, R. N., Kurtz, S. K., Vink, A. T., Peters, R. C.:* Phys. Rev. Letters 27 (1971) 183.
[29] *McFarlane III, S. H., Wang, C. C.:* J. Appl. Physics 43 (1972) 1724.
[30] *Giesecke, G., Pfister, H.:* Acta Cryst. 11 (1958) 369.
[31] *Chase, B. D., Holt, D. B.:* J. Materials Sci. 7 (1972) 265—278.
[32] *Saul, R. H.:* J. Appl. Phys. 40 (1969) 3273.
[33] *Stringfellow, G. B.:* J. Appl. Phys. 43 (1972) 3455.
[34] *Stewart, C. E. E.:* J. Cryst. Growth 8 (1971) 269—275.
[35] *Kishino, S., Ogirima, M., Kurata, K.:* J. Electrochem. Soc. 119 (1972) 617.
[36] *Feder, R., Light, T. B.:* J. Appl. Phys. 43 (1972) 3114.
[37] *Rozgonyi, G. A., Saul, R. H.:* J. Appl. Phys. 43 (1972) 1186.

2.4. SOME QUESTIONS CONCERNING THERMODYNAMICS OF CRYSTAL GROWTH

Yu. D. Čistyakov and *Yu. A. Baikov*

The Moscow Institute of Electronic Engineering, Moscow, USSR

2.4.1.   INTRODUCTION

In the first part of this section, the possibility for the existence of a thin compact film on the heated surface of a solid material is discussed. For the investigation of the processes of epitaxial growth many authors study the case of the appearance of some thin liquid films on the heated surface of the solids [1].

Here, the existence of liquid thin films is explained on the basis of the *van der Waals* interaction theory. Several quantitative characteristics of the films, which will be given below, are connected with the laws of classical thermodynamics. On the other hand, some simple questions of phase formation will be discussed according to the principles of phase equilibrium. However, only such results of the *van der Waals* interaction theory which are necessary for an understanding of the main conclusions regarding our problem have been considered.

2.4.2.   SOME QUESTIONS CONCERNING THE EXISTENCE OF THIN LIQUID FILMS ON THE HEATED SURFACE OF A SOLID

Here, a thermodynamic system consisting of three physical media is considered (Fig. 1): vapor, a thin liquid film, and a solid body. The liquid film has a thickness, or height, h. A mechanism explaining the existence of such a thin liquid film on the heated surface of a solid is not discussed here. However, it is necessary to emphasize that this mechanism, based on the *van der Waals* interaction of all three phases, is very complicated. Therefore, only the results of this mechanism and the general laws of classical thermodynamics for a case of phase equilibrium of all three media are applied. Here, we consider only a one-component system of all phases which are in the state of thermodynamic equilibrium. This condition of thermodynamic equi-

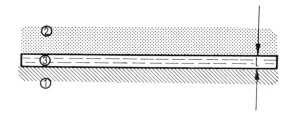

Fig. 1. The scheme of three physical phases which are in the state of thermodynamic equilibrium. (1) is a crystal, (2) represents the vapor phase, and (3) is a thin liquid film with a thickness h

librium requires the equalization of the temperatures, pressures, and chemical potentials of all three phases:

$$T^v = T^l = T^S = \text{const},\tag{1a}$$

$$p^v = T^l = p^S = \text{const},\tag{1b}$$

$$\mu^v = T^l = \mu^S = \text{const}.\tag{1c}$$

The equality of the chemical potentials enables an estimate of the thickness h of the thin liquid film on the heated surface of a solid. It is assumed that an interaction between the atoms of all phases is caused by the so-called *van der Waals* forces. This *van der Waals* interaction gives rise to some difference between the chemical potentials of a bulky liquid and those of a thin liquid film. This difference of the chemical potentials caused by the *van der Waals* forces may be expressed as a function of h

$$\Delta\mu \equiv F(h) = \frac{\hbar\bar\omega}{8\pi^2 h^3}.\tag{2}$$

Here, only the case $h \ll \lambda_0$ is considered, where λ_0 is a mean wavelength of the absorption spectrum of the liquid film; that is why this difference is proportional to h^{-3} [2].
In eq. (2) $\bar\omega$ is a function characterizing the electrodynamic properties of all three phases. On the other hand, there is the potential energy of an interaction between two atoms with a distance $R \approx 10^{-7}$ cm, $U(R)$, proportional to R^{-6}, *i.e.*

$$U(R) = -aR^{-6},\tag{3}$$

where a is *van der Waals'* constant. If all atoms are thought to be situated in two semi-spaces divided by a slit of width h (Fig. 1) and any atom of semi-space 1 is thought to be interacting with one partner atom of semi-space 2 (and vice versa) then the mean potential energy of all atoms interacting in pairs is equal to:

$$\bar U = -\frac{a\pi N_1 N_2}{2h^2}\tag{4}$$

hence

$$F(h) = -\frac{\partial \bar U}{\partial h} = -\frac{a\pi N_1 N_2}{h^3}\tag{5}$$

where N_1 and N_2 are the numbers of atoms per cm^3 of semi-space 1 and 2, respectively
The chemical potential of a thin liquid film may be assumed to be:

$$\mu_0^1 = \mu^l + \frac{\Delta\mu}{\rho_e},\tag{6}$$

160

where $\lim\limits_{h\to\infty} \Delta\mu = 0$ and μ_0^l is the chemical potential of a bulky liquid. The chemical potential of vapor is:

$$\mu^v = \mu_0^l + \frac{kT}{m^v}\ln\frac{p}{p_0},\tag{7}$$

the constant μ_0^l has the same meaning as in eq. (6) because $\mu^v = \mu^l$ when $p = p_0$ and $h \to \infty$. m^v is the mass of one atom in the vapor phase, T is the vapor temperature, $\dfrac{p}{p_0}$ is the so-called supersaturation of vapor, ρ_e is the density of the thin liquid film, and k is *Boltzmann*'s constant. According to eqs. (6) and (7) the condition (1c) yields:

$$\Delta\mu = \frac{kT}{m^v}\rho_e\ln\frac{p}{p_0}\tag{8}$$

hence:

$$h^3 = \frac{\hbar\bar\omega\, m^v}{8\pi^2 kT\rho_e \ln\dfrac{p}{p_0}}\tag{9}$$

or taking into account that:

$$N_1 m^v = \rho_s,\tag{10}$$

one obtains, with eqs. (2), (3), and (5)

$$h^3 = \frac{\pi N_2\,\rho_s\, R^6\, U(R)}{kT\rho_e \ln\dfrac{p}{p_0}}\tag{11}$$

Assuming usual values for the parameters

$N_2 = 10^{20}\ \text{cm}^{-3}$,

$R = 10^{-7}\ \text{cm}$,

$U(R) \cong 10^{-1}\ \text{eV}$,

$kT \cong 13\cdot 10^{-2}\ \text{eV}\ (T \approx 1\,500\ \text{K})$,

$\dfrac{\rho_s}{\rho_e} \approx 1$, and

$\ln\dfrac{p}{p_0} \cong 10^{-3}$,

one obtains

$$h = \sqrt[3]{\frac{\pi N_2\, R^6\, \rho_s\, U(R)}{\ln\dfrac{p}{p_0}\,\rho_e\, kT}} \sim 100\ \text{Å}.$$

11 Epitaxy

161

This estimation is in good agreement with the available experimental results from the investigations of the Si homoepitaxy [1]

2.4.3. **SOME THERMODYNAMICAL PRE-CONDITIONS FOR THE FORMATION OF LIQUID PHASES ON THE SURFACE OF A GROWING CRYSTAL**

In the previous paragraph, we have considered some applications of the *van der Waals* interaction theory which explains the existence of thin liquid films on the heated surface of a solid. When the thickness of the liquid films increases, the influence of the *van der Waals* forces reduces, and a subsequent absorption of substance from the vapor leads to the formation of dome-shaped protuberances on the surface of thin films with very small contact angles Θ (Fig. 2). This formation is conditioned by the surface tension acting along the boundary between the liquid and the vapor phases. Such formations have been observed by many investigators [1].

There is also a simple model (see Fig. 2) which explains this effect. The effective coefficient of the surface tension between solid (1) and vapor (2), taking into account the existence of the liquid film (3), may be expressed by the integral:

$$\alpha(h_{max}) = \int_{h_{max}}^{\infty} h \frac{\partial \mu}{\partial h} \, dh + \alpha_{13} + \alpha_{23} \tag{12}$$

where h_{max} is a maximum value of the film thickness when a roughness first appears, μ is the chemical potential of a thin liquid film, α_{13} and α_{23} are the coefficients of surface tension acting along the boundaries of phases (1) and (3) or (2) and (3), respectively. The condition for mechanical equilibrium of this roughness on the film surface yields

$$\cos \Theta = \frac{\alpha(h_{max}) - \alpha_{13}}{\alpha_{23}}. \tag{13}$$

After expanding $\cos \Theta$ (we assume that the angle Θ is sufficiently small) one obtains:

$$\cos \Theta \cong 1 - 1/2 \, \Theta^2 \tag{14a}$$

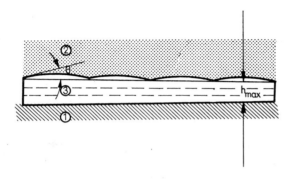

Fig. 2. The same scheme as depicted in Fig. 1 where h is the thickness of a liquid film when its surface is perturbed by some roughnesses which appear due to an influence of the surface tension forces

with

$$\cos \Theta = 1 + \frac{h\mu}{\alpha_{23}} \bigg|_{h_{max}}^{\infty} - \frac{1}{\alpha_{23}} \int_{h_{max}}^{\infty} \mu \, dh \, . \tag{14b}$$

Taking into account that the chemical potential μ may be written as a function of h as:

$$\mu = \frac{\hbar \overline{\omega}}{8\pi^2 h^3} \tag{2a}$$

when the thickness h is very small in comparison with the mean wavelength of the absorption spectrum for the substance considered in our system we obtain

$$\Theta \cong \frac{\sqrt{3}}{\sqrt{8} \, \pi h_{max}} \sqrt{\frac{\hbar \overline{\omega}}{\alpha_{23}}} \cong \frac{1}{5 h_{max}} \sqrt{\frac{\hbar \overline{\omega}}{\alpha_{23}}} \tag{15}$$

$\overline{\omega}$ is a *frequency* of the vapor–film–crystal system determined by the dielectric permeabilities of all three physical media. Furthermore, we have assumed with eq. (2a) that for $h \to \infty$ $\mu(h) \to 0$ more quickly than $h^{-1} \to 0$. This assumption holds when the following condition is fulfilled:

$$L \cdot \Theta \gg h_{max} , \tag{16}$$

where L is the diameter of a base of the dome-shaped protuberance, Θ is the above-mentioned contact angle (see Fig. 2). We consider here, again, the Si-system with the following available values for the appropriate parameters: $\hbar \overline{\omega} \cong 35$ cV,

$$\alpha_{23} = 2 \cdot 10^{14} \text{ eV} \cdot \text{cm}^{-2},$$

$$h_{max} = 5 \cdot 10^{-6} \text{ cm}.$$

This yields $\Theta \cong 1°$. It is necessary to emphasize that we considered only a one-component system consisting of Si, besides, the chemical composition of a roughness was the same as that of a film. Furthermore, the minimum work required for the formation of a new phase (3) on top of the surface of a thin liquid film (1) may be written as follows:

$$W = \Delta\Omega_s + \Delta\Omega_v , \tag{17}$$

where $\Delta\Omega_s$ is the change in the thermodynamic potential's surface part of the system bounded by the two surfaces (A B C) and (A D C) which have appeared instead of the primary plane boundary between a vapor and a thin liquid film.
$\Delta\Omega_v$ is the change in the thermodynamic potential's volume part of the system related to the so-called surface pressure caused by the difference of pressures in adjacent phases.
The common expression for $\Delta\Omega_v$ may be written as:

$$\Delta\Omega_v = -(p_3 - p_1)V_{13} - (p_3 - p_2) V_{23} \tag{18}$$

where:

$$V_{13} = \frac{\pi d^3}{24} (\operatorname{cosec} \Theta_1 - \cot \Theta_1)^2 \, (2 \operatorname{cosec} \Theta_1 + \cot \Theta_1)$$

$$V_{23} = \frac{\pi d^3}{24} (\operatorname{cosec} \Theta_2 - \cot \Theta_2)^2 \, (2 \operatorname{cosec} \Theta_2 + \cot \Theta_2).$$

Here, d is the diameter of the base of the phase-expressed spherical segments (Fig. 3); $(p_3 - p_1)$ and $(p_3 - p_2)$ are the differences of the pressures in the phases (3) and (1) or (3) and (2), respectively; and Θ_1 and Θ_2 are the contact angles depicted in Fig. 3. Now we shall consider the magnitude of $\Delta\Omega_s$. Mechanical equilibrium of the entire system requires that the following equation is fulfilled:

$$\vec{\alpha}_{12} + \vec{\alpha}_{13} + \vec{\alpha}_{23} = 0. \tag{19}$$

The components of these vectors along the two axes "X" and "Y" (see Fig. 3) are:

$$-\alpha_{12} + \alpha_{13} \cos \Theta_1 + \alpha_{23} \cos \Theta_2 = 0$$
$$-\alpha_{13} \sin \Theta_1 + \alpha_{23} \sin \Theta_2 = 0. \tag{20}$$

Furthermore, the change of $\Delta\Omega_s$ may be written as:

$$\Delta\Omega_s = \alpha_{13} S_1 + \alpha_{23} S_2 - \alpha_{12} S_{gr} \tag{21}$$

where

$$S_{gr} = \pi r_1^2 \sin \Theta_1 = \pi r_2^2 \sin \Theta_2$$
$$S_1 = 2\pi r_1^2 (1 - \cos \Theta_1)$$
$$S_2 = 2\pi r_2^2 (1 - \cos \Theta_2).$$

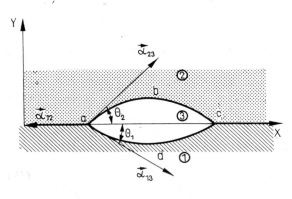

Fig. 3. The scheme of a physical system consisting of three phases. (1) is the thin liquid film, (2) is the vapor phase, and (3) is the phase-expressed drop of a liquid

164

Here, r_1 and r_2 are the radii of curvature for the surfaces (A B C) and (A D C), respectively. Conditon (20) yields:

$$\cos \Theta_1 = \frac{\alpha_{12}^2 + \alpha_{13}^2 - \alpha_{23}^2}{2\alpha_{12}\alpha_{13}}$$

$$\cos \Theta_2 = \frac{\alpha_{12}^2 - \alpha_{13}^2 + \alpha_{23}^2}{2\alpha_{12}\alpha_{13}} \tag{22}$$

where the coefficients α_{12}, α_{13}, α_{23} are interrelated by the following conditions:

$$\alpha_{ij} \leqq \alpha_{kl} + \alpha_{mn},$$

$$\alpha_{ij} \leqq |\alpha_{kl} - \alpha_{mn}|, \tag{23}$$

i, j, k, l, m, n = 1, 2, 3.

Finally, utilizing the results of the previous formulae, we obtain:

$$\Delta\Omega_s = \frac{\pi d^2}{4} \alpha_{eff} \tag{24}$$

with

$$\alpha_{eff} = 2\left[\frac{2(\alpha_{12}\alpha_{13})^2}{(2\alpha_{13}\alpha_{23})^2 - (\alpha_{12}^2 - \alpha_{13}^2 - \alpha_{23}^2)^2} \times (\alpha_{13} + \alpha_{23} - \alpha_{12}) - \alpha_{12}\right].$$

2.4.4.   CONCLUSIONS

The main objective of this section was an attempt to demonstrate a new method for the understanding of some questions connected with crystal growth and epitaxy. Questions of merely physical character were not considered here as this would lead to very complicated and unnecessary problems. The main issue of this part of the book was concentrated on the application of commonly used laws of classical thermodynamics. The practical side of such a consideration is obvious, though some estimations of the physical magnitudes discussed here were not quite correct due to the lack of experimental data. Nevertheless, the authors consider this section as an inception for subsequent investigations of a similar kind. The relative simplicity of the formulae offered is very convenient for a general discussion of the results. The simple models used here, and the corresponding common formulae which have been given in this section, may be examined and checked by the available experimental data.

2.4.5.   REFERENCES

[1] *Čistyakov, Yu. D.*: Collection of articles on microelectronics, vol. 4 (Epitaxy). MIET Moscow (1969).
[2] *Dzyaloshinsky, I. E., Lifshicz, E. M., Pitaevsky, L. P.*: U.F.N. 73 (1961) 381.

3. FORMATION OF EPITAXIAL SEMICONDUCTOR LAYERS BY EVAPORATION AND INTERDIFFUSION UNDER ISOTHERMAL CONDITIONS

3.1. THE ROLE OF MERCURY VAPOR PRESSURE IN THE ISOTHERMAL EPITAXY OF HgTe AND CdTe*

F. Bailly, *L. Svob* and *G. Cohen-Solal*

Laboratory of Solid Physics, National Center of Scientific Research (CNRS), Bellevue, France

3.1.1. KINETICS OF EQUILIBRATION BETWEEN ISOLATED HgTe AND MERCURY VAPOR

3.1.1.1. Introduction

Previously, the interdiffusion [1], [2] between HgTe and CdTe, as well as the growth of HgTe on CdTe under isothermal conditions (EDRI) [3], [4], had been investigated. The effect of various parameters such as temperature, duration of treatment, nature of initial materials, distance between source and substrate, and different temperature gradients on interdiffusion or growth was studied systematically. The results obtained provided a better understanding of the phenomena involved and revealed how these phenomena might be used for a controlled preparation of materials, with variable forbidden band widths, having electronic characteristics of special interest [5], [6].

During these investigations we qualitatively observed the remarkable effect that an excessive vapor pressure of mercury or tellurium [7] in the ampulla can have on interdiffusion and growth. Subsequently, *Tufte* and *Stelzer* [8], [9] experimentally studied the effect of excessive vapor pressure on layer growth under the conditions of EDRI (Evaporation Diffusion en Régime Isotherme) and arrived at certain conclusions with respect to the mode of preparing layers which *a priori* are expected to have definite characteristics regarding the composition of the surface and the composition gradients in its vicinity.

Our investigations were carried out in the same direction attempting to give a theoretical interpretation — based on a simple model — of the experimental effects observed during growth and interdiffusion as well.

In the course of these studies, we became aware of the fact that it was easy to correlate the characteristics of the phenomenon with the excessive vapor pressure p_0 of mercury in the isothermal enclosure, but that it was difficult to give a theoretical interpretation of such correlations when this pressure p_0 was not identical with the real pressure p of mercury, as our knowledge of the kinetics of establishing an equilibrium between HgTe and its vapor is restricted. Therefore, we undertook an ex-

---

* Chapter 3.1 continues Chapter 4 (by *G. Cohen-Solal, Y. Marfaing, F. Bailly*) in "Epitaxie — Endotaxie" (Ed.: H. G. Schneider), VEB Deutscher Verlag für Grundstoffindustrie, Leipzig (1969) pp. 187–212.

tensive study of HgTe itself, which was placed into an isothermal enclosure both in the presence of an excess of mercury and with no excess of mercury. This study is the subject of this section, whereas Section 3.1.2 deals with the analysis of the effect exerted by the mercury pressure on the growth itself, *i.e.* in the presence of CdTe and in the aforementioned disposition EDRI.

The parameters whose influences on the kinetics of equilibration between HgTe and the mercury vapor will be studied are:

a) the ratio between the volume v of the specimen and the volume V of the enclosure,
b) the excessive vapor pressure p_0 of mercury in the enclosure, and
c) the duration t of the treatment.

All experiments mentioned here, and the theoretical results obtained, refer to a treatment temperature of 550°C.

In addition, it is well known [10] that the vapor over HgTe almost merely consists of two elements, *i.e.* mercury and tellurium. As there is a large difference between their partial pressures — p_{Hg} is several times higher than p_{Te_2}, — it seems justified to consider the mercury pressure only.

First and foremost, we are presenting the principle of the method and the apparatus used for the experiments. This will be followed by a theoretical analysis, its comparison with the experiments, and a discussion of the results obtained.

3.1.1.2.   Experimental conditions

3.1.1.2.1.   *Principle of the method*

The experimental values of p_{Hg} are indirectly determined from the loss of weight after the hardening of the specimens treated at a certain temperature in a sealed enclosure. The knowledge of the quantity of material available in the form of gas in a measured free volume enables the determination of the pressure p_{Hg} prevailing in the enclosure at the temperature considered by applying the ideal gas law.

The mercury pressure depends upon the composition of the respective material. Therefore, the specimens are previously subjected to a thermal treatment at a temperature of 200°C in the presence of saturated mercury vapor ($T = 175$°C) for 8 days, a period that is sufficiently long to render the specimens stoichiometrically [11]. Furthermore, since the volatilization of the material possibly changes the composition of the solid phase, which may proceed to such an extent, that is, leaves a zone where the compound between HgTe (Hg saturated) and HgTe (Te saturated) exists, it is necessary to carry out the tests in a restricted volume. In our study, we modified the experimental conditions before sealing the enclosure by introducing a certain quantity of pure mercury. This quantity, which was calculated before each experiment taking into account the free volume of the enclosure, gives a known excessive vapor pressure $p_{0\,Hg}$ at the experimental temperature, assuming the mercury to behave like an ideal gas. Practically, we proceeded in the following manner: The free volume V of the ampulla was measured. From the curves of Fig. 1 the quantity of Hg was determined which was assumed to evolve the specified dry vapor pressure $p_{0\,Hg}$ at the

168

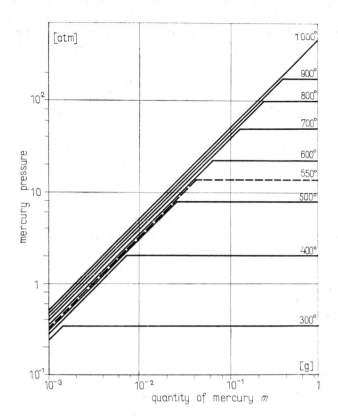

Fig. 1. Determination of mercury vapor pressure (in atm) obtained for a mercury sample of m (g), in an enclosure of 1 cubic centimeter exposed to various temperatures; the horizontal sections of the curves represent the pressures of saturated mercury vapor, Hg (g), in equilibrium with solid mercury, Hg (c)

chosen temperature. The horizontal sections of the curves correspond to saturated pressures and are not considered in this case. Then, the quantity of Hg, thus determined, was introduced into the ampulla containing the specimen, and finally the ampulla was sealed.

3.1.1.2.2. *Experimental apparatus*

The enclosure employed for the experiments had to meet two necessary requirements: It had to withstand the high mercury vapor pressures expected to evolve at the chosen temperatures (sometimes ranging up to 15 atmospheres at 600°C), and it had to be designed so that the free volume over the specimen could be reduced and precisely measured. In addition, sealing the ampulla was not to be allowed to result in an uncontrollable modification of the material due to warming-up.

The apparatus consists of two concentric silica tubes with flat bottoms. The inner tube, representing the ampulla, permits the free volume to be fixed. Such a device had already successfully been used by *Tufte* [9]. The development of a semi-automatic sealing device in our laboratory enabled us, relatively easily, to seal the system at a precisely chosen point. During the sealing operation, which was carried out in an argon residual vacuum of approximately 10^{-4} torr, the part containing the specimen and, possibly, the free mercury, is maintained at an ambient temperature by a

169

water circulation system. Before the experiment, each specimen is decreased by means of boiling trichloroethylene and subjected to a chemical treatment with bromoethyl alcohol followed by repeated rinsing with deionized water. The apmullas are cleaned with acid solutions (HF and HNO_3) and carefully rinsed with deionized water.

The free volumes are determined by filling-up with distilled water and weighing the quantity of liquid thus obtained. The quantities of mercury, ranging from 1 to 10 mg, are obtained from small drops emerging from a capillary having a diameter of about 10^{-2} mm. All weighing operations are carried out with a chemical balance.

The analysis of the accuracy of this method, given in the appendix, reveals that under certain precautions it is possible experimentally to determine the pressure occurring in the enclosure at the moment of hardening, at about ± 0.1 atmospheres.

3.1.1.3. Theoretical analysis

The material exchange between solid HgTe (c) and gaseous Hg (g) in a sealed iso-thermal enclosure is considered next.

3.1.1.3.1. *The state of equilibrium*

A specimen of solid HgTe (c) (volume v) is placed into an isothermal enclosure (volume V) at a temperature T in the presence of n_0 Hg (g) atoms in excess. At the temperature considered ($T = 550°C$) it is assumed that the main defects created in HgTe (c) are Hg vacancies with the concentration N_f^v and Hg interstitials with the concentration N_f^i, the initial concentrations being N_1^v and N_1^i, respectively.

After heat treatment at the temperature T we obtain:

$$N_f^i = N_1^i + \Delta N_f^i \tag{1}$$

$$N_f^v = N_1^v + \Delta N_f^v. \tag{2}$$

The equation of state for Hg (g) is:

$$p_f V = n_f \frac{RT}{N_0} \tag{3}$$

where p_f is the mercury vapor pressure in equilibrium with HgTe (c).

The Hg gas is composed of atoms from different sources:

a) atoms escaped from the solid leaving vacancies (ΔN_{1f}^v);
b) evaporation and subsequent dissociation of HgTe according to HgTe (g) Hg (g) + + 1/2 Te_2 (g) served as another source of atoms (n_f');
c) excess atoms (n_0).

Taking into account the atoms lost by the gas phase and incorporated as interstitials in the solid HgTe (c) (Δn_{if}) we obtain the equation of equilibrium:

$$n_f = \Delta n_{1f} - \Delta n_{if} + n_f' + n_0 \tag{4}$$

170

where

$$\Delta n_{1f} = v \cdot \Delta N_f^v \tag{5}$$

$$\Delta n_{if} = v \cdot \Delta N_f^i. \tag{6}$$

From the relations (1) and (2) and the equations describing the equilibrium between the respective phases

$$N_f^i = K_i p_f \tag{7}$$

$$N_f^v = K_1 p_f^{-1} \tag{8}$$

we derive the expressions

$$\Delta N_f^i = K_i p_f - N_1^v \tag{9}$$

$$\Delta N_f^v = K_1 p_f^{-1} - N_1^v. \tag{10}$$

Finally the dissociation of HgTe provides the relation:

$$p_f p_{Te_2 f}^{1/2} = K. \tag{11}$$

By substituting into eq. (4) the expressions derives from eqs. (5), (6), (9)–(11) we obtain an equation in which the real equilibrium vapor pressure, p_f, is related to the ratio between the volume of the specimen and that of the enclosure $\left(\dfrac{v}{V}\right)$, and to the excess pressure (p_0):

$$p_f^3 \left(1 + \frac{v}{V}\frac{RT}{N_0} K_i\right) + p_f^2 \left(\frac{v}{V}\frac{RT}{N_0} N_1^v - \frac{v}{V}\frac{RT}{N_0}\cdot N_1^i - p_0\right) -$$

$$- \frac{v}{V}\frac{RT}{N_0} K_1 p_f - 2K^2 = 0. \tag{12}$$

This equation can numerically be solved in respect to $\dfrac{v}{V}$, $p_0 = \dfrac{n_0 RT}{N_0 V}$, the various constants of equilibrium, and to the characteristic parameters N_1^v and N_1^i of the initial HgTe (c).
It is, however, possible to simplify eq. (12) employing some approximations:

(a) K and N_1^v are small quantities, and
(b) the influence of the interstitials is negligible, at least for not too high pressures p_f.

This leads to a second-order equation for p_f:

$$p_f^2 - p_0 p_f - \frac{v}{V}\frac{RT}{N_0} K_1 = 0. \tag{13}$$

171

For the case $p_0 = 0$ this is reduced to:

$$p_{f0} = \left(\frac{v}{V} \frac{RT}{N_0} K_1 \right)^{1/2}.$$ (14)

Equation (14) can be verified by experiment when p_f is not too large (Section 3.1.1.3.3). The same applies to eq. (13). Taking into account eq. (14), the solution of eq. (13)

$$p_f = \frac{1}{2} \left[p_0 + \left(p_0^2 + 4 \frac{v}{V} \frac{RT}{N_0} K_1 \right)^{1/2} \right]$$ (15)

can be expressed as

$$p_f = \frac{1}{2} [p_0 + (p_0^2 + 4p_{f0}^2)^{1/2}].$$ (16)

When the pressures p_f are approaching a value corresponding to the stability limit of HgTe (c), the approximations discussed above do not apply and the pressure p_f must be obtained from eq. (12) by means of a computer.* The parameters p_0 and $\frac{v}{V}$ are chosen such that p_0 varies from 0 to 1 in steps of 0.05. The values of the constants have been derived from results published elsewhere, as listed in the references:

$K = 2 \cdot 10^{-2} \, \text{atm}^{3/2}$ [10];

$K_i = 1.13 \cdot 10^{18} \, \text{cm}^{-3} \, \text{atm}^{-1}$

$K_1 = 2.5 \cdot 10^{20} \, \text{cm}^{-3} \, \text{atm}$ [2];

$N_1^i = 8.3 \cdot 10^{16} \, \text{cm}^{-3}$ [2] and

$N_1^v = 6.75 \cdot 10^{16} \, \text{cm}^{-3}$ [2], [11].

The molar volume of HgTe was assumed to be $40 \, \text{cm}^3$. Under these conditions, the curves shown in Fig. 2 are obtained. It should be noted that for all values of $\frac{v}{V}$ the curves $p(p_0)$ are always above the straight line $p = p_0$ except for those cases corresponding to high mercury vapor pressures in excess ($p_0 \cong 15$ atm). For these cases, the curves are below the straight line. This fact, revealing the predominant influence of Hg interstitials in comparison to Hg vacancies, may be related to the phase limit of HgTe (c) which occurs at about 15 atm at this temperature. Likewise, a limiting mercury vapor pressure of 0.7 atm is obtained in accordance with the other limit of the existence of HgTe (c) if, into the enclosure, tellurium, instead of mercury, is added in excess.

* For the program and the calculations we express our gratitude to Mr. *Frédéric Raymond*.

172

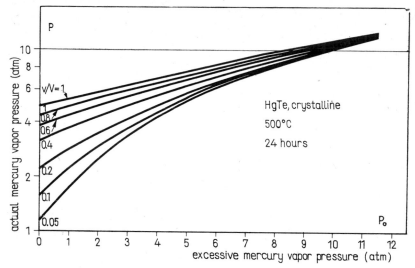

Fig. 2. Theoretical curves of mercury vapor pressure in equilibrium with solid HgTe (c) as a function of the excessive mercury vapor pressure for various values of the parameter $\dfrac{v}{V}$, the ratio between the volumes of HgTe (c) and the enclosure

3.1.1.3.2. *Kinetics of equilibrium*

In the foregoing paragraph, we ascertained the relations between the mercury vapor pressure in equilibrium with HgTe (c) and the main parameters of the system. However, it is known [11] from experiments that equilibration is a slow process. The kinetics of equilibration should be analyzed.

In the following, a simplified analysis of the kinetics is carried out using the following fundamental hypothesis: Equilibration proceeds by means of diffusion of Hg vacancies (diffusion of mercury atoms from the solid into the gas). For the sake of simplicity, we assume the diffusion to be uni-directional. The extension of this analysis to a three-dimensional specimen (real specimen) is a straightforward one. The same applies for the application to simultaneous diffusion of Hg interstitials or any other defects. We consider a specimen with thickness $2h$ and cross-section S. The diffusion of Hg vacancies occurs in the direction x perpendicular to S. At the end of a time interval t, the concentration N^v of vacancies satisfies the diffusion equation:

$$D_v \frac{\delta^2 N^v}{\delta x^2} = \frac{N^v}{t} \qquad -h < x < h \tag{17}$$

with the boundary conditions

$$N^v(x, \sigma) = N_1^v$$

$$N^v(x, \infty) = N_f^v \qquad -h < x < h. \tag{18}$$

Considering the conditions (18) the solution of eq. (17) is obtained:

$$N^{\mathrm{v}}(x, t) = N_1^{\mathrm{v}} + N_2^{\mathrm{v}}\left[\frac{2}{\sqrt{\pi}} \int_{-\infty}^{\alpha} \exp\left(-u^2\right) \mathrm{d}u + \frac{2}{\sqrt{\pi}} \int_{\beta}^{+\infty} \exp\left(-u^2\right) \mathrm{d}u\right] \qquad (19)$$

where

$$\alpha = \frac{x - h}{2\sqrt{D_{\mathrm{v}}t}} \qquad -h < x < h \qquad (20)$$

$$\beta = \frac{x + h}{2\sqrt{D_{\mathrm{v}}t}} \qquad -h < x < h. \qquad (21)$$

D_{v} is the diffusion coefficient of Hg vacancies into HgTe (c). Furthermore, the equation

$$N_1^{\mathrm{v}} + 2N_2^{\mathrm{v}} = N_{\mathrm{f}}^{\mathrm{v}} \qquad (22)$$

is satisfied. The solution given by eq. (19) is interpreted as the superposition of two diffusion processes in the semi-infinite medium from the surfaces represented by $x = +h$ and $x = -h$ towards the interior of the specimen.* Let us now consider the flow of vacancies at the surface, $J_{\mathrm{S}}^{\mathrm{v}}$. The mass of mercury, ΔM, which the material has lost at the end of the interval t is:

$$\Delta M = -S \int_0^t J_{\mathrm{S}}^{\mathrm{v}} \, \mathrm{d}t = D_{\mathrm{v}}S \int_0^t \left(\frac{\partial N^{\mathrm{v}}}{\partial x}\right)_h \mathrm{d}t \qquad (23)$$

or, according to (19):

$$\left(\frac{\partial N^{\mathrm{v}}}{\partial x}\right)_h = \frac{N_2^{\mathrm{v}}}{\sqrt{\pi D_{\mathrm{v}}t}}\left[1 - \exp\left(-\frac{h^2}{D_{\mathrm{v}}t}\right)\right]. \qquad (24)$$

Hence, after integration by parts and appropriate transformation of the variable one obtains:

$$\Delta M = 2N_2^{\mathrm{v}}\sqrt{\frac{D_{\mathrm{v}}}{\pi}}\, S\sqrt{t}\left[1 - \exp\left(-\frac{h^2}{D_{\mathrm{v}}t}\right)\right] + 2vN_2^{\mathrm{v}}\left(1 - \frac{2}{\sqrt{\pi}}\int_0^{\frac{h}{\sqrt{D_{\mathrm{v}}t}}} \exp\left(-u^2\right) \mathrm{d}u\right) \qquad (25)$$

where

$$v = hS. \qquad (26)$$

---

* The extension to three dimensions is obtained by superposing three solutions of this kind, x replaced by y and z, respectively. The extension to other types of defects, *e.g.* Hg interstitials, is carried out by replacing D_{v}, N_1^{v}, N_2^{v}, N^{v} by D_{I}, N_1^{i}, N^2, and N_t^{i}, respectively.

The second term of eq. (25) is the *Gaussian error function*. In connection with the hypothesis that the Hg vapor pressure in the enclosure arises essentially as a result of the formation of vacancies in the specimen, eq. (25) describes the vapor pressure as a function of time (kinetics of equilibration) since λ is a constant, and hence:

$$p = \lambda \, \Delta M. \tag{27}$$

For the initial stage of the equilibration process (*i.e.* if t is small) the law of pressure variation may be expressed by the simpler equation:

$$p_{t \to 0} \simeq \frac{2N_2^v D_v}{\pi} S \sqrt{t} \left[1 - \frac{D_v t}{2h^2} \exp \left(-\frac{h^2}{D_v t} \right) \right]. \tag{28}$$

In this case, the pressure is established to be proportional to the square root of time (with a corrective term) and proportional to the surface of the specimen. However, in the case of a prolonged treatment, the expression for the pressure obtained by an asymptotic development of the *Gaussian error function*

$$p_{t \to \infty} \simeq \frac{2N_2^v}{\lambda} v \left(1 - \frac{h}{\sqrt{\pi D_v t}} \right) \tag{29}$$

eveals that the quantity of mercury lost by the specimen in the gas tends to approach a limit which is defined by the conditions of equilibrium, $2N_2^v = \Delta N_f^v v$.

The geometrical quantity playing the dominant role is, in this case, not the surface S, but the volume v of the specimen.

We express the various terms of the relation yielding the pressure as a function of the main parameters in accordance with eq. (22) as

$$pV = \frac{n}{N_0} RT \tag{30}$$

with

$$n = \eta \Delta M \qquad \eta = C^{ste}. \tag{31}$$

As a simplification, we consider the case $p_0 = 0$:

$$p = \eta \left(\frac{v}{V} \frac{RT}{N_0} K_1 \right)^{1/2} \left[\left(\frac{D_v t}{\pi h^2} \right)^{1/2} \left\{ 1 - \exp \left(-\frac{h^2}{D_v t} \right) \right\} + \text{erfc} \frac{h}{\sqrt{D_v t}} \right]. \tag{32}$$

At a given instant in time, the pressure is a square root function of $\frac{v}{V}$ (eq. (14)).

The importance of the term $\frac{h^2}{D_v}$, which has the dimensions of time, should be emphasized. It actually characterizes the rate of vacancy diffusion.

In all the cases characterized by $p_0 \neq 0$ and in those cases where other kinds of defects are supposed to exist as interstitials, a relation is obtained where time is given explicitly. This expression is similar to eq. (12) with $p_f = p_{t \to \infty}$.

175

3.1.1.3.3.   *Discussion of the results*

Figure 3 represents the function $p = \mathrm{f}\left(\dfrac{v}{V}\right)$, for $p_0 = 0$ and $t = 24$ h, which was obtained experimentally using the method described in Section 3.1.1.2.1. The agreement with the theoretical curve (in dotted lines) calculated according to eq. (12) is satisfactory. It seems to be proved that if p is not too high the simplified eq. (12) correctly describes the experimental results. The deviations are due to the choice of numerical values for the equilibrium constants. The agreement can also be seen by considering the percentage of weight loss of the specimens, Δ, as a function of $\dfrac{v}{V}$. From eq. (14) we obtain:

$$\Delta[\%] = 100\,\Delta m/m = \frac{\gamma}{\dfrac{v}{V}} \tag{33}$$

with

$$\gamma = m_{\mathrm{Hg}}\,\frac{V_{\mathrm{m}}}{M}\,\sqrt{\frac{K_1}{N_0\,RT}} \tag{34}$$

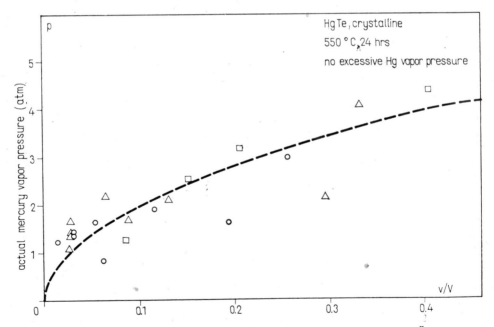

Fig. 3. Mercury vapor pressure in equilibrium with solid HgTe (c) as a function of $\dfrac{v}{V}$, the ratio of the volumes of HgTe (c) and the enclosure for the case of the excessive mercury vapor pressure zero

176

where m is the mass of the specimen, $m = \dfrac{M_v}{V_m}$. M is the molar mass of HgTe, V_m its molar volume, and m_{Hg} the atomic mass of Hg. From the results shown in Fig. 4 it is evident that the analysis is well founded.

In Fig. 5 the Hg vapor pressure over HgTe (c) is given as a function of the excessive vapor pressure for defined values of $\dfrac{v}{V}$ and t. The agreement with the theoretical curve (in dotted lines) is still satisfactory. In the case of equal excessive vapor pressures p_0, the final pressure depends upon the duration of treatment, which actually confirms the existence of a kinetic process of establishing the equilibrium vapor pressure. According to the observed variation of p with time (Fig. 6) the pressure is found to be proportional to $\sqrt{t}$ in the initial stage, as predicted by theory, then it begins to vary less and less rapidly until it attains its value of equilibrium. The existence of a critical time that is required to establish this equilibrium pressure, *i.e.* about 160 hours, substantiates the hypothesis of a relatively slow process. From the experimental curve, the value of $\dfrac{h^2}{D_v}$ can be determined as 50 hours at a temperature of 550°C. With an

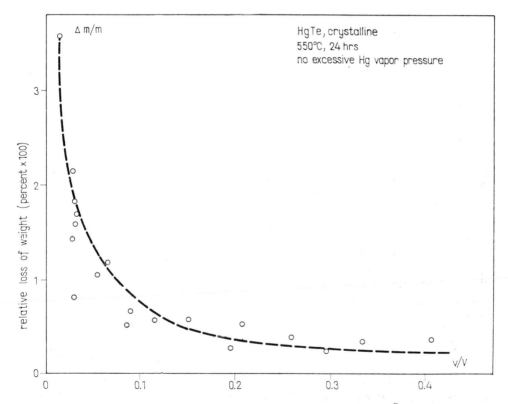

Fig. 4. The relative loss of weight of the HgTe (c) specimen as a function of $\dfrac{v}{V}$ for the case of a 24-hour treatment at a temperature of 550°C and without any excessive mercury vapor pressure

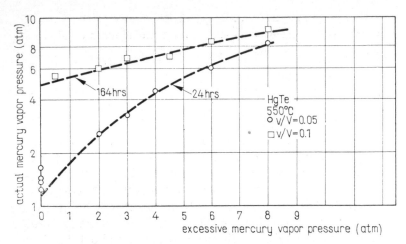

Fig. 5. Vapor pressure of mercury in equilibrium with HgTe crystals as a function of the excessive mercury vapor pressure for a 24-hour, and for a 164-hour treatment (theoretical curves in dashed lines)

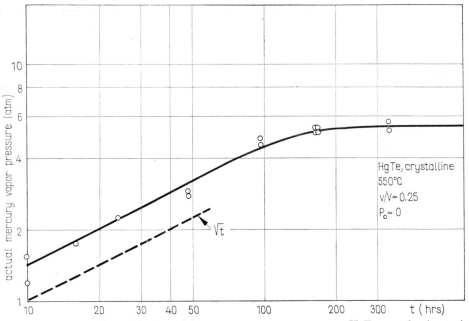

Fig. 6. The process of equilibration of the mercury vapor pressure over HgTe crystals; the straight line represents the theoretical function $t^{1/2}$. The excessive Hg vapor pressure is zero

average value of $h \cong 2 \cdot 10^3 \ \mu m$, the value of the diffusion coefficient for Hg vacancies in HgTe (c) is calculated to be $D_v \cong 8 \cdot 10^4 \ \mu m^2/h$, which is considerable, since the interdiffusion coefficient at the limit of pure HgTe attains a value of only some tens of μm^2 per hour at the same temperature.

178

INFLUENCE OF THE MERCURY VAPOR PRESSURE ON THE
        PARAMETERS AND THE KINETICS OF GROWTH IN THE
        HgTe–CdTe SYSTEM

3.1.2.1.   Introduction

The point of view adopted in this Section differs slightly from the one assumed in
Section 3.1.1. Because of the additional deposition of CdTe specimens into the enclosure, which previously contained only HgTe — a condideration of equilibrium,
or of equilibration kinetics, similar to Section 3.1.1 is impossible. One can only argue
within the framework of a system developing according to certain permanent conditions. This important modification is due to the fact that the phenomenon of layer
growth closely links the characteristics of vapor with those of volume diffusion in
the solid substrate, the interaction between these two processes constituting the base
of growth itself [4].

It is the existence of such an interaction which enables the characteristics of growth
to be modified by varying the excessive vapor pressures occurring in the enclosure [7].
Briefly, a certain number of previous results obtained on layer growth in the absence
of an excess vapor pressure will be considered and these will be generalized for an
interpretation in terms of microscopic mechanisms.

However, it should be noted that former studies of interdiffusion between CdTe and
HgTe [1], [2] have revealed how the nature of the initial CdTe (content of impurities,
deviations from stoichiometry) influences the coefficient of interdiffusion. Since the
growth process is largely dependent on the interdiffusion between the deposited layer
and the CdTe substrate, it is obvious that the influence of the substrate properties
on the characteristics of this growth cannot be neglected. This is to be verified in
the following sections, especially in Section 3.1.2.5, where it will be shown that, to
a certain extent, the experimental results depend upon the nature of the CdTe used.
Therefore, we shall designate the CdTe material according to its origin either with
the index "SAT" (made available by SAT) or with the index "CNRS" (prepared in
the laboratory by *R. Triboulet*). The two materials mainly differ in their tellurium
content, the first being richer in tellurium (which influences the kinetics of growth
by accelerating it, leading to thicker layers as is shown below), whereas the latter is
much purer.

3.1.2.2.   Generalization of previous results

Since there was no excessive vapor pressure of mercury ($p_0 = 0$), it is reasonable,
according to our previous results to distinguish between two temperature ranges in
the analysis of growth.

At high temperatures ($580°C > T >$ about $450°C$) the surface of the layers was
formed by pure HgTe, their thickness being sufficiently high so as not to be neglected
(for example, $10 \ \mu m$). Therefore, an attempt was made to explain the isothermal
growth of HgTe on CdTe by means of mercury vacancies produced in the interdiffu-

179

sion zone (at the *Kirkendall interface*) and migrating towards the surface of the layer [4], thus providing a mediator between the vapor and interdiffusion.

At low temperatures ($T < 450°C$), however, the surface of the deposited layers was formed by the solid solutions $Cd_cHg_{1-c}Te$ and this played an active role in the interdiffusion process, thus easily explaining the development of isothermal growth.

In the following (Section 3.1.2.5), it will be seen that, even at high temperatures, the presence of excessive mercury vapor may result in the appearance of a solid solution on the surface of the deposited layers (instead of pure HgTe, if $p_0 = 0$.)

In addition, the experiments carried out by *Tufte* and *Seltzer* at 600°C with larger distances between the source and the substrate [9] have shown that under these conditions the surface of layers may still be composed of solid solutions.

Also, for a swift analysis of growth, we shall distinguish between two distinct cases rather than between two temperature ranges, *i.e.* the case where the superficial layer is *not directly* involved in the interdiffusion process occurring underneath, and the case where the superficial layer participates *directly* in the interdiffusion process (solid solution $Cd_cHg_{1-c}Te$).

3.1.2.2.1.   *Pure HgTe on the surface* [4], [7]

The intermediate region at the surface contains mercury vacancies produced by volume interdiffusion and which diffuse towards the surface of the layer, thus giving rise to a concentration gradient of vacancies which impedes the re-evaporation of the deposited HgTe.

During the time dt the layer thickness grows by $d\xi$ which is proportional to the vacancy flux J_v and the probability of evaporation P:

$$d\xi = PJ_v \, dt. \tag{35}$$

The flux J_v is expressed as follows (D_v being the diffusion coefficient for the migration of vacancies at the concentration C_v):

$$J_v = \frac{\Delta C_v}{\xi} D_v. \tag{36}$$

Hence, we obtain by integration of eq. (35), taking into account the initial conditions $\xi(t = 0) = 0$:

$$\xi = \left(\frac{2P\Delta C_v}{\Gamma} D_v \right)^{1/2} \sqrt{t}. \tag{37}$$

In this expression, the influence of the diffusion of mercury vacancies in the substrate is evident (from the term $D_v \Delta C_v$).

3.1.2.2.2. *Solid solution $Cd_cHg_{1-c}Te$ on the surface*

In this case, it is not necessary to refer to the vacancy mechanism in order to explain the growth.

Growth is liable to two different models: the first is similar to the preceding one, where, however, the probability of evaporation depends on the surface composition of the layer, and the second results directly from interdiffusion laws. In the first case, we have:

$$\Gamma d\xi = \left(PD \frac{\partial C_{Hg}}{\partial x} \right)_S dt .\qquad(38)$$

Here, D is the coefficient of interdiffusion and $\dfrac{\partial C_{Hg}}{\partial x}$ is the concentration gradient of mercury. P and D depend on C_{Hg}. The index S indicates that the values refer to the surface of the layer.

The study of interdiffusion [2] reveals that C_{Hg} depends upon the parameter

$$y = xt^{1/2}.$$

Hence, we obtain:

$$\frac{\partial C_{Hg}}{\partial x} = \frac{dC_{Hg}}{dy} \frac{1}{\sqrt{t}}\qquad(39)$$

and eq. (38) is transformed into:

$$d\xi = \left(\frac{PD}{\Gamma} \frac{dC_{Hg}}{dy} \right)_S \frac{dt}{\sqrt{t}} .\qquad(40)$$

When $(C_{Hg})_S$ is constant (permanent range) it follows that D_S and $\left(\dfrac{dC_{Hg}}{dy} \right)_S$ are also constant.

Consequently, integration of eq. (40) can be carried out in order to obtain the law of growth:

$$\xi = 2 \left(\frac{PD}{\Gamma} \frac{dC_{Hg}}{dy} \right)_S \sqrt{t} .\qquad(41)$$

It should be noted that the dependence expressed as a function of time is always the same, viz.

$$(t^{+1/2})$$

as is usual for a phenomenon controlled by diffusion. However, dependence expressed as a function of the interdiffusion coefficient D and the probability P changes according to Section 3.1.2.2.1 $\left(\text{the power becomes 1 instead of } \dfrac{1}{2}\right).$

The second growth model results directly from the equation of diffusion:

$$\frac{d}{dy}\left(D\frac{dC_{Hg}}{dy}\right) = \frac{y}{2}\frac{dC_{Hg}}{dy} \tag{42}$$

or from

$$\frac{dD}{dC_{Hg}}\left(\frac{dC_{Hg}}{dy}\right)^2 + D\frac{d^2C_{Hg}}{dy^2} = \frac{y}{2}\frac{dC_{Hg}}{dy} \tag{43}$$

as D depends on C_{Hg} only. If we apply eq. (43) to the surface S, the origin of y being at the initial surface of the substrate (compared with the interface of interdiffusion) [2], we obtain:

$$\left[\frac{dD}{dC_{Hg}}\left(\frac{dC_{Hg}}{dy}\right)^2\right]_S + \left(D\frac{d^2C_{Hg}}{dy^2}\right)_{S'} = \frac{\xi}{2\sqrt{t}}\left(\frac{dC_{Hg}}{dy}\right)_S \tag{44}$$

which gives evidence of the proportionality of δ to $t^{1/2}$, the other quantities being independent of t (permanent range). In addition, the curvature of the function $C_{Hg}(y)$ in the vicinity of the surface is very low as shown by experiment [3]. Hence, we can write as a good approximation:

$$\xi \approx 2\left[\frac{dD}{dC_{Hg}}\frac{dC_{Hg}}{dy}\right]_S\sqrt{t}. \tag{45}$$

3.1.2.3.   Experimental equipment

The same device is used as for the investigation of HgTe only, and the same precautions are taken with regard to the necessary preparations and arrangements. The only difference is that one or two CdTe specimens were placed opposite the HgTe specimen from which they were separated by silica plates with a well-defined thickness as depicted in Fig. 7. The growth of a HgTe layer on the CdTe substrate does not modify the characteristics of the device, at least as far as the measurements of vapor pressures of the elements are concerned and which are carried out as described above (the vapor pressure of Cd has the same order of magnitude as the vapor pressure of Hg).

3.1.2.4.   Theoretical study of the real vapor pressure p_0. Comparison with experiments

As already pointed out in Section 3.1.2.1 one cannot speak of an equilibrium between HgTe (c) and Hg (g) if CdTe is present in the enclosure since CdTe intervenes and fosters the growth mechanism of layers consisting of Cd HgTe solid solutions, thus permitting a continuous dissolution of the gaseous mercury. The theoretical descrip-

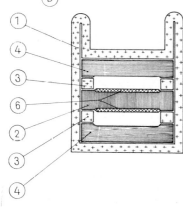

Fig. 7. Schematic representation of the apparatus
employed

(1) Silica

(2) CdTe substrate

(3) Silica spacer

(4) HgTe source

(5) Silica holder

(6) Epitaxial layers

experimental arrangement
for the isothermal
growth method EDRI.

(A) one layer
(B) two layers

tion of such a situation is rather complicated. Nevertheless, a simple approach to the problem is possible by only considering the diffusion mechanism of gaseous Hg in the layer when p_{Hg} is to be determined. This hypothesis is based on the fact that the growth process involves a simultaneous displacement of mercury and tellurium atoms and, therefore, does not sensibly change the balance of mercury loss from the solids. This hypothesis is valid only if the amount of HgTe displaced remains low compared to the remaining quantities of material, in other words, if the amount of solid solution formed by interdiffusion can be neglected in comparison with the amount of HgTe (c), and if the respective vapor pressures are different from those of HgTe (c). Under these circumstances, it is easier to verify that the previously established laws have generally to subsist in the case of HgTe (c), although they may quantitatively be modified by the growth process.

A comparison of the experimental results for HgTe (c) and for the HgTe–CdTe system is given in Fig. 8. It should be noted that the variation of p_{Hg} as a function of $p_{0\,Hg}$ proceeds in the same way for both cases and that the vapor pressure of the sys-

183

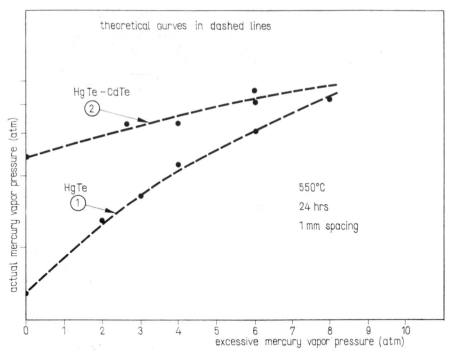

Fig. 8. The actual mercury vapor pressure over crystalline HgTe (curve 1) and over a solid solution of HgTe–CdTe (curve 2) as a function of the excessive mercury vapor pressure

tem is always higher than that for HgTe only. (This is obvious from the effective values of $\dfrac{v}{V}$ or K_1 which are higher than the real values.) According to Fig. 9 the agreement with the theoretical curves (in dotted lines) remains good, even if the duration of treatment is changed.

Nevertheless, the most interesting aspect in studying the effect of CdTe is the analysis of the kinetics leading to the Hg vapor pressure as it appears already in the curve of Fig. 10. This kinetics actually results from the interaction between two processes. The first process is the evaporation of Hg, creating vacancies in the source as already described (Section 3.1.1.3.2). But for the evaluation of the constant, the existence of CdTe and the growth resulting from it has to be considered here. The second process is the diffusion of some of the gas atoms generated in this manner in the solid solutions resulting from the process of interdiffusion. Under these conditions, the study of the Hg (c) balance for the system gives rise to a function p_{Hg}, which does not vary monotonously with time (as in the case of HgTe (c) alone), but passes through a maximum. Considering the amount of Hg which is lost by the formation of vacancies in the source, it will be found that this function can be expressed by the relation (25), since, according to the aforementioned hypothesis, the "coupling effect" between HgTe and CdTe due to layer growth influences only the magnitude of the values to be given as h and N_2^v.

184

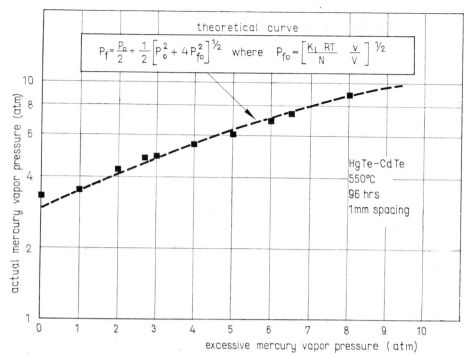

Fig. 9. The actual mercury vapor pressure as a function of the excessive mercury vapor pressure during epitaxial growth of HgTe

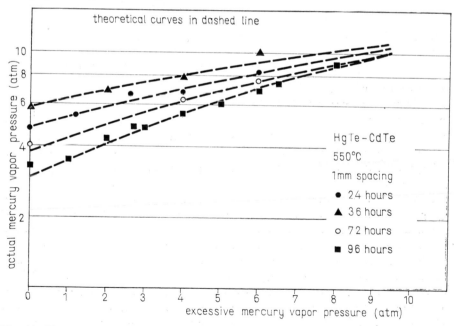

Fig. 10. The actual mercury vapor pressure as a function of the excessive vapor pressure for different growth times

185

It is possible to calculate the amount of Hg, Δm, which diffuses from the gas into the solid solutions formed at the end of a certain time t. As before, we assume unidirectional diffusion with a coefficient D_{Hg}. The three-dimensional generalization is, however, always possible.

The Hg concentration in the material, C_{Hg}, satisfies the equation of diffusion

$$D_{Hg} \frac{\partial^2 C_{Hg}}{\partial x^2} = \frac{\partial C_{Hg}}{\partial t} \tag{46}$$

in the range $-k < x < +k$ with the supplementary conditions

$$C_{Hg}(x, 0) = 0$$
$$\text{for } -k < x < +k. \tag{47}$$
$$C_{Hg}(x, \infty) = C_{Hg_1}$$

The final state of the material characterized by the value D_{Hg_1} is, to a certain extent, a fictitious state, for it implies:

(a) the final Hg concentration phase only is due to the phenomenon of diffusion out of the gas phase whereas the phenomenon of growth has not been taken into account,
(b) the amount of source material is sufficiently large so as to attain the final state without the composition of source material being sensibly changed by the amount of Hg that is lost by the formation of vacancies.

In general, our experiments were carried out so as to meet these two conditions. Nevertheless, it should be emphasized that, on carrying out certain lengthy experiments, the preparation of Hg specimens induced us to presume that they have already a high concentration of vacancies (evidenced by the presence of tellurium microprecipitates). We observed superficial fusion after the treatment, both on the source and on the layer itself. These fusion processes indicate that the degree of vacancy formation is so high as to initiate a phase segregation process. Beyond these borderline cases, a solution of eq. (46) that satisfies eq. (47) is:

$$C_{Hg} = C_{Hg_1} \left[\frac{2}{\sqrt{\pi}} \int_{-\infty}^{\gamma} \exp\left(-u^2\right) du + \frac{2}{\sqrt{\pi}} \int_{\delta}^{+\infty} \exp\left(-u^2\right) du \right] \tag{48}$$

with:

$$\gamma = \frac{x - k}{2\sqrt{D_{Hg}t}}$$
and
$$\delta = \frac{x - k}{2\sqrt{D_{Hg}t}} . \tag{49}$$

Concerning the physical interpretation of eq. (48), the same arguments apply as in the case of HgTe alone (eq. (19)). The diffused amount of Hg, Δm, is:

$$\Delta m = D_{Hg} S_{Hg} \int_0^t \left(\frac{\partial C_{Hg}}{\partial x} \right)_{x=k} dt \tag{50}$$

where, as before, the integral refers to the time of flow through the surface S_{Hg}.

$$\Delta m = 2C_{Hg_1} \left[\sqrt{\frac{D_{Hg}}{\pi}} S_{Hg} \sqrt{t} \left(1 - \exp \left(\frac{-k^2}{D_{Hg}t} \right) \right) + \right.$$

$$\left. + 2V_{Hg} C_{Hg_1} \left(1 - \frac{2}{\sqrt{\pi}} \int_0^{\frac{k}{\sqrt{D_{Hg}t}}} \exp (-u^2) \, du \right) \right] \tag{51}$$

where

$$V_H = kS_H. \tag{52}$$

Considering the balance of the amount of mercury in the gas phase, ΔQ, we obtain

$$\Delta Q = \Delta M - \Delta m \tag{53}$$

and

$$\frac{\partial \Delta Q}{\partial t} = \frac{\partial \Delta M}{\partial t} - \frac{\partial \Delta m}{\partial t}. \tag{54}$$

The relation (54) yields the variation of p_{Hg} since η is a constant

$$p_{Hg} = \eta \Delta Q \tag{55}$$

$$\frac{\partial \Delta Q}{\partial t} = N_2^v S \sqrt{\frac{D_v}{\pi t}} \left(1 - \exp \left(-\frac{h^2}{D_v t} \right) \right) - C_{Hg_1} S_{Hg_1} \sqrt{\frac{D_{Hg}}{\pi t}} \left(1 - \exp \left(\frac{-k^2}{D_{Hg}t} \right) \right). \tag{56}$$

It is found that eq. (56) becomes zero, if $t = t_0$, leading to

$$\frac{1 - \exp (-k^2/D_{Hg}t_0)}{1 - \exp (-h^2/D_v t_0)} = \frac{N_2^v S}{C_{Hg_1} S_{Hg}} \sqrt{\frac{D_v}{D_{Hg}}}. \tag{57}$$

For $t < t_0$ then $\dfrac{\partial \Delta Q}{\partial t} > 0$, and when $t > t_0$ then $\dfrac{\partial \Delta Q}{\partial t} < 0$, which proves that $\Delta Q(t)$ and, hence, the pressure $p_{Hg}(t)$ passes through a maximum for $t = t_0$. The value t_0 can be evaluated approximately as follows: First, only specimens with $h \approx k$ and $S \approx S_{Hg_2}$ were considered. On the other hand, it was determined that $D_v = 8 \cdot 10^4 \, \dfrac{\mu}{h}$. According to the study of growth [1], $D_{Hg} \approx 4 \cdot 10^2 \, \mu m \, h^{-1}$ and $\dfrac{D_v}{D_{Hg}} \approx 2 \cdot 10^2$. If t_0 is

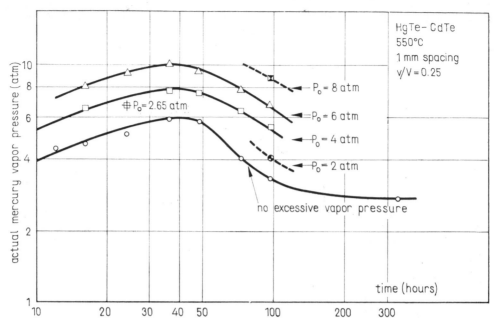

Fig. 11. Kinetics of equilibration of the actual mercury vapor pressure over the HgTe–CdTe system as a function of the excessive mercury vapor pressure

small, eq. (56) reduces to:

$$\frac{1}{1 - \exp\left(-\dfrac{h^2}{D_v t_0}\right)} = 14 \frac{N_2^v}{C_{Hg_1}}.$$ (58)

The evaluation of the second term is more delicate. It is, however, possible to conclude:

a) The conditions of equilibrium of HgTe furnish $N_2^v V_m / N_0 \approx 10^{-3}$.
b) The "final" concentration C_{Hg_1} is likely to be of the same order of magnitude as the concentration of vacancies in HgTe, presumably $2 \cdot C_{Hg_1} \approx 3 \cdot 10^{-2}$ at its limit of stability.

Hence, it is possible to calculate the time the system requires to attain its maximum pressure, and t_0 is found to be approximately 40 hours.
The experimental results are in good agreement with theoretical expectation. Actually, according to Fig. 11 the Hg vapor pressure has a maximum regardless of the specific value of the excessive vapor pressure p_0.
Furthermore, this maximum is attained in any case at the end of a period ranging between 30 and 40 hours. In addition, according to the curve for $p_0 = 0$, after a period of 200 hours the vapor pressure over the system is virtually stabilized at a value having the same order of magnitude as observed in the case of HgTe alone.

188

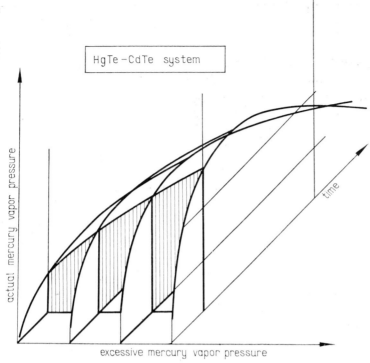

Fig. 12. Schematic diagram showing the dependence of the actual mercury vapor pressure on excessive mercury vapor pressure and time for the HgTe–CdTe system

Summarizing, the results concerning the dependence of the Hg vapor pressure, over the HgTe–CdTe system, upon the excess vapor pressure of Hg (g), p_0, and the time, may be readily illustrated by the scheme depicted in Fig. 12.

3.1.2.5. Effect of the excessive vapor pressure p_0 on growth characteristics

The rates of the interdiffusion and evaporation processes determining the growth of layers depend upon the amount of vacancies in the substrate and, consequently, upon the excessive vapor pressure in the gas phase (even if equilibrium is not realized). Hence, we presume that the excessive vapor pressure p_0 not only influences the real vapor pressure p (condition for *evaporation*), but also the coefficient of interdiffusion, D, (condition for *interdiffusion*) and, finally, the thickness e of the layers, resulting from the interaction of the two phenomena and from the surface composition C_S.

In this Section, the results are given with regard to these quantities. The interpretation of these results based on microscopic and thermodynamic models will follow in Section 3.1.2.6.

3.1.2.5.1　*The interdiffusion coefficient D*

3.1.2.5.1.1.　*Experimental results and discussion*

Recently we have shown that interdiffusion at the initial stage of the growth process is impeded by an excessive Hg vapor pressure [7]. This suppression can be interpreted simply as a considerable reduction of the interdiffusion coefficients, assuming that diffusion is initiated by a mechanism of vacancy formation. The excessive Hg vapor pressure tends to reduce the number of vacancies in the layer by means of the solid/gas exchange. The resulting diminution of interdiffusion entails a reduction of vacancy formation beyond equilibrium. As a result, the mechanism of mass transfer is greatly

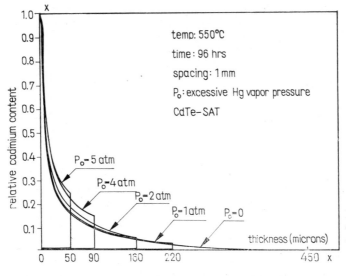

Fig. 13. The composition-distance curves for different excessive mercury vapor pressures (CdTe—SAT)

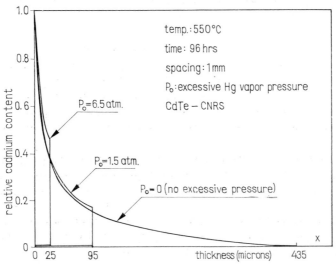

Fig. 14. The composition-distance curves for different excessive mercury vapor pressures (CdTe–CNRS)

190

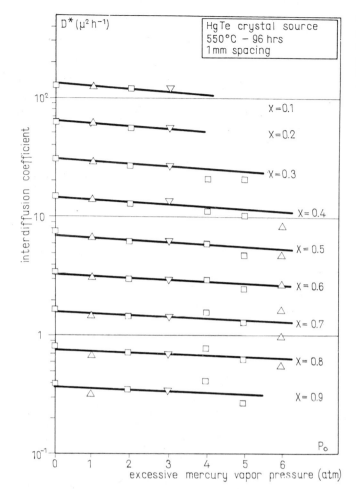

Fig. 15. The interdiffusion coefficient as a function of the excessive mercury vapor pressure for solid solutions of a composition $0.1 < x < 0.9$; $T = 550°C$

diminished. A remarkable consequence of this complex process is that, on the one hand, the composition *versus* distance curve remains unchanged whatever the pressure p_0 might be, and, on the other hand, the composition of the specimen at the surface only depends upon p_0, as is shown below. These results depicted in Figs. 13 and 14, agree well with the results obtained by *Tufte* [9] at a higher temperature (600°C) and with sources of a different nature (crystal or sintered powder of mercury telluride).

D is obtained from the analysis of the composition-distance curves employing the *Matano method*. The curves depicted in Fig. 15 represent the interdiffusion coefficient as a function of the pressure p_0 for various values of the mole fraction X of the solid solution. Note that D, which ranges from 0.5 $\mu m^2 h^{-1}$ for solid solutions rich in cadmium to 100 $\mu m^2 h^{-1}$ for solid solutions rich in mercury ($X = 0.1$), decreases with p_0. These results agree well with those derived from the composition-distance curves given by *Tufte* [9] and confirmed by our own experiments (Fig. 16) with the excep-

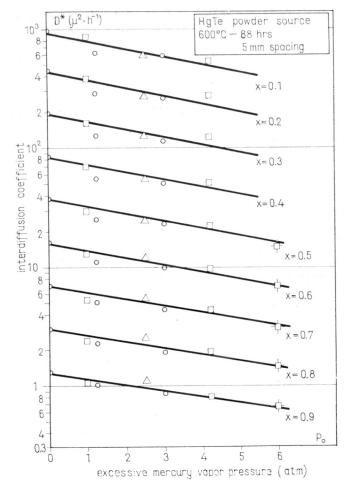

Fig. 16. The interdiffusion coefficient as a function of the excessive mercury vapor pressure for solid solutions of the composition $0.1 < x < 0.9$; $T = 600°C$ (according to *Tufte, Stelzer* and our own results)

tion that D is higher at 600°C, being between 1 μm^2 h^{-1} and 1 000 μm^2 h^{-1}, and that it decreases with p_0 more rapidly.

The invariability of the composition-distance curves in respect to p_0 and, in addition, the invariability of the curves obtained when the composition is plotted versus the reduced distance

$$(y = xt^{-1/2})$$

in respect to p_0 and t reveals that the interdiffusion coefficient D does not change with time. This implies the existence of well-defined relations concerning the balance of vacancies in the substrate on a microscopic level.

This item will be discussed in Section 3.1.2.6. All these results may be expressed (for

192

a given composition C of solid solution $Cd_cHg_{1-c}Te$) in a general form:

$$D(p_0, t) = f_1(p_0)$$
$$\left. \frac{\partial f_1}{\partial p_0} < 0 \right\} . \tag{59}$$

This result is schematically illustrated in parts C and D of Fig. 21a and summarized in the scheme of Fig. 21b.

3.1.2.5.1.2. *Complementary verification*

The characteristics of the interdiffusion coefficient (especially its invariance in respect to time) may be verified by the following experiment:
Two layers are prepared in an identical way. One of which serves as a reference, and the other is subjected to an annealing treatment under conditions of excessive pressure, the overall composition of vacancies being maintained. If our hypotheses concerning the evaluation of D in respect to t and the influential significance of p_0 are exact, we should obtain identical results when analyzing the interdiffusion coefficient prior to, and after, the annealing process. This should be verified under the following conditions:

(a) Preparation of two layers at $T = 550°C$, $t = 72$ hours, $p_0 = 1$ atm. At the end of the growth process the real pressure p over the layers was found to be about 8 atm.

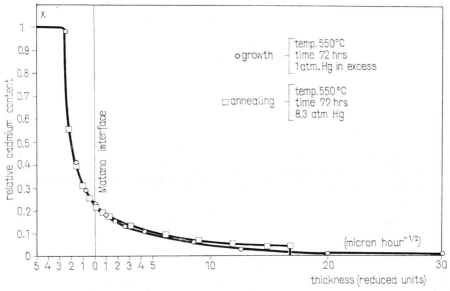

Fig. 17. The composition of an epitaxial layer subjected to an annealing treatment at a high temperature and exposed to different mercury vapor pressures

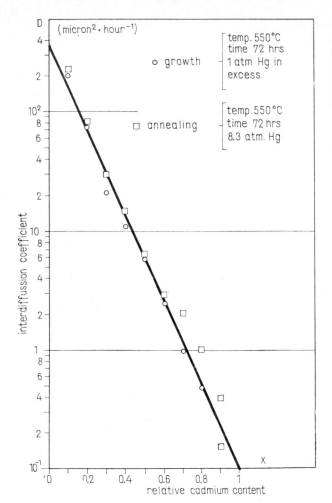

Fig. 18. The interdiffusion coefficient as a function of the composition for two specimens prepared under equal conditions, one of them, however, having been annealed exposed to a mercury vapor atmosphere

(b) Annealing of one of the layers at $T = 550°C$, $t = 72$ hours, the saturated mercury pressure being $p_{sat} = 8$ atm. This was established by means of mercury placed at the cold point of the annealing ampulla ($T = 500°C$).

Figure 17 shows that the composition *versus* reduced distance curves prior to, and after, the annealing process coincide almost completely (taking into account the variation corresponding to the *total* time of treatment being the total time of growth and annealing = 144 hours).

This accounts for the fact that, under such conditions, the interdiffusion coefficient is solely related to the composition of the solid solution and to the temperature, as is illustrated in Fig. 18.

The independence of D on time and the effect of p_0 on the variation of D are, therefore, confirmed in this way.

194

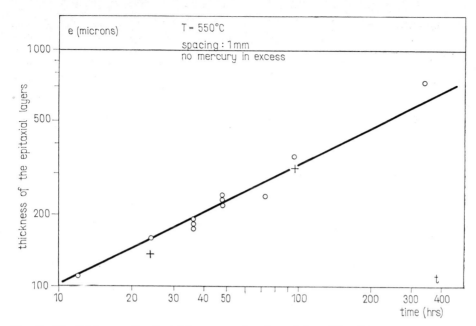

Fig. 19. The thickness of epitaxial layers as a function of deposition time

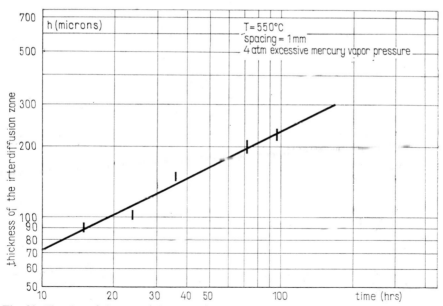

Fig. 20. The size of the interdiffusion zone as a function of time for an excessive mercury vapor pressure of 4 atmospheres

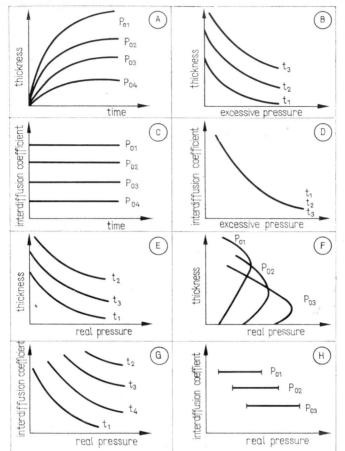

Fig. 21a. Schematic representation of the layer thickness and the interdiffusion coefficient as functions of p_0 and t. In each case, one variable (p_0 or t) is eliminated and the other is applied as a parameter ($t_{1,2,3}$ or $p_{0\,1,2,3}$)

3.1.2.5.2. *Thickness of layers and surface composition*

From the properties of the interdiffusion coefficient D certain characteristics of the variation $e(p_0, t)$ can be derived, especially the fact that the thickness of the layers (cf. Fig. 19) varies as $t^{1/2}$, no matter what value the excessive mercury pressure p_0 has (an example is given in Fig. 20 where $p_0 = 4$ atm). Consequently, we write quite generally:

$$\left.\begin{aligned} e(p_0, t) &= f_2(p_0)\,\sqrt{t} \\ \frac{df_2}{dp_0} &< 0 \end{aligned}\right\}. \tag{60}$$

The variations in t and in p_0 are schematically illustrated in Figs. 21a and 21b, respectively, and can be summarized by the schemes depicted in Figs. 22a and 22b, giving evidence of the roles of the two parameters.

The theoretical justification of eq. (60) was given in the references of Section 3.1.2.2.

196

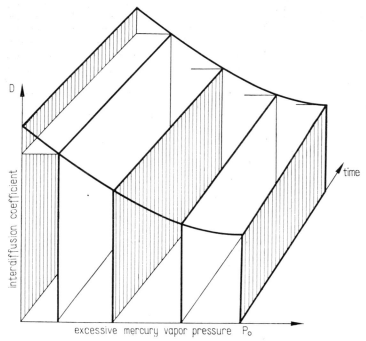

Fig. 21b. Schematic representation of the function D (p_0, t)

The study of the dependence of the composition C_S upon p_0 (in $Cd_{C_S} Hg_{1-C_S}Te$) at the surface of the layers makes possible the shape of the function $f_2(p_0)$, as revealed experimentally in Figs. 23 and 24, to be considered.

First, it should be noted that the composition C_S of cadmium at the surface of the layers does not depend upon time, but only upon p_0 as may be concluded from the study of the preceding section concerning the interdiffusion coefficient and the composition-distance curves:

$$C_S(p_0, t) = f_3(p_0) .$$ (61)

The experiment reveals that the function $f_3(p_0)$ can be expressed in the following form:

$$f_3(p_0) \approx Bp_0 .$$ (62)

The analysis presented in the references of Section 3.1.2.2 concerning the growth law for the case of the formation of the layer surface by a solid solution, makes it possible to express the law $f_2(p_0)$ for the variation of thickness (eq. (60)) originating from $f_3(p_0)$. Indeed, according to eq. (41), we obtain:

$$\xi \equiv e = \left[\frac{2P}{\Gamma} D \frac{dC_{Hg}}{dy} \right]_S \sqrt{t} .$$ (63)

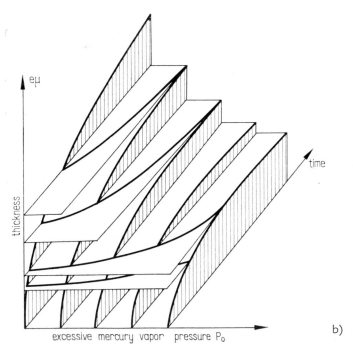

Figs. 22a and 22b. Schematic representation of the functions $e\,(p_0,\,t)$ for **EDRI** growth processes

a)

eμ

thickness of the layer

time

excessive mercury vapor pressure

eμ

thickness

time

excessive mercury vapor pressure P_0

b)

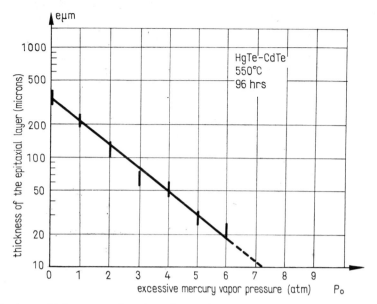

Fig. 23. The layer thickness as a function of excessive vapor pressure at $T = 550°C$

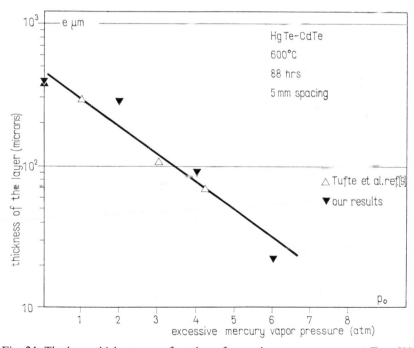

Fig. 24. The layer thickness as a function of excessive vapor pressure at $T = 600°C$

The experiment reveals that, at the surface, $\dfrac{dC_{\mathrm{Hg}}}{dy}$ depends only slightly on p_0. Consequently, the significant variations of e, result from those of D. Therefore, the study of interdiffusion [2] has shown that, at a given temperature

$$D(C) = D_0 \exp(-\alpha C) \tag{64}$$

and consequently

$$e = \left[\frac{2P}{\Gamma} \frac{dC_{\mathrm{Hg}}}{dy} \right]_S \sqrt{t}\, D_{0S} \exp(-\alpha C_S). \tag{65}$$

If eq. (61) is taken into account, we obtain, according to eq. (60):

$$f_2(p_0) = \left[\frac{2P}{\Gamma} \left(\frac{dC_{\mathrm{Hg}}}{dy} \cdot \right) \right]_S D_{0S} \exp(-\alpha f_3(p_0)) = e_0 \exp(-kp_0) \tag{66}$$

with

$$k = \alpha B. \tag{67}$$

Knowing the values of α and B (with p_0 in atm and C in per cent) with $\alpha \approx 5 \cdot 10^{-2}$ and $B \approx 11$, we determine — in accordance with eq. (67) — the value $k \approx 0.55$, which agrees well with the experimental value $k \approx 0.5$.
The law of variation, eq. (60), is now precisely expressed as

$$e(p_0, t) = e_0 \sqrt{t} \exp(-kp_0). \tag{68}$$

Considering now the curves depicted in Figs. 25–28, we can verify that the law $f_3(p_0)$ concerning the variation of C_S in dependence upon p_0, is represented by a linear re-

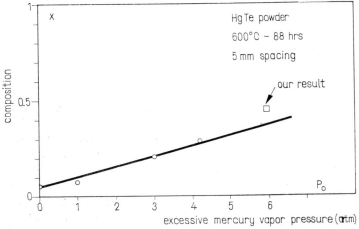

Fig. 25. The surface composition as a function of excessive mercury vapor pressure at $T = 600°$C and $\delta = 5$ mm (according to *Tufte*, *Stelzer*, and our own results)

200

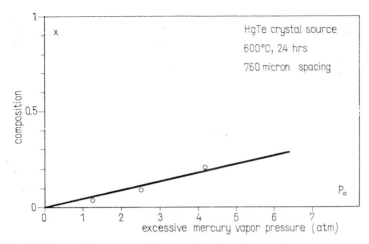

Fig. 26. The surface composition as a function of excessive mercury vapor pressure at $T = 600°C$ and $\delta = 760$ μm (according to *Tufte et al*)

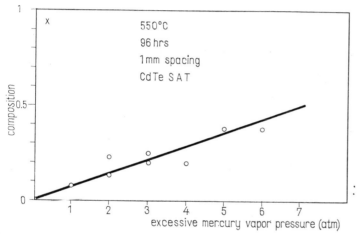

Fig. 27. The surface composition as a function of excessive mercury vapor at $T = 550°C$ and $\delta = 1$ mm; SAT material

lation, irrespective of the temperatures or the lengths of time chosen for the growth. Some remarks of special interest should be emphasized:

Figure 28, obtained from the experiments for various values of t, reveals the independence of B on t. It should be noted that the dispersion of the experimentally obtained points for $p_0 > 4$ atm is closely related to the fact that, under these conditions, the growth of layers is totally inhibited and the C_S values obtained must be attributed to an interdiffusion mechanism for which the surface condition seems to have a significant influence. The comparison of the straight lines obtained at a temperature of 500°C, but for different durations of treatment (88 and 24 hours) (Figs. 25 and 26) reveals the already mentioned fact that the variation of C_S as a function of p_0 does

201

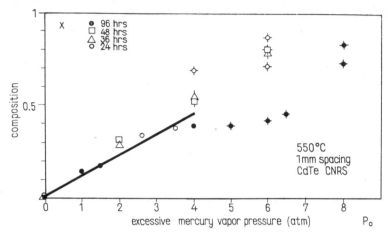

Fig. 28. The surface composition as a function of excessive mercury vapor pressure at $T = 550°C$ and $\delta = 1$ mm; CNRS material. (The crosses refer to interdiffusion processes with no measurable layer thickness)

not depend upon time, (the slope of the two straight lines is virtually the same). The two straight lines of Figs. 25 and 26 cannot be superposed since the one shown in Fig. 25 does not pass through zero. This provides evidence for the effect of the parameter δ, the distance between the source and the substrate, on C_S. This could partly explain the law concerning the variation of thickness of the layers with δ as indicated in Fig. 29. Moreover, within this hypothesis, certain apparent disagreements with the results obtained by *Tufte* could be explained as well, especially the fact that *Tufte* never obtained pure HgTe at the surface of his layers.

3.1.2.5.3. *Influences of the real pressure p*

In the foregoing, the following relation was established both experimentally and theoretically (see Section 3.1.2.4):

$$p = f_4(p_0, t).\tag{69}$$

Principally, it is possible to obtain numerous desired relations between e and p as well as between D and p by graphic evaluation of the functions $e(p_0, t)$ and $p(p_0, t)$. For the discussion, it should be noted that two independent variables, p_0 and t, which can be checked by experiments are associated with the system and that p appears as a dependent variable according to eq. (69). In order to make p an independent variable, it is necessary to eliminate p_0, or t, employing eqs. (59), (60), and (69). Depending on whether the eliminated variable is p_0 or t, *different* functions $e(p)$ and $D(p)$ are obtained, because the faces $e(p, p_0)$ and $e(p, t)$ differ from each other, as do the faces $D(p, p_0)$ and $D(p, t)$. In these cases, the variations of e and D are qualitatively depicted in Fig. 21, E and F refer to e, and G and H refer to D.

In Figs. 21 E and G, p_0 was eliminated and t appears as a parameter, whereas in

202

F and H it is eliminated, and p_0 appears as a parameter. It is noted that the inversion of the order of the parameter t in Figs. 21 E and G expresses only the existence of a maximum in the kinetics of establishing the pressure p over the HgTe–CdTe system as already referred to above. Hence, we obtain the important result which is implicitly contained in the mathematical formulation of eqs. (59), (60), and (69), that there are no unique relations $e(p)$ and $D(p)$. These relations, are, however, derived from specified experimental conditions (the system has been studied as a function of time at a certain pressure p_0, or the system has been studied for a chosen time as a function of various values of p_0).

According to Fig. 21H D seems to be independent of p when t has been eliminated. This observation is important, for it allows the variations $p(t)$ to be clarified in a particular case: the independence of D upon p indicates that, as far as the solid solutions are concerned, nothing is changed if, during growth, p_0 varies in such a way that the interdiffusion mechanism is not affected.

This is expressed exactly by the hypothesis established above in order to take theoretically into account the variations $p(t)$. We supposed that Hg atoms diffuse from the gas into the formed layers thus counterbalancing the effect of the formation of vacancies due to interdiffusion. More precise details concerning this point could be determined by a quantitative study of the balance of vacancies in the layers which takes into consideration the creation and annihilation of vacancies associated with the various processes.

The conclusions from this study, described in the following Section, indicate that all the studied phenomena may be completely interpreted in those cases where the Hg vacancies play a decisive role in the process of growth; in other words, when the surface of the deposit is composed of pure HgTe.

3.1.2.6.   Interpretation of the results

In order to justify, or verify, the conclusions drawn in the preceding Section, we consider the balance of vacancies in the substrate; we then attempt to find the thermodynamic reasons involved in the compensation between the creation and annihilation of vacancies as it results from this balance.

3.1.2.6.1.   *Balance of vacancies in the substrate*

In the substrate, the *Kirkendall effect* intimately connected with interdiffusion between HgTe and CdTe appears to be the principal source of vacancies [4].

As D'_{Hg} and D'_{Cd} are the *individual diffusion coefficients* of mercury and cadmium, respectively, the flow of vacancies initiated by their difference at the *Kirkendall interface* is approximately [12]:

$$(I_v)_{\mathrm{K}} = \left[(D'_{\mathrm{Hg}} - D'_{\mathrm{Cd}}) \frac{\partial C_{\mathrm{Hg}}}{\partial x} \right]_{\mathrm{K}} . \tag{70}$$

Hence, the quantity σ_v of the vacancies generated between the onset of interdiffusion and the time t is:

$$\sigma_v = S \int_0^t (I_v)_K \, dt = S \int_0^t \left[(D'_{Hg} - D'_{Cd}) \frac{\partial C_{Hg}}{\partial x} \right] dt \tag{71}$$

where S is the cross-sectional area of the specimen perpendicular to the direction of diffusion, x.

However, the composition C_{Hg} at the *Kirkendall interface* remains constant. When taking the quantity

$$y = x \cdot t^{-1/2}$$

as a characteristic parameter, we obtain

$$\left[\frac{\partial C_{Hg}}{\partial x} \right]_K = \left[\frac{d C_{Hg}}{dy} \right]_K \cdot t^{-1/2},$$

where $\left[\dfrac{d C_{Hg}}{dy} \right]_K$ is a constant. As a result, we can rewrite eq. (71) as

$$\sigma_v = 2 \left[(D'_{Hg} - D'_{Cd}) \frac{d C_{Hg}}{dy} \right]_K S \sqrt{t}. \tag{72}$$

The vancancies thus generated by interdiffusion then evolve according to two possible mechanisms: some of the vancancies diffuse from the *Kirkendall interface* into the surface of the deposited layer establishing a concentration gradient at the origin of growth, the rest of the vacancies is annihilated due to diffusion of mercury from the vapor. The respective contributions of these two mechanisms are estimated in the following.

In respect of the first mechanism, we remember that a constant vacancy concentration gradient was assumed, taking into account the kinetics of growth [4], *i.e.*

$$C_v = \frac{C_{vS} - C_{vK}}{\xi} x + C_{vK} \tag{73}$$

where C_{vS}, C_{vK}, and ξ represent the concentration of vacancies at the surface of the layer, the saturated concentration at the *Kirkendall interface*, and the thickness of the deposited layer, respectively, (x being counted here from the *Kirkendall interface*). The growth law of the layer [4] was derived as:

$$\xi = \left[2 D_{Hg} \frac{C_{vK} - C_{vS}}{\Gamma} P \right]^{1/2} \sqrt{t}. \tag{74}$$

The number of vacancies, n_v, involved in this process may be expressed in two ways:

$$n_v = S \int_0^\xi C_v \, dx = S D_{Hg} \int_0^t \frac{\partial C_v}{\partial x} \, dt. \tag{75}$$

204

This leads to

$$n_v = \frac{C_{vK} + C_{vS}}{2} S\xi = \frac{\Gamma}{P} S\xi. \tag{76}$$

Considering the identification from eq. (76):

$$\frac{\Gamma}{P} = \frac{C_{vK} + C_{vS}}{2} \tag{77}$$

and

$$\xi = \left[4D_{Hg} \frac{C_{vK} - C_{vS}}{C_{vK} + C_{vS}} \right]^{1/2} \sqrt{t} \tag{78}$$

we finally obtain

$$n_v = [(C_{vK} + C_{vS})(C_{vK} - C_{vS}) D_{Hg}]^{1/2} S\sqrt{t}. \tag{79}$$

Regarding the second mechanism (annihilation of vacancies by diffusion of mercury from the vapor) it is sufficient for evaluating the number of vacancies involved, N_v, to refer to eq. (51) which provides it exactly.

According to this relation, we find that N_v is not a simple function (in $t^{1/2}$) of time as the other quantities (σ_v and n_v). Nevertheless, it is possible to reduce N_v to that type of function provided that the time t is not too long:

$$N \approx 2C_{Hg_1} \left(\frac{D_{Hg}}{\pi} \right)^{1/2} S' \sqrt{t}. \tag{80}$$

In order to justify the conclusions drawn from Section 3.1.2.5 regarding the independence of D as a function of time it should be finally verified that all of these processes are compensated exactly, i.e.:

$$\sigma_v = n_v + N_v. \tag{81}$$

For σ_v the various constants involved in eq. (72) are given in [2]. With

$$D'_{Hg} \approx 33 \ \mu m^2 \ h^{-1} \tag{81a}$$

$$D'_{Cd} \approx 15 \ \mu m^2 \ h^{-1}$$

$$\frac{dC_1}{dy} \approx 4.5 \cdot 10^{20} \text{ at } \mu m^{-1} h^{1/2} \ (i.e. \ 30\% \text{ per } 10 \ \mu m \ h^{-1/2})$$

we obtain (S in units of cm^2)

$$\sigma_v \approx 1.6 \cdot 10^{18} \ S\sqrt{t}. \tag{81b}$$

For n_v, experience furnishes (by the study of growth, *i.e.* the study of the variation of ξ as a function of t [4]):

$$4 D_{Hg} \frac{C_{vK} - C_{vS}}{C_{vK} + C_{vS}} \approx 30 \mu m \, h^{-1/2} . \tag{81c}$$

From this we derive

$$D_{Hg} \approx 4 \cdot 10^{-6} \, cm^2 \, h^{-1} \tag{81d}$$

which represents a mean value somewhere between D'_{Hg} and D_v, as expected. Hence, the value of

n_v is approximately $\dfrac{10^{18} \, S}{\sqrt{t}}$.

Finally, with this value of D_{Hg} and with $C_{Hg_1} \approx 2.25 \cdot 10^{20} \, cm^{-3}$ we obtain:

$$N_v \approx 5.5 \cdot 10^{17} \, S \sqrt{t} . \tag{81e}$$

Consequently, the relation (81) is approximately verified. However, the validity of eq. (81), which enables us to understand how the process of composition in the deposited layer takes place, does not indicate why it occurs.

The process of compensation is originated by the thermodynamic conditions of the system considered. This is discussed in the subsequent Section.

3.1.2.6.2. *Thermodynamic conditions for the onset of the compensation process*

In the following, we again assume that the surface of the deposited layer is always composed of pure HgTe. The extension to the more general case where this surface consists of a CdTe–HgTe solid solution is discussed later.

It is essential for our analysis that we determine the distribution of the molecular free energy in the system and, consequently, the distributions of the potentials and forces resulting from it, taking into account the concentration distributions of the various chemical species as schematically illustrated in Fig. 30.

In order to study the free energy, we refer to the results published elsewhere [4], thus simplifying the problem by the important restriction that we confine ourselves to studying the two ternary systems, (Hg, Cd, Te) and Hg, (V_{Hg}, Te). For the case of low cadmium concentrations, however, interaction between mercury vacancies and cadmium atoms in the deposited layer has to be considered. This would require the complex treatment of the quaternary system (Hg, Cd, (V_{Hg}, Te). In the following, it will be shown that the simplification is acceptable and that no major difficulties arise by joining the two ternary systems.

Under these circumstances, and designating all quantities referring to mercury, tellurium, and cadmium (or to the mercury vacancies) by the indices A, B, and C,

206

respectively, the results of Refs. [2] and [4] enable us to schematically illustrate the distribution of the potentials

$$\frac{dG_m}{dC_A} = \mu_A - \mu_c \quad \text{and of the forces} \quad \partial(\mu_A - \mu_c)\,\partial x$$

as depicted in Figs. 31 and 32, respectively (where any intermediate region for which the analysis of the quaternary system would be required is distinctly marked). These schemes, and notably Fig. 32, indicate clearly the tendency of the mercury to migrate from the source towards the substrate.

The forces acting on the mercury atoms are of great interest, especially for the intermediate region of connection, but also for the surface near regions of the source and for the deposited layer.

If F_{HgCd} is the force acting on a mercury atom in the ternary system (Hg, Cd, Te) and if F_{Hgv} L is the corresponding force in the ternary system (Hg, V_{Hg}, Te) then, employing the numerical expressions and values of [2] and [4], we obtain:

$$F_{HgCd} = \left(\frac{RT}{C_{Cd}} + 4.4\right)\frac{\partial C_{Hg}}{\partial x} \tag{82}$$

$$F_{Hgv} = \left(\frac{RT}{C_v} - 60.4\right)\frac{\partial C_{Hg}}{\partial x}. \tag{83}$$

(RT is expressed in kcal/mole and C in molar percent.)
For a substrate temperature of 550°C these two forces are equal under the condition:

$$\frac{C_{Cd} - C_v}{C_{Cd}\, C_v} = 39.4. \tag{84}$$

Taking into account the distribution of cadmium atoms and of vacancies in the substrate, this condition is fulfilled in the region close to the *Kirkendall interface* (where $C_{Cd} = 0.09$, *i.e.* 18 per cent cadmium, and $C_v \approx 0.025$, *i.e.* 5% vacancies represent the limit of solubility).

The continuity of forces is ensured and contained in the scope of our approximation. Therefore, the two domains may be connected without any difficulty.

Equation (83) reveals that the force due to the vacancies and acting on the mercury atoms is the more effective the smaller C_v and the larger the corresponding gradient is.

For the present case, this force decreases as C_v increases, to become negative passing the limit of solubility, *i.e.* beyond the *Kirkendall interface*. Thus, compensation may arise according to the mechanism discussed above. When the forces acting at the surface of the layer and of the substrate are written as functions of time the change of the vapor pressure in the enclosure as a function of time may be explained from a point of view supplementary to the aforementioned. Designating all quantities referring to the properties of the source and the substrate by the indices 1 and 2, respectively, and all those referring to the surfaces by S and taking into account eq.

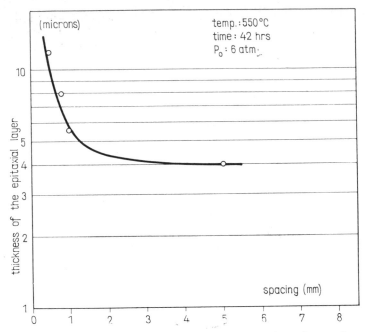

Fig. 29. The epitaxial layer thickness as a function of the distance between source and substrate for the case of an excessive mercury vapor pressure of 6 atm.

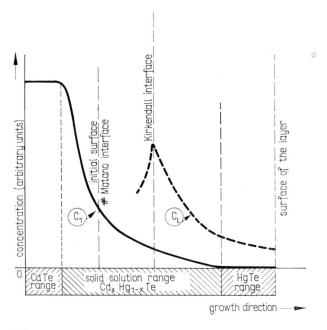

Fig. 30. Schematic distribution of various species (Hg, Cd, N_{Hg}^{v}) in the system

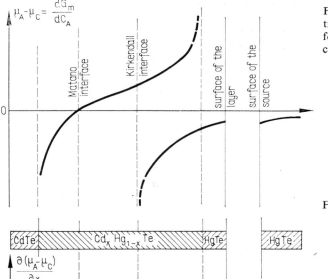

(83) we obtain, according to eqs (19) and (48) established for the distribution of the mercury atoms (or vacancies) in HgTe,

$$(F_{\mathrm{Hgv}})_{\mathrm{S}_1} = \left(\frac{RT}{C_{\mathrm{vS}_1}} - 60.4 \right) \frac{N_2}{\sqrt{\pi D_{\mathrm{v}} t}} \left[1 - \exp \left(- \frac{h^2}{D_{\mathrm{v}} t} \right) \right] \tag{85}$$

$$(F_{\mathrm{Hgv}})_{\mathrm{S}_2} = \left(\frac{RT}{C_{\mathrm{vS}_2}} - 60.4 \right) \frac{C_{\mathrm{Hg}_1}}{\sqrt{\pi D_{\mathrm{Hg}} t}} \left[1 - \exp \left(- \frac{k^2}{D_{\mathrm{Hg}} t} \right) \right] \tag{86}$$

where:

$$\Delta F_{\mathrm{S}} = (F_{\mathrm{Hgv}})_{\mathrm{S}_1} - (F_{\mathrm{Hgv}})_{\mathrm{S}_2} . \tag{87}$$

On comparing $(\Delta F)_{\mathrm{S}}$ with the expression for the quantity ΔQ of eq. (53) we obtain, (supposing that $C_{\mathrm{vS}_1} \simeq C_{\mathrm{vS}_2} = C_{\mathrm{vS}}$, which is conceivable)

$$\Delta F_{\mathrm{S}} = \frac{\partial \Delta Q}{\partial t} \left(\frac{RT}{C_{\mathrm{vS}}} - 60.4 \right) . \tag{88}$$

Thus, the difference between the forces acting between the two surfaces appears to be proportional to the derivative compared with the time of mercury vapor pressure exposure. Therefore, the function $p(t)$ can be interpreted in terms of the evolution of forces on the surfaces of the layer and the substrate.

When t is small, the force tending to extract mercury from the source is superior to that tending to incorporate mercury into the substrate. Hence, the vapor enriches in mercury and the vapor pressure increases. Then, the forces become equalized followed by an inversion of their difference, *i.e.* the substrate tends to capture more mercury than the source is able to supply; the vapor becomes poorer in its mercury content, and the pressure decreases.

This statement explains the experimental observation of a maximum of the function $p(t)$, which was interpreted above from a more fundamental point of view.

3.1.2.7.   Conclusion

A satisfying interpretation of the effect of excessive mercury vapor pressure on the growth according to the EDRI mechanism in the HgTe–CdTe system has been provided. A theoretical model has been proposed that takes into account, both qualitatively and quantitatively most phenomena observed in the experiments.

In addition, the obtained results may be useful for the development of a flexible and controllable method for the preparation of layers with well-defined characteristics for application, especially in the field of photoelectricity, notably the detection ([13], [14], [15]).

3.1.2.8.   Appendix. Determination of mercury pressures and the reliability of the method

First and foremost, we note that on employing this method we have neglected the loss of weight due to tellurium and have supposed that the vapor pressure in the enclosure is established only by the mercury during the treatment. This approximation is justified because of the above-mentioned enormous difference between the partial pressures of these two constituents (4 to 5 orders of magnitude).

The weighing method implies a primary source of errors causing an estimated error of 10^{-4} g. The inaccuracy resulting from the chosen value for the water density, which is considered to be systematic, has only little effect on the formulation of the experimental relations. Another source of errors results from the evaluation of the volume between the walls of the two concentric tubes when measuring the free volume of the ampulla; this volume is estimated to be 0.02 cubic centimeters on average. Taking into consideration these various sources and having carried out all calculations we find that $| \Delta p | \leq 10^{-1}$ atmospheres for an excessive vapor pressure of 2 atmospheres, and $| \Delta p | \leq 2 \cdot 10^{-1}$ atmospheres for an excessive vapor pressure of 6 atmospheres.

Nevertheless, it is necessary to consider a source of error that is inherent in the method, *i.e.* the possibility that the mercury gas will condense at the specimen surface

210

during hardening. As a result, wrong weighing values will be found. In view of the difficulties of evaluating this source of error, we try to restrict its effect as far as possible by quenching the specimen in such a way that the walls cool down more rapidly in direct contact with water and, thus, ensure condensation of the entire gas.

This can be seen from the fact that a deposit appears at the walls, which does not occur in the case of a slow cooling-down procedure.

3.1.3. REFERENCES

[1] *Bailly, F.:* C. R. Acad. Sci. 262 (1966) 635.
[2] *Bailly, F.:* Thèse, Orsay (1967).
[3] *Cohen-Solal, G.:* Thèse, Orsay (1967).
[4] *Bailly, F., Marfaing, Y., Cohen-Solal, G., Melngailis, J.:* J. Phys. 28 (1967) 273.
[5] *Cohen-Solal, G., Marfaing, Y., Bailly, F., Rodot, M.:* C. R. Acad. Sci. 261 (1965) 931.
[6] *Marfaing, Y., Cohen-Solal, G., Bailly, F.:* Int. Conf. Crystal Growth. Boston (1966).
[7] *Cohen-Solal, G., Marfaing, Y., Bailly, F.:* Rev. Phys. Appl. 1 (1966) 11.
[8] *Tufte, O. N., Stelzer, E. L.:* Intrinsic infrared detector development, Honeywell Corporate Research Center, Mai 1967.
[9] *Tufte, O. N., Stelzer, E. L.:* J. Appl. Phys. 40 (1969) 4559.
[10] *Brebrick, R. F., Strauss, A. J.:* J. Phys. Chem. Sol. 26 (1965) 989.
[11] *Rodot, H.:* J. Phys. Chem. Sol. 25 (1964) 85.
[12] *Darken, L. S.:* Trans A.I.M.E. 175 (1948) 184.
[13] *Cohen-Solal, G., Marfaing, Y.:* Sol. Stat. Elect. 11 (1968) 1131.
[14] *Cohen-Solal, G., Riant, Y.:* Appl. Phys. Lett., 19, (10) (1971) 436.
[15] *Cohen-Solal, G., Janik, E., Castro, E., Marfaing, Y., Svob, L.:* IX. Photovoltaïc Specialists Conference. Philadelphia, May 1972.

4. FORMATION OF EPITAXIAL SEMICONDUCTOR FILMS BY CRYSTAL GROWTH FROM THE MELT

4.1. BINARY AND TERNARY PHASE DIAGRAMS AS A BASIS OF THE LIQUID PHASE EPITAXY OF $A^{III}B^V$ COMPOUNDS

G. *Kühn* and *A. Leonhardt*

Department of Chemistry, Karl-Marx University, Leipzig, German Democratic Republic

4.1.1. INTRODUCTION

$A^{III}B^V$ compounds have attracted great interest as materials for light emitting semiconductor elements (light emitting diodes and LASERS). The most important substances are specified in Table 1.

Table 1. $A^{III}B^V$ compounds with a zinc blende structure [1]

Compound	Lattice Constant (Å)	Eg (eV) at 300 K
AlP	5.451	2.45
AlAs	5.6622	2.16
AlSb	6.1355	1.5
GaP	5.45117	2.261
GaAs	5 65321	1.435
GaSb	6.09593	0.72
InP	5.86875	1.351
InAs	6.0584	0.35
InSb	6.47937	0.180

The range of the variants of material is considerably extended due to the intensive formation of solid solutions between the compounds. In Fig. 1 a survey of the most interesting combinations with the exception of nitrides is given.

Different aspects of application can be seen from Fig. 2. Thus, for instance, IR emitters of about 0.9 μm are well suited for switching purposes since as receivers in connection with silicon detectors they constitute an ideal couple for optical coupling elements. Epitaxial layers have proved to be most suitable for the preparation of such component parts. As the real structure of these layers is of great importance for the efficiency of *pn* junctions, the lattice constants and expansion coefficients of the layer and the substrate are allowed to differ only insiginificantly. Ideal configurations are:

n-GaAs : Si–p-GaAs : Si

n-GaAs–p-Ga$_x$Al$_{1-x}$As .

Luminescence components having exceptionally high external quantum efficiencies are prepared by means of Liquid Phase Epitaxy (LPE). In the case of the crystallization from gallium melts, this effect is attributed to a decrease in the gallium vacancy

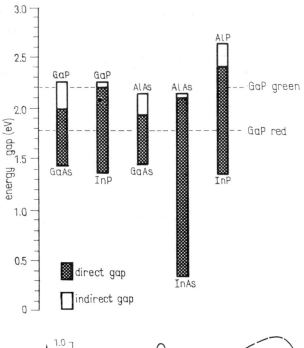

Fig. 1. Energy gaps of some ternary $A^{III}B^V$ compounds for visible light emission. The energy range for direct transitions is represented by vertical lines [1]

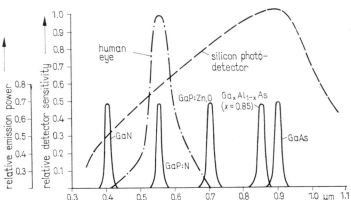

Fig. 2. The region of the optical spectrum and its coverage by known light emitting diodes [2]

concentration. It is for this reason and because of the simplicity of apparatus employed that LPE is often preferred to gaseous phase deposition (chemical transport reaction, pyrolysis and vacuum deposition processes).

Since the composition of the melt determines quality and composition of the layers, the choice of the epitaxy technique employed and the temperature regime during crystal growth are important. Knowledge of the respective phase diagrams is of particular interest. Therefore, in the following Section 4.1.2 experimental and computational ways of determining phase diagrams will be considered. Subsequently, a survey of the most common LPE processes and a discussion on the relations between layer characteristics and growth parameters (Section 4.1.3) are given.

The determination of these phase diagrams is often considerably hampered by experimental difficulties because of the high temperatures required and, consequently, by the high vapor pressures of the substances. In addition, it is extremely difficult to obtain precise measured values of the respective solidus curves because the diffusion coefficients of A^{III}B^{V} systems are low and a noticeable homogenization often requires several months. Therefore it seems to be advantageous to calculate solidus and liquidus curves from thermodynamic data. The main problem is the determination of the activity coefficients. Various methods of approximation have been utilized; some of them are well known, *e.g.* the *"Regular Solution* (RS) *Model"*, *Darken*'s *quadratic formalism*, and the *"Quasi-chemical Equilibrium* (QCE) *Model"*. However, these methods have in common that they can be applied successfully only if some experimental points of the respective system are known.

4.1.2.1.   **Determination by experiments**

Phase diagrams of A^{III}B^{V} compounds are mainly determined according to the following methods:

a) Differential thermal analysis (DTA)
b) Solubility determinations
c) Epitaxial layer analysis
d) Tempering between the liquidus and the solidus curves.

The first two possibilities mainly serve to trace liquidus curves whereas the last two are used to determine solidus curves. Vapor pressure measurements will not be considered in this connection.

4.1.2.1.1.   *Differential thermal analysis*

As most A^{III}B^{V} compounds melt incongruently, the experiments have to be carried out in closed ampoules [3], [4]. Such an arrangement is shown in Fig. 3. As for liquidus curves, the cooling-down rates are comprised between 1 and 2 degrees/min. As a rule, undercooling effects are negligible. The accurate determination of solidus curves, however, is rather problematic. Very low heating rates (< 2 degrees/min) are required due to the low diffusion coefficients of A^{III}B^{V} compounds. Likewise, careful annealing before the experiment proves to be very favorable. Neglect of this fact may lead to incorrect results. For example, the pseudo-binary phase diagram GaSb–AlSb had been identified as a simple eutectic system with a degenerated eutectic point [5]. *Woolley* and *Smith* [6], however, were able to prove by annealing experiments that these two compounds form a continuous series of solid solutions.

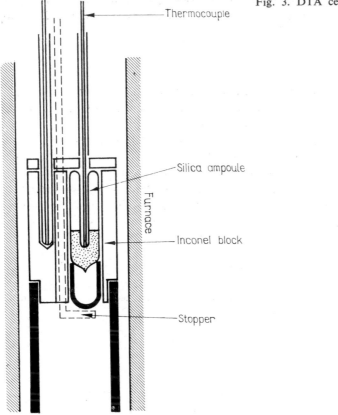

Fig. 3. DTA cell

Thermocouple

Silica ampoule

Furnace

Inconel block

Stopper

4.1.2.1.2. *Solubility determination and epitaxial layer analysis*

This method is based upon the measurement of the loss of weight of a polycrystalline ingot of an $A^{III}B^V$ compound during exposure to a certain temperature in a slightly undersaturated solution. The liquidus curve is reached when, after repeated immersions of the ingot in the melt, no further loss of weight can be detected. Stirring the molten solution accelerates the process of establishing equilibrium. The liquidus temperatures thus ascertained can be determined with an accuracy better than $\pm 5°$ and the composition of the melt with an accuracy of ± 0.1 mole per cent. When the immersion technique depicted in Fig. 18a is modified for these measurements by arranging a second crystal mounting it is possible to immerse a substrate into the molten solution at this equilibrium temperature and to deposit an epitaxial layer by means of a cooling process. In this case, the composition of the layer close to the substrate corresponds to the solidus composition at the temperature T. This method has been utilized, for instance, by *Wu* and *Pearson* [7] for the investigation of the Ga–In–As system. Solubility experiments based on similar methods have been carried out earlier by *Hall* [8], *Rupprecht* [9], *Ilegems* and *Pearson* [10]. *Panish* determined the liquidus curves for the In–P [1], Ga–In–P [12], and Ga–As–P [3] systems by directly observing the melt surface in a closed quartz ampoule during the heating and cooling processes, making use of the fact that the $A^{III}B^V$ compounds have densities lower than those of their saturated solutions.

4.1.2.1.3. *Annealing experiments*

Woolley and coworkers [13], [14] applied the following method to determine the solidus curve of a pseudo-binary system: For the purpose of prehomogenization a mixture of compounds was annealed slightly above the melting point of the higher melting component, quenched and powdered. Subsequently, the annealing process was repeated near the melting point of the lower melting compound until a monophase behavior could be observed by means of *Debye–Scherrer micrographs* (application of *Vegard's rule*). The annealing periods often last up to several months. *Foster* and *Scardefield* [15] improved this method by annealing sintered powder samples at a temperature between the solidus and the liquidus curves. Replica had been obtained from the quenched samples, molten and crystalline zones were identified by microscopy, and the composition of the respective phase was determined by means of an electron-beam microprobe analyzer. After only one week of annealing treatment mono-crystalline zones were found sufficiently large to permit a reliable evaluation. Thus the solidus limit had been determined in the pseudo-binary systems GaP–InP and InAs–AlAs [15], [16].

4.1.2.2. **Methods of calculation**

4.1.2.2.1. *Calculations from binary phase equilibria according to Panish*

Panish et al. [3], [12], [17]–[19] applied their method to various ternary $A^{III}B^V$ systems. It constitutes a simple extrapolation of binary phase equilibria to ternary systems. In the case of a $A^{III}B^{III}C^V$ a solid solution $A_xB_{1-x}C$ is in equilibrium with a liquid phase along a liquidus curve. Consequently, considering the system as a "quasibinary" one we can write

$$AC_{solid} \rightleftharpoons A_{liquid} + C_{liquid}$$
$$BC_{solid} \rightleftharpoons B_{liquid} + C_{liquid}$$

and

$$K_{AC_{solid}} = \frac{N_A \cdot \gamma_A \cdot N_C \cdot \gamma_C}{X_{AC} \cdot \gamma_{AC}} \tag{1a}$$

$$K_{BC_{solid}} = \frac{N_B \cdot \gamma_B \cdot N_C \cdot \gamma_C}{X_{BC} \cdot \gamma_{BC}} \tag{1b}$$

N_i and X_{ij} being the atom or mole fractions of the liquid and solid phases, respectively, γ_i and γ_{ij} the corresponding activity coefficients, and K_{AC} and K_{BC} the equilibrium constants. If we assume the mixture to behave as an ideal solution — which applies to some $A^{III}B^V$ solid solutions in a first approximation especially if one component predominates (e.g. $N_A \gg N_B$) — then all activity coefficients $\gamma_i = 1$, and eqs. (1a) and (1b) are simplified:

$$K'_{AC} = \frac{N_A N_C}{X_{AC}} \cdot \qquad (2a)$$

$$K'_{BC} = \frac{N_B N_C}{X_{BC}} \cdot \qquad (2b)$$

If $\sum_i N_i = 1$ we obtain:

$$X_{AC} = \frac{N_A(1 - N_A - N_B)}{K'_{AC}} \qquad (3a)$$

$$X_{BC} = \frac{N_B(1 - N_A - N_B)}{K'_{BC}} \qquad (3b)$$

and with

$$X_{AC} + X_{BC} = 1$$

$$\frac{N_A(1 - N_A - N_B)}{K'_{AC}} + \frac{N_B(1 - N_A - N_B)}{K'_{BC}} = 1 \qquad (4)$$

and finally

$$K'_{BC}(N_A - N_A^2 - N_A N_B) + K'_{AC}(N_B - N_B^2 - N_A N_B) - K'_{AC} K'_{BC} = 0. \qquad (5)$$

From eq. (5) the liquidus area of a ternary $A^{III}B^V$ mixture can be calculated, K'_{AC} and K'_{BC} being the ideal equilibrium constants of the binary systems.

In order to determine one equilibrium constant, at least one binary phase diagram — or points of this diagram — must be known. Then the second equilibrium constant will be obtained by an adjustment to experimentally determined values of the ternary system.

However, this method leads to sufficient agreement with experimental results only if the following restrictions are taken into account:

a) The interaction between the A–B pairs in the liquid phase and the AC–BC pairs in the solid phase are neglected in the ternary system A–B–C, i.e. those systems are assumed to behave as ideal solutions.
b) The equilibrium constants K'_{AC} and K'_{BC} are determined from the binary phase diagrams and used for ternary systems.
c) $N_A(N_B) \gg N_B(N_A) + N_C$.
d) The difference of lattice constants of the solid compounds must be small.

218

On this basis *Panish* and *Sumski* [3], [17], [19] calculated liquidus isotherms for the Ga–Al–As, Ga–As–P, and Ga–In–As systems.

Especially the Ga–Al–As system revealed a good agreement with experimental results, *i.e.* this system shows nearly ideal behavior (Fig. 4a). This result agrees well with calculations of ΔG^E (free excessive enthalpy) by *Kühn* and *Leonhardt* [20].

But such a simple assumption is not sufficient in every case, as some systems remarkably differ from the ideal behavior. In particular, mixtures containing phosphorus and arsenic tend to display major deviations. This is taken into consideration by the fact that the activity coefficient of the B constituent (phosphorus, arsenic) is assumed not to be constant in the liquid phase, but varying with the temperature according to the model of regular solutions (RS model).

According to the RS model we have [21]:

$$\ln \gamma_{C(liquid)} = \frac{\Omega_{AC}}{RT} (1 - N_C)^2 . \tag{6}$$

Ω_{AC} is the parameter of interaction between the components A and C (Section 4.1.2.2.2.4). Equation (5) then becomes:

$$K''_{BC}(N_A - N_A^2 - N_A N_B) + K''_{AC}(N_B - N_B^2 - N_A N_B) - \frac{K''_{AC} K''_{BC}}{\gamma_C} = 0 \tag{7}$$

and with eq. (6):

$$K''_{BC}(N_A - N_A^2 - N_A N_B) + K''_{AC}(N_B - N_B^2 - N_A N_B) - K''_{AC} K''_{BC} \cdot e^{-\frac{\Omega_{AC}}{RT}(1 - N_C^2)} = 0. \tag{8}$$

This correction results in an improved agreement with experimental values, as has been shown for the Ga–In–P [12], Ga–Al–P [18], and In–Al–P [22] systems.

In addition to the determination of ternary liquidus isothermal curves by means of eqs. (5) and (8), respectively, it is possible to determine solidus curves (Fig. 4b) by means of eqs. (1a), (1b), and (6).

The great advantage of this method is its simplicity with regard to the required calculations. The general course of the liquidus and solidus isothermal curves, which provides valuable information for the carrying out of crystal growth experiments, can be rapidly determined.

4.1.2.2.2.   *The method according to Vieland*

In a binary $A^{III}B^V$ system, the equilibrium between the melt (A + B) and the binary compound AB is represented by the liquidus curve (the eutectic points in the vicinity of the final points of the phase diagram have been neglected):

$$\mu_A(T) + \mu_B(T) = \mu^0_{AB}(T) . \tag{9}$$

The chemical potentials of the liquid phase are expressed by:

$$\mu_A(T) = \mu^0_A(T) + RT \ln \gamma_A N_A \tag{10a}$$

$$\mu_B(T) = \mu^0_B(T) + RT \ln \gamma_B N_B \tag{10b}$$

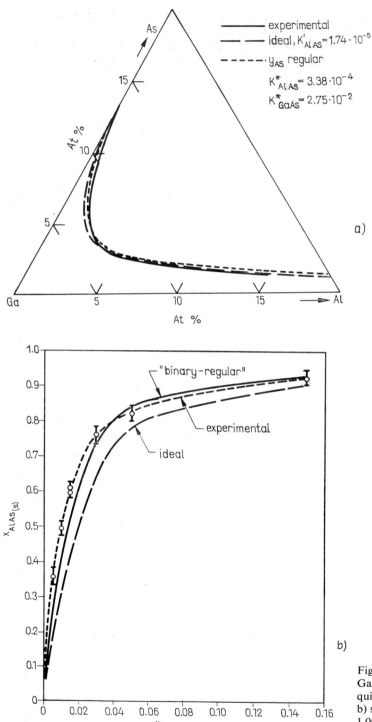

Fig. 4. Ga rich corner of the Ga–Al–As system [17] a) Liquidus isotherm for 1000°C; b) solidus isotherm for 1000°C

where μ_A^0, μ_B^0, and μ_{AB}^0 are the chemical potentials of the pure components, $\gamma_{A,B}$ the activity coefficients, and $N_{A,B}$ the atom fraction.

According to *Vieland* [23], the potential μ_{AB}^0 of the solid phase can be expressed by the following equation (when the variation of the specific heat is neglected):

$$\mu_{AB}^0(T) = \mu_A^{s.l.}(T) + \mu_B^{s.l.}(T) - \Delta S_{AB}^F(T_{AB}^F - T) \tag{11}$$

where $\mu_A^{s.l.}$ and $\mu_B^{s.l.}$ are the chemical potentials of A and B in the liquid phase for a stoichiometric composition (*i.e.* $N_A = N_B = 0.5$), ΔS_{AB}^F the entropy of melting, and T_{AB}^F the melting point of the respective binary compound AB.

According to eq. (9), we obtain:

$$\mu_A^0(T) + RT\ln(\gamma_A N_A) + \mu_B^0(T) + RT\ln(\gamma_B N_B) = \mu_A^{s.l.}(T) + \mu_B^{s.l.}(T) - \Delta S_{AB}^F(T_{AB}^F - T) \tag{12}$$

$$\mu_A^{s.l.}(T) = \mu_A^0(T) + RT\ln\left(\frac{\gamma_A^{s.l.}}{2}\right) \tag{13a}$$

$$\mu_B^{s.l.}(T) = \mu_B^0(T) + RT\ln\left(\frac{\gamma_B^{s.l.}}{2}\right). \tag{13b}$$

Substituting eqs. (13a) and (13b) into eq. (12) yields

$$\ln\frac{1}{4 N_A N_B} + \ln\frac{\gamma_A^{s.l.} \cdot \gamma_B^{s.l.}}{\gamma_A \cdot \gamma_B} = \frac{\Delta S_{AB}^F}{R}\left(\frac{T_{AB}^F}{T} - 1\right). \tag{14}$$

Equation (14) is *Vieland*'s expression for the calculation of the liquidus curve of a binary $A^{III}B^V$ system. *Ilegems* and *Pearson* [10] as well as *Stringfellow* and *Greene* [24] applied this equation to ternary systems.

In a ternary system, a liquid phase comprising the three components A, B, and C and a solid phase comprising the compounds AC and BC are in equilibrium, with AC and BC forming a continuous series of solid solution. Consequently:

$$\mu_A(T) + \mu_C(T) = \mu_{AC}(T) \tag{15a}$$

$$\mu_B(T) + \mu_C(T) = \mu_{BC}(T). \tag{15b}$$

μ_A, μ_B, and μ_C can be expressed according to eqs. (8a) and (8b); the chemical potentials of the solid phase by

$$\mu_{AC}(T) = \mu_{AC}^0(T) + RT\ln(\gamma_{AC} X_{AC}) \tag{16a}$$

$$\mu_{BC}(T) = \mu_{BC}^0(T) + RT\ln(\gamma_{BC} X_{BC}). \tag{16b}$$

For μ_{AC}^0 and μ_{BC}^0, *Vieland*'s expression (eq. (11)) can be employed.

Employing the equilibrium condition (eqs. (15)) and including eqs. (13) we obtain the following system of equations:

$$\ln(\gamma_{AC} X_{AC}) = \ln\frac{4 N_A \gamma_A N_C \gamma_C}{\gamma_A^{s.l.} \gamma_C^{s.l.}} + \frac{\Delta S_{AC}^F}{RT}(T_{AC}^F - T) \tag{17a}$$

$$\ln(\gamma_{BC} X_{BC}) = \ln\frac{4 N_B \gamma_B N_C \gamma_C}{\gamma_B^{s.l.} \gamma_C^{s.l.}} + \frac{\Delta S_{BC}^F}{RT}(T_{BC}^F - T). \tag{17b}$$

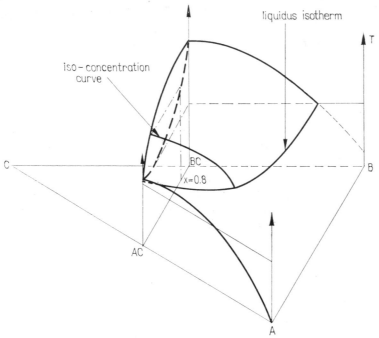

Fig. 5. Liquidus isotherm and iso-concentration curves for $A_{1-x}B_xC$ systems

Because of

$$N_A + N_B + N_C = 1 \tag{18a}$$

$$X_{AC} + X_{BC} = 1 \tag{18b}$$

it is possible to determine the ternary liquidus areas by means of computer techniques. The system of eqs. (17) and (18) yields two series of curves within a ternary system:

1. Liquidus curves for a constant temperature T,
2. Iso-concentration curves, $i.e.$ curves representing the same content of AC and BC in the respective solid phases. Such curves connect points of the liquidus area with the solid being at equilibrium.

These curves are depicted in Fig. 5.
The main advantage of this method of calculating liquidus isotherms would be that the only experimental data required are entropies of melting, ΔS_{ij}^F, and the melting points T_{ji}^F of the $A^{III}B^V$ compounds.
The primary problem is how to determine the activity coefficients. Various conceivable methods will be discussed in the following.

4.1.2.2.2.1.   *Calculation of activity coefficients according to the model of a regular solution*

The model of a regular solution (RS model) devised by *Hildebrand* [25] applies only when the following criteria are fulfilled:

1. Low polarity of the components.
2. Enthalpy of mixture not equal to zero.
3. The energy in respect to the nearest neighbors depends upon the distance only.
4. The configuration of the components in the solution is merely statistical (ideal entropy of mixture).
5. The volume change corresponding to the mixing is equal to zero.

In order to determine the activity coefficients it is necessary to know the partial molar quantities of the mixture. These are defined according to the RS model as follows [26]:

1. Partial molar enthalpy of mixing

$$\Delta \bar{H}_{M,A} = \Omega_{AB} N_B^2 + \Omega_{AC} N_C^2 + (\Omega_{AB} + \Omega_{AC} - \Omega_{BC}) N_B N_C \qquad (19)$$

where $\Delta \bar{H}_{M,A}$ is the partial molar enthalpy of mixing of component A in the ternary A–B–C system.
The Ω_{ij} are the interaction parameters as defined by *Guggenheim* [21] (Section 4.1.2.2.2.4).

2. Partial molar entropy of mixing

$$\Delta \bar{S}_{M,A} = -R \ln N_A . \qquad (20)$$

This expression is identical with the partial entropy of mixing of the component A of an ideal solution.

3. The partial molar free enthalpy of mixing of the component A is according to eqs. (19) and (20) [26]:

$$\Delta \bar{G}_{M,A} = \Delta \bar{H}_{M,A} - T \Delta \bar{S}_{M,A}$$
$$\Delta \bar{G}_{M,A} = \Omega_{AB} N_B^2 + \Omega_{AC} N_C^2 + (\Omega_{AB} + \Omega_{AC} - \Omega_{BC}) N_B N_C + RT \ln N_A . \qquad (21)$$

With the well-known relation between $\Delta \bar{G}_{M,A}$ and the activity coefficient

$$\Delta \bar{G}_{M,A} = RT \ln \gamma_A + RT \ln N_A \qquad (22)$$

an expression for the calculation of the activity coefficients is derived

$$RT \ln \gamma_A = \Omega_{AB} N_B^2 + \Omega_{AC} N_C^2 + (\Omega_{AB} + \Omega_{AC} - \Omega_{BC}) N_B N_C . \qquad (23a)$$

Analogous considerations for the B and C components lead to

$$RT \ln \gamma_B = \Omega_{AB} N_A^2 + \Omega_{BC} N_C^2 + (\Omega_{AB} + \Omega_{BC} - \Omega_{AC}) N_A N_C \qquad (23b)$$
$$RT \ln \gamma_C = \Omega_{AC} N_A^2 + \Omega_{BC} N_B^2 + (\Omega_{AC} + \Omega_{BC} - \Omega_{AB}) N_A N_B . \qquad (23c)$$

223

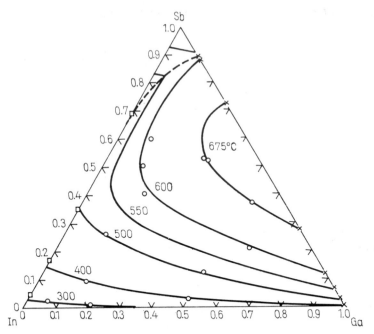

Fig. 6. Liquidus isotherms of the In–Ga–Sb system [31] (———— according to the RS model and ○ representing experimental values)

According to the criteria of the RS model the parameter of interaction is postulated as a quantity not depending on temperature and concentration ("strictly regular solution"). *Thurmond* [27] and *Arthur* [28], however, proved that Ω_{ij} approximately linearly depends on temperature. Hence, interaction parameters depending on temperature are frequently used for the calculation of activity coefficients according to the RS model ("quasi-regular solution"). If eqs. (23) are substituted in *Vieland*'s expression, eqs. (17), the respective values of N_A, N_B, and N_C can be determined at a desired temperature T depending on the composition of the solid phase X. According to this method, the following $A^{III}B^V$ systems have been calculated: Ga–Al–As [10], Ga–In–P [26], [29], Ga–In–Sb [30], Ga–As–P [4, 30], and Ga–In–As [7]. The agreement with experimental results is good. In Fig. 6 the phase diagram of the In–Ga–Sb system as calculated by *Blom* and *Plaskett* [31] is shown.

4.1.2.2.2.2. *Calculation of activity coefficients according to Darken's "quadratic formalism"*

Another possibility of calculating the activity coefficients has been devised by *Darken et al.* [32]–[34]. These investigators determined the activity coefficients of binary metallic systems for concentrations up to 60 mole% and achieved sufficient agreement with experimental results:

$$RT \ln \gamma_A = \Omega_{AB} N_B^2 \tag{24}$$

$$RT \ln \left(\frac{\gamma_B}{\gamma_B^0} \right) = \Omega_{AB} \left(-2 N_B + N_B^2 \right). \tag{25}$$

Component A is assumed to be a solvent, component B a solved component. γ_B^0 represents the activity coefficient of component B in case of infinite dilution. This "quadratic formalism" by *Darken* becomes identical with the regular model if

$$RT \ln \gamma_B^0 = \Omega_{AB}. \tag{26}$$

Antypas and *James* [35] applied this model to $A^{III}B^V$ mixtures. They extrapolated eqs. (24) and (25) to a ternary system (*e.g.* $Ga_{1-x}In_xAs$).

$$\ln \gamma_A = \alpha_{AB} N_B^2 + \alpha_{AC} N_C^2 + (\alpha_{AB} + \alpha_{AC} - \alpha_{BC}) N_B N_C \tag{27}$$

$$\ln \left(\frac{\gamma_B}{\gamma_B^0} \right) = -2 \alpha_{AB} N_B + (\alpha_{BC} - \alpha_{AB} - \alpha_{AC}) N_C +$$
$$+ \alpha_{AB} N_B^2 + \alpha_{AC} N_C^2 + (\alpha_{AB} + \alpha_{AC} - \alpha_{BC}) N_B N_C \tag{28}$$

$$\ln \left(\frac{\gamma_C}{\gamma_C^0} \right) = -2 \alpha_{AC} N_C + (\alpha_{BC} - \alpha_{AB} - \alpha_{AC}) N_B +$$
$$+ \alpha_{AB} N_B^2 + \alpha_{AC} N_C^2 + (\alpha_{AB} + \alpha_{AC} - \alpha_{BC}) N_B N_C \tag{29}$$

with

$$\alpha = \frac{\Omega}{RT}.$$

In the ternary system, too, A represents the solvent, B and C are the solved components.

By substituting eqs. (27), (28), and (29) into eqs. (17a) and (17b) one value of the activity coefficients in case of infinite dilution (γ_B^0 or γ_C^0) is eliminated. The other value is found by adjusting eqs. (17) to experimentally obtained points of the liquidus isotherms. The system of eqs. (27)–(29) attains those of the RS model, if

$$\ln \gamma_B^0 = \alpha_{AB} \tag{30a}$$

$$\ln \gamma_C^0 = \alpha_{AC}. \tag{30b}$$

For calculating the activity coefficients in the solid phase *Antypas et al.* [35]–[38] applied the RS model, *i.e.*

$$\ln \gamma_{AC} = \alpha_{AC-BC} X_{BC}^2 \tag{31a}$$

$$\ln \gamma_{BC} = \alpha_{AC-AB} X_{AC}^2. \tag{31b}$$

When eqs. (27)–(29), and (31) are substituted into eq. (17) the ternary phase diagram can be calculated numerically (Fig. 7).

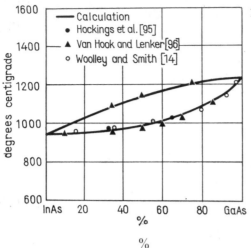

Fig. 7. The Ga–In–As system [35]

The advantage of *Darken*'s "quadratic formalism" is that it has an additional correction factor with $\ln \gamma_i^0$ which allows a better adjustment to experimental values. However, some experimental values are required to determine $\ln \gamma_i^0$.

4.1.2.2.2.3.   *Calculation of activity coefficients according to the QCE model*

The quasi-chemical equilibrium model (QCE) is a modification of the RS model. The number of A–A, B–B, and A–B pairs is assumed to depend exponentially on the interaction parameter Ω_{AB}^Q in a binary solution [21]:

$$\frac{n_{AA}\, n_{BB}}{(n_{AB})^2} = \frac{1}{4}\, e^{\frac{2\,\Omega_{AB}^Q}{Z\,RT}} \tag{32}$$

$$2\,n_{AA} + n_{AB} = Zn_A \tag{33a}$$

$$2\,n_{BB} + n_{AB} = Zn_B, \tag{33b}$$

where n_{ij} is the number of i–j pairs (i, j = A, B), R the gas constant, Z the coordination number, n_A and n_B are the numbers of A and B atoms of the solution, respectively.
A consequence of eq. (32) is that $\Delta S_M^E \neq 0$ for the QCE model. The molar excessive quantities are calculated on the basis of this model as:

$$\Delta H_M = N_A N_B \left(1 - \frac{2\,N_A\,N_B\,\Omega_{AB}^Q}{Z\,RT}\right) \tag{34}$$

$$\Delta S_M^E = - \frac{N_A^2\,N_B^2\,\Omega_{AB}^{Q^2}}{Z\,RT^2} \tag{35}$$

$$\Delta G_M^E = N_A N_B \,\Omega_{AB}^Q \left(1 - \frac{N_A\,N_B\,\Omega_{AB}^{Q^2}}{Z\,RT}\right). \tag{36}$$

226

The coefficients of activity may be derived from eqs. (34), (35) and (36) as

$$\gamma_A = \left\{ \frac{W - 1 + 2Y}{Y(W + 1)} \right\}^{Z/2} \tag{37a}$$

$$\gamma_B = \left\{ \frac{W + 1 - 2Y}{(1 - Y)(W + 1)} \right\}^{Z/2} \tag{37b}$$

with

$$W = [1 + 4Y(1 - Y)(S^2 - 1)]^{1/2} \tag{38a}$$

$$Y = \frac{N_A}{N_A + N_B} \tag{38b}$$

$$S = \exp\left(\frac{\Omega_{AB}^Q}{Z RT}\right). \tag{38c}$$

Stringfellow and *Greene* [24] extended the system of eqs. (32) and (33) for a ternary mixture:

$$\frac{n_{AA}\, n_{BB}}{(n_{AB})^2} = \frac{1}{4}\, e^{\frac{2\Omega_{AB}^Q}{Z RT}} \tag{39a}$$

$$\frac{n_{BB}\, n_{CC}}{(n_{BC})^2} = \frac{1}{4}\, e^{\frac{2\Omega_{BC}^Q}{Z RT}} \tag{39b}$$

$$\frac{n_{AA}\, n_{CC}}{(n_{AC})^2} = \frac{1}{4}\, e^{\frac{2\Omega_{AC}^Q}{Z RT}} \tag{39c}$$

$$2\,n_{AA} + n_{AB} + n_{AC} = Z n_A \tag{40a}$$

$$2\,n_{BB} + n_{AB} + n_{BC} = Z n_B \tag{40b}$$

$$2\,n_{CC} + n_{AC} + n_{BC} = Z n_C. \tag{40c}$$

By combining eqs. (39) and (40) we obtain a system of equations, which can be solved with a computer and furnishes n_{AC}, n_{BC}, and n_{AB} for a certain temperature:

$$(n_A - Y_{AB} - Y_{AC})(n_B - Y_{AB} - Y_{BC}) - g_{AB}^2\, Y_{AB}^2 = 0 \tag{41a}$$

$$(n_B - Y_{AB} - Y_{BC})(n_C - Y_{AC} - Y_{BC}) - g_{BC}^2\, Y_{BC}^2 = 0 \tag{41b}$$

$$(n_A - Y_{AB} - Y_{AC})(n_C - Y_{AC} - Y_{BC}) - g_{AC}^2\, Y_{AC}^2 = 0 \tag{41c}$$

with

$$Y_{ij} = \frac{n_{ij}}{Z}; \quad g_{ij} = \exp\frac{\Omega_{ij}^Q}{Z RT}; \quad i, j = A, B, C. \tag{42}$$

If the enthalpy of mixing is defined by

$$\Delta H_M = Y_{AB}\,\Omega^Q_{AB} + Y_{AC}\,\Omega^Q_{AC} + Y_{BC}\,\Omega^Q_{BC} \tag{43}$$

the activity coefficients of the ternary system can be determined by means of the free excessive enthalpy of mixing which is obtained by integrating the *Gibbs–Helmholtz equation* as

$$\ln \gamma_A = \frac{1}{R}\left[\frac{\partial}{\partial n_A}\left(\int_0^{T/1} (Y_{AB}\,\Omega^Q_{AB} + Y_{AC}\,\Omega^Q_{AC} + Y_{BC}\,\Omega^Q_{BC})\,\mathrm{d}\frac{1}{T}\right)\right]_{n_B,\,n_C}. \tag{44}$$

An analogous expression also applies to γ_B and γ_C. As can be seen from eq. (44), the coefficients of activity cannot be determined explicitly, but only by a numerical evaluation for each individual point of the liquidus isotherm. This, however, leads to a lower convergence of the procedure of solving the system of eqs. (17). This is an obvious disadvantage because the expenditure of the required calculation increases considerably. However, the QCE model is presently regarded as the most precise approximation for the calculation of activity coefficients.

Stringfellow and *Greene* [24], [39] and *Huber* [40], [41] determined some ternary phase diagramms of $A^{III}B^V$ systems (Fig. 8) by means of this model.

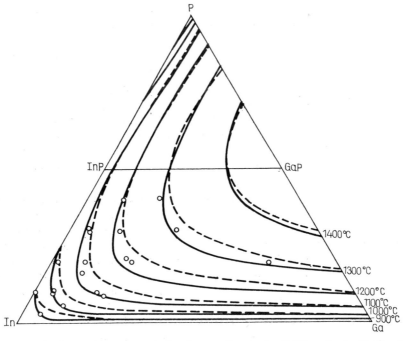

Fig. 8. The Ga–In–P system calculated according to the QCE method, with Ω values independent on temperature (———), with Ω values depending on temperature (- - - -), and with experimental values (○)

The different methods of calculation yield results with relatively good agreement with experimental results. There are, however, gradual differences. In all methods, simple extrapolations of binary data for the ternary system are made. For example, *Panish et al.* use the equilibrium constants of the respective binary systems for the calculation in a ternary system. Likewise, an extrapolation from the binary system to the ternary one is done in calculating liquidus isotherms according to *Vieland's* method (eqs. (17)).

Panish assumes that at least two main components (*i.e.* the components forming the major portion) of the liquid phase and the entire solid phase display an ideal behavior. However, this is a very rough approximation. Good agreement with experimental results will therefore be achieved only in those systems which differ only slightly from ideality. According to *Van Vechten* [42], *Foster, Woods* [43], *Foster, Scardefield* [15], [16], and *Kühn* and *Leonhardt* [20], however, some $A^{III}B^V$ systems display non-ideal behavior and, in part, an excess entropy that cannot be neglected (*e.g.* In–As–Sb, Al–In–As, and Ga–P–Sb).

It is for this reason that *Vieland's* method (strictly regular solution) was modified taking into account the dependence of the interaction parameter upon the temperature as a linear function $\alpha = A + BT$ (A, B are constants) [27], [28]. This modification is called a quasi-regular solution and yields good results.

Likewise, Fig. 9 shows that on calculating the activity coefficient according to the RS model a better agreement with experimental values is achieved if the interaction parameter $\Omega = -3.7\,T$ is applied to the Ga As system, instead of the interaction parameter $\Omega = -4380$ cal/mole not depending on temperature and established by *Stringfellow* and *Greene* [24].

This does not mean, however, that the better agreement for the binary system entails better results for the ternary case as well.

Darken's treatment does not raise any new basic requirements for the behavior of solutions as compared with the RS model. It has merely the advantage that on calculating the phase diagrams an additional factor of correction is available with the value of $\ln \gamma_i^0$ with which a better adjustment to experimental results can be achieved. The quasi-chemical equilibrium model, on the other hand, narrows the assumption of approximation even more. The assumption of the purely incidental configuration of components in the solution is abandoned. As a result, the energetic interactions between the atoms are more precisely considered. Likewise, a molar excess entropy is postulated, which constitutes a quantity that cannot be neglected in many $A^{III}B^V$ systems. As can also be seen from Fig. 9, the activity coefficients determined from eqs. (37) agree sufficiently with experimental results. Furthermore it can be seen that γ_{As} appreciably depends upon the interaction parameter. This quantity Ω will now be studied in detail. It reflects the energetic interactions between the individual components in the phase of mixtures and is defined as follows [21]:

$2\,\Omega_{AB}$ is the energy corresponding to the transfer of an A–A pair and a B–B pair into two A–B pairs.

According to this definition, $\Omega_{AB} = 0$ applies to an ideal binary solution, as the inter-

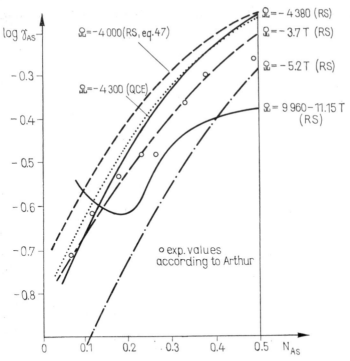

Fig. 9. The activity coefficient of As along the liquidus line of the Ga–As system; the experimental values according to *Arthur* [28] are depicted as (○)

action between different components is assumed to be equal:

$$A \longleftrightarrow A = B \longleftrightarrow B = A \longleftrightarrow B. \tag{45}$$

However, interactions will be different in case of non-ideal mixtures:

$$A \longleftrightarrow B \gtrless A \longleftrightarrow A \tag{46a}$$

$$A \longleftrightarrow B \gtrless B \longleftrightarrow B. \tag{46b}$$

If Ω_{AB} is positive, preferably A–A and B–B pairs will be present in the solution; if Ω_{AB} is negative, then preferably A–B pairs will exist.

The three interaction parameters Ω^l_{AC}, Ω^l_{BC} and Ω^l_{AB} of the liquid phase and the interaction parameter Ω^s_{AC-BC} of the solid phase are required for the calculation of activity coefficients of a ternary mixture based on the RS or QCE models (the solid phase being considered as quasi-binary).

The determination of interaction parameters of the liquid phase may be accomplished in different ways:

230

1. If the liquidus curve of the corresponding binary phase diagram or some points of it are known, then Ω_{ij}^l can be calculated according to *Vieland* [23]:

$$\Omega_{ij}^l = -\frac{RT}{2(0.5 - N_j)^2}\left[\ln 4 N_j(1 - N_j) + \frac{\Delta S_{ij}^F}{R}\left(\frac{T_{ij}^F}{T} - 1\right)\right] \tag{47}$$

where T is the liquidus temperature, T_{ij}^F the melting temperature of the binary system, ΔS_{ij}^F the entropy of melting, R the gas constant, and N_j the atom fraction of the element of the 5th main group of the periodic system of elements.

2. Ω_{ij}^l can be calculated from experimentally determined activity coefficients by means of eqs. (24), (25), and (37). The activity coefficients are determined from vapor pressure measurements [28] applying:

$$\gamma_j = \frac{1}{N_j}\left(\frac{P_{j_4}}{P_{j_4}^0}\right)^{1/4} \tag{48}$$

where P_{j4} is the partial vapor pressure of the tetra-atomic species of component j over the binary solution, $P_{j_4}^0$ the vapor pressure over the pure j melt.

For GaAs, GaP, and InP, these Ω^l values agree well with the results obtained from eq. (47) [44].

3. *Stringfellow* [45] derived the following formula:

$$\Omega_{i,j}^l \approx \left(\frac{V_i + V_j}{2}\right)(\delta_i - \delta_j) - 3\cdot 10^4(\kappa_i - \kappa_j)^2 \tag{49}$$

where $\delta_{i,j}$ are the parameters of solubility according to *Hildebrand* (square root of sublimation energy per unit volume [46]), $V_{i,j}$ are the molar volumes of the respective elements, and $\kappa_{i,j}$ the electronegativities according to *Pauling* [47].

In Table 2, experimentally determined interaction parameters are compared with those calculated with eq. (49).

Table 2. Interaction parameters Ω_{ij}^l of some binary $A^{III}B^V$ systems (in cal/mole)

System	$\Omega^l{}_j$ according to eq. (49)	Ω_{ij}^l experimentally determined
Ga–P	−4 800	−3 500 [39]
Ga–As	−4 000	−4 380 [24]
In–P	−1 020	0 [39]
In–As	−4 240	−6.070 [24]
In–Sb	−2 680	−3 980 [24]

4. Another possibility of calculating the interaction parameter Ω_{ij}^l provides the RS model if the binary system has a characteristic critical temperature (miscibility gap). In such a case, Ω_{ij}^l is calculated according to the simple relation

$$\Omega_{ij}^l \approx 2 R T_C \tag{50}$$

231

where T_C is the critical temperature [21]. From the binary systems which are of interest here numerous combinations of the elements of the IIIrd and Vth main groups of the periodic system of elements have miscibility gaps (*e.g.* As–Sb [48], Al–In [45]).

5. The interaction parameter Ω^s_{AC-BC} of the solid phase is in most cases determined by adjusting the phase diagram to known experimental points. *Kühn* and *Leonhardt* [20] suggested a computational determination of Ω^s_{AC-BC}. According to the RS model the interaction parameter is related to the integral molar enthalpy of mixing, ΔH_M, by the following simple relation:

$$\Omega^s_{AC-BC} = \frac{\Delta H_M}{X(1-X)} \tag{51}$$

where X is the mole fraction of the pseudo-binary system.

ΔH_M can be determined employing the spectroscopic theory established by *Phillips* and *Van Vechten* [42], [50], [51]:

$$\Delta H_M(X) = \Delta H_{ABC}(X) - [X\Delta H^f_{AC} + (1-X)\Delta H^f_{BC}]$$

$$\Delta H_{ABC}(X) = \Delta H_0 \left(\frac{a_{Ge}}{Xa_{AC} + (1-X)a_{BC}} \right)^3 \cdot$$

$$\cdot \left[1 - 4b \left(\frac{E_2(X)}{E_0(X) + E_1(X)} \right)^2 \right] [Xf_{iAC} + (1-X)f_{iBC}] \tag{52}$$

where ΔH_0 is a factor of proportionality equal to 68.6 kcal/mole, a_{Ge} the lattice constant of germanium, a_{AC}, a_{BC} the lattice constants of AC and BC compounds, $b = 0.0467$, E_0, E_1, E_2 are transitions at the critical points of the optical spectrum, f_{iAC}, f_{iBC} ionicities of $A^{III}B^V$ compounds [52], and ΔH^f_{AC}, ΔH^f_{BC} are the heats of formation of the binary $A^{III}B^V$ compounds.

From eq. (51) we obtain interaction parameters displaying a weak dependence on concentration and which in a strict sense apply to $T = 300$ K only. In order to calculate Ω^s_{AC-BC} the maximum value of ΔH_M was used.

In Table 3 some interaction parameters determined with eq. (51) are given. A comparison with the Ω^s_{AC-BC} values obtained by adjustment to pseudo-binary phase diagrams reveals a satisfactory agreement.

Table 3. Some interaction parameters Ω^s_{AC-BC} of ternary $A^{III}B^V$ systems (in cal/mole)

System	Ω^s_{AC-BC} according to eq. (51)	Ω^s_{AC-BC} determined by adjustment to a pseudo-binary phase diagram
GaAs$_x$Sb$_{1-x}$	4 056	4 500 [35]
Ga$_x$In$_{1-x}$Sb	1 120	1 500 [30]
Ga$_x$In$_{1-x}$As	1 328	2 000 [37]
GaP$_x$As$_{1-x}$	811	1 000 [36]

In all methods presently utilized to determine the interaction parameters in the liquid and solid phases certain errors are involved. For both, the RS and QCE model it could be shown that the activity coefficients do not seriously affect the shape of the liquids isotherms. The most crucial quantities are the entropies of melting for binary compounds and their melting points, since they are exponents in eqs. (17). They influence the shapes and the positions of the liquidus isotherms much more than the activity coefficients. The melting points are very well and accurately known especially for the low-melting $A^{III}B^V$ compounds. However, many entropies of melting have not yet been determined by experiments, or the experimental values differ widely. Moreover, the errors of the experimental determination increases with rising temperatures.

For this reason, theoretical evaluations were carried out by *Thurmond* [27] and *Sirota* [53].

Table 4. Melting points and entropies of melting of $A^{III}B^V$ compounds

System	T^F (K)	T^F (K) according to eq. (58)	ΔS^F (e. u.)	ΔS^F (e. u.) according to eq. (53)
1	2	3	4	5
AlP	2 823 [54]	2 305	15.8 [45]	17.2
AlAs	2 013 [54]	2 190	15.8 [45]	16.7
AlSb	1 323 [55]	2 180	15.14 [55]	15.2
GaP	1 740 [39]	1 683	16.25 [53]	16.7
GaAs	1 513 [55]	1 527	16.64 [55]	16.2
GaSb	985 [55]	1 075	15.8 [55]	14.7
InP	1 333 [39]	1 475	14.7 [53]	15.2
InAs	1 210 [55]	1 253	14.5 [55]	14.7
InSb	798 [55]	853	14.32 [55]	13.2

According to *Thurmond* holds

$$\Delta S^F_{3,5,i,j} = \Delta S^F_{4j} + \Delta S^F_{4i} + R \ln 4 \tag{53}$$

where i and j designate the period in which the A and B elements are located in the periodic system of elements, and the numbers designate the main group.

If the entropy of melting of a compound in a quasi-binary system is known, the other value can also be estimated from eqs. (17) when melting temperatures and the activity coefficients are known [10].

A possibility of calculating phase diagrams according to the RS model without having any experimentally determined points has been proposed by *Stringfellow* [56]:

The interaction parameters for the calculation of the activity coefficients of the liquid phase in eqs. (23) are determined according to eq. (49). The activity coefficients of the solid solution are calculated on the basis of an incidental atomic distribution in the A^{III} or B^V sublattices. In this case holds:

$$\Delta G^E_M = \Delta H_M \tag{54}$$

233

and, consequently:

$$RT \ln \gamma_{AC} = \left(\frac{\partial(\Delta H_M \, n_0)}{\partial n_{AC}} \right)_{n_{BC}, T, P} \tag{55a}$$

$$RT \ln \gamma_{BC} = \left(\frac{\partial(\Delta H_M \, n_0)}{\partial n_{BC}} \right)_{n, AC, T, P} \tag{55b}$$

where $n_0 = n_{AC} + n_{BC}$; n_{AC} and n_{BC} are the numbers of moles of AC and BC in the alloy, respectively. $\Delta H_M(X)$ follows from eq. (52), and ΔS_{ij}^F from eq. (53). T_{ij}^F is obtained from a cycle via ΔS_{ij}^F:

$$A_{solid} + C_{solid} \xrightarrow{-\Delta H_{AC}^{l}} AC_{solid} \xrightarrow{\Delta H_{AC}^{F}} AC_{liquid}$$

$$\xrightarrow{-(\Delta H_A^F + \Delta H_C^F)} A_{liquid} + C_{liquid} \xrightarrow{\Delta H_M}$$

where $\Delta H_{f, AC}$ is the heat of formation and ΔH_{AB} the enthalpy of melting. and

$$\Delta H_M = \frac{\Omega_{AC}}{4} .$$

$$\Delta H_{AC}^F = \Delta H_{AC}^f - \frac{\Omega_{AC}^l}{4} + \Delta H_A^F + \Delta H_C^F . \tag{57}$$

The melting temperature is given by

$$T_{AC}^F = \frac{\Delta H_{AC}^F}{\Delta S_{AC}^F} . \tag{58}$$

The values calculated for T_{AC}^F and ΔS_{AC}^F are also given in Table 4. Thus, all the parameters required for solving the eqs. (17) have been obtained by calculations.

4.1.2.3.   Further possibilities of calculating phase equilibria

4.1.2.3.1.   Pseudo-binary systems

Steininger [57] derived a relation for the liquidus and solidus curves, which can also be attributed to *Prince* [58]:

$$RT \left[\ln \left(\frac{X_{AC}}{X_{BC}} \right) - \ln \left(\frac{N_{AC}}{N_{BC}} \right) \right] = \Delta H_{AC}^F \left[1 - \left(\frac{T}{T_{AC}^F} \right) \right] - \Delta H_{BC}^F \left[1 - \left(\frac{T}{T_{BC}^F} \right) \right] - D \tag{59}$$

where X_{AC} and X_{BC} define the composition of the solid phase and N_{AC} and N_{BC} the

234

composition of the liquid phase, ΔH_{AC}^F and ΔH_{BC}^F are the enthalpies of melting, and T_{AC}^F and T_{BC}^F the melting points,
D is a term indicating the deviation from ideal behavior:

$$D = \left(\frac{\partial\,\Delta G_{M, solid}^E}{\partial X_{AC}}\right) - \left(\frac{\partial\,\Delta G_{M, liquid}^E}{\partial N_{AC}}\right) = RT \ln\left[\left(\frac{\gamma_{AC, solid}}{\gamma_{AC, liquid}}\right)\left(\frac{\gamma_{BC, solid}}{\gamma_{BC, liquid}}\right)\right]. \qquad (60)$$

In the case of ideal solutions D becomes 0, according to its definition, as $\Delta G_{M, solid}^E = \Delta G_{M, liquid}^E = 0$. *Steininger* calculated D values directly by means of eq. (59) for some $A^{III}B^V$ systems determined by experiments (*e.g.* InSb–GaSb, InAs–GaAs). These systems revealed an approximately ideal behavior as D took relatively low values over the entire concentration range.
This method can be applied only if either liquidus or solidus curves were experimentally determined. Furthermore, the enthalpies of melting of the respective binary compounds have to be known.

4.1.2.3.2.   *Calculation of systems of the $A^{III}B^V$ metal type*

This type of system is of particular interest for doping and contact purposes. Calculations are based on the RS model.
Method according to *Furukawa* and *Thurmond* [59]. These authors calculated liquidus isotherms for the Ga–As–Cu system within the region of primary precipitation of GaAs.
They considered GaAs to be stoichiometric and the solubility of Cu in GaAs to be negligible.
For this reason, only the following equilibrium along the liquidus isotherms is considered:

$$GaAs \rightleftharpoons Ga_{liquid} + As_{liquid}. \qquad (61)$$

When regular behavior is assumed, the expression for the calculation of liquidus isotherms is found to be

$$RT \ln\left[\frac{K_{GaAs}}{N_A\,N_C}\right] = \{(\Omega_{AB}\,N_B + \Omega_{AC}\,N_C)\,(N_B + N_C) + (\Omega_{AB}\,N_A + \Omega_{BC}\,N_C)\cdot$$

$$\cdot\,(N_A + N_C) - \Omega_{BC}\,N_B\,N_C - \Omega_{AC}\,N_A\,N_C\}, \qquad (62)$$

where K_{GaAs} is the constant of equilibrium for the equation of reaction (61), the Ω_{ij} are the respective interaction parameters, and N_i the molar fractions of the liquid phase. *Furukawa, Thurmond* [59], and *Panish* [60], [61], [62] applied this method to the Ga–As–Cu, Ga–As–Zn, Ga–As–Sn, and Ga–As–Ge systems. They achieved satisfactory agreement with experimental results.
Method according to *Jordan* [63], [64].
The premise is the same as that of *Furukawa* and *Thurmond* [59]. The low solubility of the metallic component in the $A^{III}B^V$ compound is also neglected in this case.

Hence, we obtain the following expression for the liquidus isotherm:

$$\Delta S_{AB}^F(T_{AB}^F - T) + RT \ln(4 N_A N_B) = \Omega_{AB}[0.5 - N_B(1 - N_A) - N_A(1 - N_B)] +$$
$$+ \Omega_{AC} N_C(2 N_A - 1) + \Omega_{BC}(2 N_B - 1). \tag{63}$$

The Ga–As–Zn and Ga–P–Zn systems [64] were calculated according to this method.

4.1.3. LIQUID PHASE EPITAXY (LPE)

For all $A^{III}B^V$ compounds the solubility in liquid metals like Ga, In, Sn, Bi and Pb greatly depends upon temperature. The temperature coefficient is always positive, and the solubility for the same solvent decreases within a homologous series with equal A^{III} constituents with increasing E_g and rising ionicity of the chemical bond using the definition by *Phillips* [52]. Gallium is the metal most used as solvent because of its low melting pointand vapor pressure, good wetting properties and its electrical inactivity (compared with Sn and Pb).

The principle of the LPE is that from an oversaturated molten solution a mono-crystalline layer is deposited onto a mono-crystalline substrate of a definite orientation. As substrates, $A^{III}B^V$ compounds or element semiconductors are exclusively used. Coherent layers on insulating substances such as $MgAl_2O_4$ and sapphire can be produced only if a pre-deposition had been applied by means of pyrolysis or a chemical transport reaction [65].

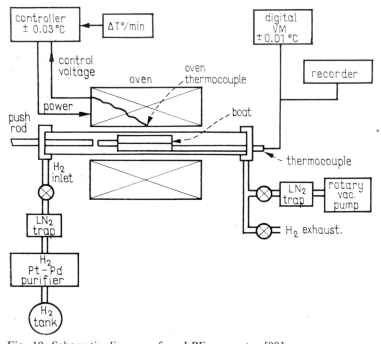

Fig. 10. Schematic diagram of an LPE apparatus [80]

Figure 10 shows a schematic diagram of an LPE apparatus. The individual variants differ mainly in the position and dimensions of the furnace (vertical or horizontal position), the temperature range and the growth vessel. The apparatus is made of high-purity quartz, can be evacuated, and is rinsed with H_2 or noble gas. The growth vessel is made of high-purity graphite. All materials which are in the growth zone and in its vicinity influence the result of epitaxy, partly by their contents of impurities [66], [67]. However, the coefficients of impurity distribution are much lower compared to the stoichiometric melt ("getter action of the solvent").

The LPE processes may be classified as follows:

1. According to the method of establishing an oversaturation of the melt (Table 5).
2. The method of bringing the substrate into contact with the melt (tilting, shifting, rotary shifting, rolling, dipping, dripping).

We use the notion of item 2 for the classification of the LPE methods in Table 5.

Table 5. Classification of LPE processes according to the way of producing an over-saturation

Process	Remarks
Programmed cooling	Thickness of layers depends upon: (i) solubility of the compound; (ii) ratio between the substance crystallized at the melt surface and the quantity grown on the substrate. If amphoteric doping elements are applied, it is possible to produce *pn* transition with one melt; accurate temperature control required.
Heat dissipation by the crystal holder (relaxation method, see Section 4.1.3.1).	Low thickness of layers (up to 10 μm), only fair reproducibility, accurate temperature control not necessary.
Deposition at a constant temperature: a) temperature gradient between the growth and solution zones:	Source material required; suitable for compounds with low solubility; growth at low temperatures (used to study the incorporation of impurities and to produce homogeneous solid solutions), layer thickness at choice.
b) "Isothermal Solution Mixing Growth" (ISM)	Preferred deposition of very thin layers (Å range), accurate temperature control necessary (see Section 4.1.3.2).

4.1.3.1.   Experimental arrangements to deposit LPE layers

4.1.3.1.1.   *Tilting*

Figure 11 shows an experimental arrangement according to *Nelson* [68]. This arrangement is reminiscent of the decanting process known from metallurgy. On one side of a graphite vessel there is a Ga solution saturated with GaAs, on the other side the GaAs substrate. On tilting the furnace by an angle of about 10 degrees the Ga solu-

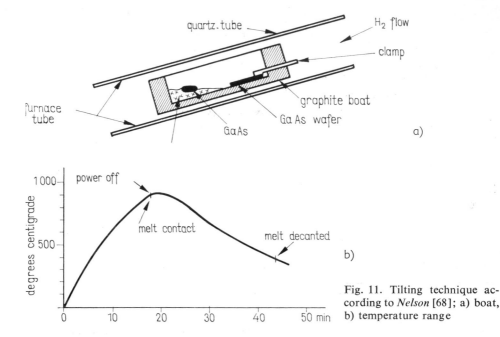

Fig. 11. Tilting technique according to *Nelson* [68]; a) boat, b) temperature range

tion flows onto the substrate. After the temperature is decreased the melt is tilted back to its initial position. If a quartz tube closed either at one end or at both ends is used, melts containing volatile components (GaP, InP with Zn-, ZnO-, and N-doping) can also be used [69], [70].

The tilting around an axis perpendicular to the furnace axis, however, has some disadvantages. Firstly, it disturbs the temperature homogeneity by convection, and secondly, the radial gradient of temperature causes the substrate to be always warmer than the surface of the melt. Due to these temperature differences a large portion of the solved compound is crystallized on the surface of the melt and prevents the melt from rolling off after tilting. Furthermore, the temperature gradient perpendicular to the phase boundary between substrate and layer often attains a value permitting constitutional undercooling [71]. This effect and the crystallization from the remaining melt still adhering at the layer account for the insufficient surface quality of the layers which prohibits the instantaneous growth of further layers (multilayer structures). Therefore, the tilting technique is modified, and other processes are developed in order to remedy such shortcomings. For instance, *Rosztoczy* [72] suggested providing the vessel with a quartz slide and placing the substrate into a bottom recess (Fig. 12a). This arrangement permits complete removal of the melt from the substrate. In Fig. 12b, however, it is the substrate which is moved. This arrangement was used by *Panish* [17] for the deposition of $Ga_{1-x}Al_xAs$ onto GaAs substrates. Since the melt always tends to form oxides due to its Al content, the slide serves above all for the removal of the oxide layer.

A remarkable improvement of the tilting technique is achieved by placing the tilting axis parallel to the axis of the vessel. These variants have preferably been used for the formation of high-purity GaAs for Gunn elements. *Donahue* and *Minden* [73]

238

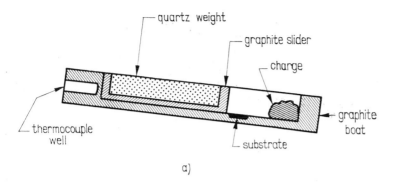

quartz weight
graphite slider
charge
graphite
boat
substrate
thermocouple
well

a)

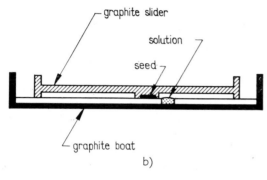

graphite slider
solution
seed
graphite boat

b)

Figs. 12a and 12b. Modifications of the vessels for the *Nelson technique* [17], [72]

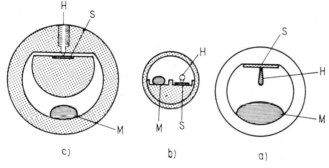

H   S
M

c)

H
M   S

b)

S
H
M

a)

Figs. 13a–13c. Variants of the vessel for the tilting process (cross-section) with the tilting axis parallel to the axis of the vessel. H — the substrate holder, M — the melt, and S — the substrate

discussed the arrangements depicted in Fig. 13. In Fig. 13a, a turn of 180 degrees by means of a rod leads to a contact between substrate and melt. According to *Maruyama*, this rotation can also be achieved by rolling the cylindrical epitaxy vessel in a quartz vessel with a length $L = \dfrac{\pi D}{(2 + D)}$ (D = outer diameter of the graphite vessel). It is advantageous that the required tilting angle measures only a few degrees and, hence, does not disturb the temperature field [74]. A similar kind of vessel was suggested by *Grobe* and *Salow* [75] (Fig. 14). The angle of rotation is 90°. Figure 13b shows a vessel which during the entire epitaxy process is rotated in one direction by an angle of 360°. The vessel depicted in Fig. 13c is preferably used to produce very thin layers. The width of the slot, however, has to be larger than 0.12 inches; otherwise, the melt is prevented from entering the slot due to surface tension. The authors suggest that 5 percent of indium should be added in order to improve the wetting properties of a Ga–GaAs solution. This shape of vessel has some advantages which can be seen both from the surface quality and the homogeneous thickness of the layer. The radial temperature gradients in the vessel shown in Figs. 13a and 13b are equal to those prevailing in the *Nelson process*. The melt is closer to the furnace axis than the substrate. As a result, crystallization on the melt surface and constitutional undercooling resulting in a deterioration of the quality of layer surfaces are favored. In contrast, in Fig. 13c an inversion of the temprature gradient between the melt and the substrate is obtained which largely eliminates such disadvantages and increases the reproducibility. Three-zone diffusion furnaces with a very homogeneous temperature distribution were used for these experiments. The vessels were turned one minute after the cooling programme (0.2°/min) had been initiated.

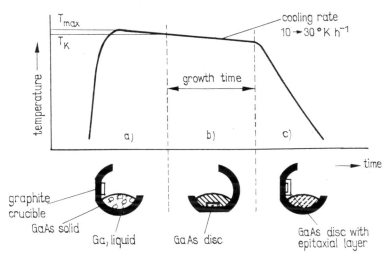

Fig. 14. The substrate/melt arrangement and temperature according to *Grobe* and *Salow* [75]

4.1.3.1.2. *Shifting technique*

This arrangement can be used both for cooling processes and isothermal deposition. It is interesting from the technological point of view since it allows the preparation of multilayer structures in a single process. A vessel with which $Ga_{1-x}Al_xAs$ layers were grown for luminescence and laser diodes is shown in Fig. 15. As indicated by the arrow, the substrate is shifted under one of the four differently composed melts and is exposed to a special cooling program. However, strict care has to be taken that the substrate surface is not damaged when shifted. A similar arrangement had been suggested by *Panish* [77]. The main advantage of the shifting process is that the substrate is completely separated from the adhering Ga melt. This guarantees good surface qualities and a homogeneous layer thickness. *Blum* and *Shih* prepared $Al_xGa_{1-x}As$ layers for monolithic planar structures by means of the improved arrangement depicted in Fig. 16 [78], [79] .When the required growth parameters are observed −

240

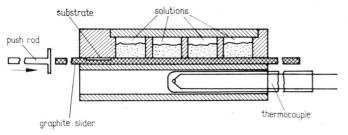

push rod

graphite slider

thermocouple

Fig. 15. Vessel for the shifting technique according to *Beneking* [76]

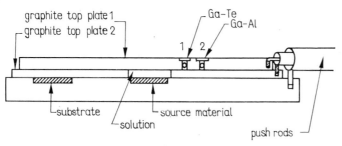

graphite top plate 1
graphite top plate 2

Ga-Te
Ga-Al

substrate   source material

solution

push rods

Fig. 16. Apparatus preparing $Ga_{1-x}Al_xAs$ layers for monolithic planar structures [78], [79]

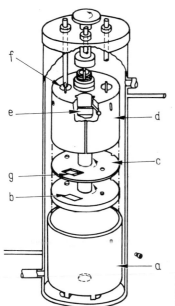

Fig. 17. Apparatus for producing "superlattice structures" (rotation in the direction of the arrow) (according to *Woodall* [81])

especially a precise temperature control — the shifting technique ensures the reproducibility required for technological purposes [80]. The contact between the melt and the substrate can also be achieved by a rotation (Fig. 17). The apparatus is also made of graphite and consists of the following parts: a) crucible, b) substrate holder, c) spacing washer for the delimination of the volume of the melt. Part d) is mounted on (a) and (b); e) melting chambers with source material; f) quartz tubes to inject doping material and g) a vacancy to keep back a small volume of melt Vg above the substrate. The entire equipment is located in a quartz recipient and surrounded by high-purity hydrogen.

4.1.3.1.3.   *Immersion technique*

Originally this technique was applied by *Rupprecht* to GaAs [9]. In Fig. 18a growing equipment for doped $Al_xGa_{1-x}As$ crystals is depicted. Figure 18b shows the temperature profile of the furnace and Fig. 18c the temperature as a function of the time during the growth [82]. $In_xGa_{1-x}As$ [7] and GaP [83] were also grown by means of this method. Since the crystal mounting can be rotated and oscillated and since it also serves for the heat diversion, the procedure can be varied widely. However, it is impossible to produce multilayer sequences by one step of the process.

The "Transient Mode Liquid Epitaxy" suggested by *Deitch* [84] can be realized by means of this equipment. In this method a substrate, *e.g.* a GaAs substrate with a temperature about 50° lower than the bath, is immersed in a saturated Ga solution with a constant temperature. Due to the heat dissipation through the substrate and the substrate holder the solution becomes locally oversaturated for a certain time. This oversaturation vanishes during layer growth ("relaxation process", see also Table 5). In Fig. 19 the temperature of the melt, T, the saturation parameter, α, and the layer thickness, x, are schematically depicted as functions of time t. T_m is the initial temperature. The saturation parameter is given by

$$\alpha = \frac{C_A(x, t, T_E)}{C_E(x, t, T_A)} \tag{64}$$

where C_A and C_E are the instantaneous actual concentration and the equilibrium concentration of GaAs in the solution, respectively. Hence C_A is determined by the equilibrium temperatures of the solution, T_E, while C_E is the equilibrium concentration corresponding to the instantaneous temperature T_A. T_1 is the temperature finally obtained due to the lost of heat by diversion.

Figure 19c represents the conditions for a GaAs–Ga solution at 800°C and a substrate temperature of 750°C. The maximum layer thickness is attained after 60 seconds (growth rate about 10 μm/min).

As the temperature gradient at the substrate is very high constitutional undercooling effects are avoided. A layer thickness of about 1 to 10 μm may be achieved by this method. A special source material is required for layer growth at constant temperature. In Fig. 20 an apparatus suggested by *Butter et al.* [85] is shown. The transport of substance occurs by convection and diffusion. By means of the inside cooling of the substrate holder the temperature gradient perpendicular to the substrate can be adjusted so that any constitutional undercooling is repressed. Under optimum conditions layers up to 400 μm with high surface qualities can be obtained. This method is suitable for the growth of epitaxial layers at low temperatures where methods based on cooling processes fail because of the low solubility of some $A^{III}B^V$ compounds.

242

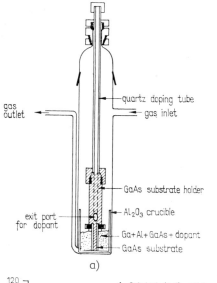

gas outlet

gas inlet

quartz doping tube

GaAs substrate holder

Al₂O₃ crucible

exit port for dopant

Ga+Al+GaAs+ dopant

GaAs substrate

a)

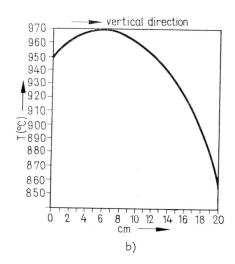

b)

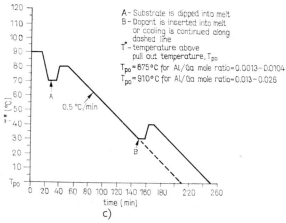

A – Substrate is dipped into melt
B – Dopant is inserted into melt
    or cooling is continued along
    dashed line
T* – temperature above
    pull out temperature, T_{po}
$T_{po} \approx 875°C$ for Al/Ga mole ratio = 0.0013 – 0.0104
$T_{po} \approx 910°C$ for Al/Ga mole ratio = 0.013 – 0.026

0.5 °C/min

c)

Fig. 18. Immersion technique. a) Apparatus, b) temperature profile of the furnace, c) temperature range during layer growth

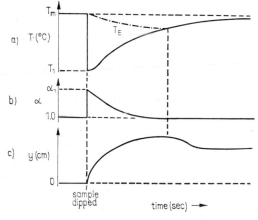

a) T (°C)

b) α

c) y (cm)

sample dipped

time (sec) →

Fig. 19. Temperature of the melt T, (a), saturation parameter α, (b), and layer thickness y, (c), as functions of time of the "Transient Mode LP" according to *Deitch* [84]

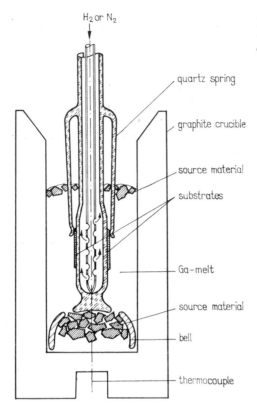

H₂ or N₂

Fig. 20. LPE apparatus for layer growth at constant temperature according to *Butter* and coworkers [85]

quartz spring

graphite crucible

source material

substrates

Ga–melt

source material

bell

thermocouple

4.1.3.2. **Relations between phase diagrams, growth parameters and layer characteristics**

The discussion of this complex field is focused on the following major aspects:

(i) Phase diagram and concentrations in mixed crystal layers,
(ii) incorporation of impurities as a function of temperature.

As examples, the Ga–Al–As, Ga–In–As, and Ga–As–X (X is the doping element) systems will be discussed.

4.1.3.2.1. *Phase diagrams and concentrations in mixed crystal layers*

The exact knowledge of phase diagrams is for liquid-epitaxial layer growth of great interest, especially for two reasons:

1. The knowledge of phase diagrams is fundamental for the choice of the composition of the initial melt and for the determination of the optimum temperature at which the substrate should be brought in contact with the melt.

244

2. When the respective phase diagrams are known the concentration profiles in growth direction of the layer could be estimated.

From the phase diagram of the Ga–Al–As system it may be concluded that for the Ga-rich region a highly preferred Al deposition occurs close to the substrate if the Al content in the melt is at the definite value. This causes a decrease of the Al concentration towards the layer surface (Fig. 21; for the apparatus employed see Fig. 18). These curves indicate that the Al distribution coefficient decreases with increasing Al concentration in the melt (Fig. 22) because X_{Al} cannot exceed the

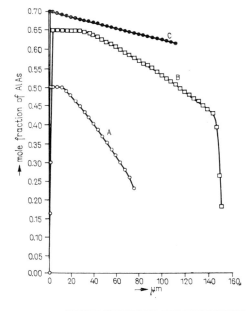

Fig. 21. Composition profiles along the growth axis. Profiles A, B, C represent layers grown from melts with $N_{Al} = 0.0088$ and an initial growth temperature of 995°C, $N_{Al} = 0.011$ and 955°C, and $A_{Al} = 0.017$ and 990°C, respectively [89]

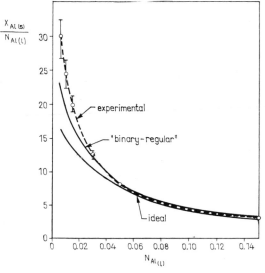

Fig. 22. Distribution coefficients of Al as a function of the Al content in the melt at 1 000°C [17]

245

value 0.5 in solid state. Furthermore, it may be concluded from the phase diagram that the distribution coefficient increases with decreasing temperatures, especially in case of low N_{Al} values. In addition, the growth rate (cooling speed) also influences the distribution coefficient.

Concerning the growth of layers, one may therefore conclude that the concentration gradient in the direction of growth can be reduced by increasing the content of Al in the melt if the temperature range chosen for the cooling process does not exceed 200°C. Based on this idea, mono-crystals can also be grown without any concentration gradient [86]. The crystals or layers will then have AlAs content ranging from 70 to 80% [10].

The following two examples show how the preferred deposition of AlAs in the initial phase of crystallization can be utilized:

The characteristic procedure of the "Isothermal Solution Mixing" (ISM) technique introduced by *Woodall et al.* [81], [87] is the mixing of two melts of compositions A and B at a constant temperature (Fig. 23). The composition of the mixture is not represented by a point on the isotherm 1, but by a point on the straight line 2 joining A and B. The exact position depends on the compositions and volumes of the initial melt. Crystal growth is possible since all points except A and B are on the straight line 2 within a region corresponding the oversaturation. The design of the employed apparatus (Fig. 17) has already been discussed in Section 4.1.3.1. After the two melts with different Al contents (A and B) have been saturated with GaAs, one melt is placed by rotation in the direction of the arrow over the hole in the disk spacer (c) and the substrate (b). If one continues to rotate the melt in the same direction, the melt enclosed in (g) remains at the substrate. The deposition of layers can now be carried out in two different ways:

a) by means of a cooling procedure at a normal LPE process,
b) by means of the ISM technique.

In case b) the second melt is mixed with the one already enclosed in volume V_g, (g), thus initiating crystal growth. In the case of repeated rotation, the concentration profile in Fig. 24 is obtained [87].

Since the Ga solution ensures a sufficient wetting of the substrate even in the case of very low layer thicknesses the melt volume may be reduced to $V_g < 0.025$ cubic centimeters by reducing the thickness of the disk spacer (c). Hence, it is possible at low temperatures to produce superlattice structures with periods of $< 1\,000$ Å. The expansion of a period is essentially determined by the following three parameters:

1. The volume of the enclosed melt V_g.
2. The composition of the melt or the difference in compositions between the two melts.
3. The growth temperature.

Furthermore, it is advantageous that the surface is protected by the enclosed melt against damages during rotation. In Fig. 25 the result of simultaneous examinations of the cathodic luminescence at 300 K and of the chemical composition along a (110) cleavage plane in the direction of growth are given. The ratios of the Al and Ga

246

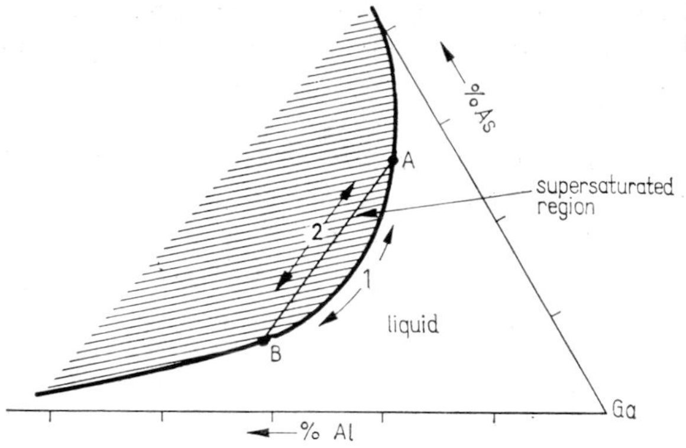

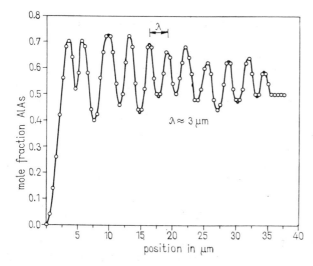

Fig. 23. Principle of the ISM technique for the Ga–Al–As system

Fig. 24. Concentration profile for layer growth at a cooling rate 0.1 °C/min and with 23 melt changes (Al/Ga: 10^{-3} and $5 \cdot 10^{-3}$). Each sub-layer is about 1.5 μm thick [87]

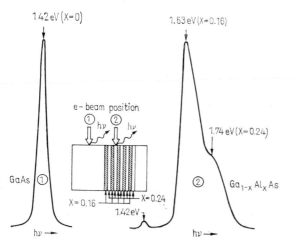

Fig. 25. Cathodic luminescence spectrum of GaAs, (1), and of Ga$_{1-x}$Al$_x$As, (2), of a "superlattice structure" (one period 7 000 Å) [81]

contents for the two melts were here $0.3 \cdot 10^{-3}$ and $1 \cdot 10^{-3}$, respectively. The analysis reveals that the structure is composed of two sub-layers having the compositions $X_{AlAs} = 0.16$ and $X_{AlAs} = 0.24$.

With the second example a method of defined interruption of crystal growth is represented based on the ISM technique. It is well known that at the application of the tilting or immersion techniques material solved in the residual melt adhering to the layer continues to crystallize during the cooling process. This causes a deterioration of surface quality. Hence, the following procedure has been suggested by *Potemski* [88] for the growth of $Ga_{1-x}Al_xAs$ layers:

After being exposed to a cooling program the substrate with the enclosed melt (by the disk spacer) is placed between two melts (apparatus as in Fig. 17, but four different melts). By further rotation the substrate is contacted with a "quenched melt". In contrast to all other melts, this melt is composed of Al and Ga only (Al/Ga = 0.1). Because of its high content of Al, the As arised from the growth melt (Al/Ga = $1.2 \cdot 10^{-3}$) is deposited as nearly pure AlAs. This layer can be dissolved by diluted HCl. By this the layer growth is interrupted in a defined way, and the occurrence of a concentration gradient of Al in growth direction is avoided.

As a further example, revealing the relations between layer quality and phase diagram, the growth of $Ga_{1-x}In_xAs$ layers will be discussed. Because of appreciable differences between the lattice constant of the two compounds ($\Delta a = 7.17\%$) complications should be expected for the formation of a defined lattice configuration. *Wu* and *Pearson* [7] found that mono-crystalline layers on (111) GaAs substrates can be deposited only if $X < 0.2$. For $X > 0.7$ InAs substrates are used. According to the phase diagram of this system GaAs is the component with the lower solubility. Consequently, the ratios representing the composition in the growth melt are inverted as well [89]:

If the In/Ga ratio is 1 in the melt, we obtain $X \approx 0.01$, and if the In/Ga ratio is 30 in the melt, we obtain $X \approx 0.2$.

The concentration profiles in growth direction of these layers are flat, analogous to the case of $Ga_{1-x}Al_xAs$ ($0.7 < X < 0.8$). As can be seen from Fig. 26, this statement applies both to solid Ga rich solutions and to those rich in In (the calculation of the

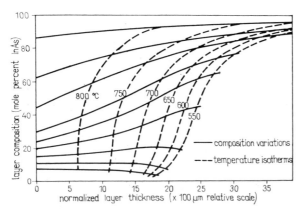

Fig. 26. Crystal composition vs. layer thickness normalized to a 5 g melt and substrate area of 1 cm². The 850°C isotherm is superimposed on the ordinate axis (according to [7])

curves is based on the assumption that the melt remains in equilibrium with the layer during growth and that crystallization takes place on the layer surface only).

4.1.3.2.2.   *Incorporation of impurities in dependence on growth parameters*

The incorporation of impurities in a semiconductor essentially depends upon four factors:

(i) doping concentration of the melt,
(ii) rate of crystallization,
(iii) temperature of crystallization,
(iv) temperature treatment after crystallization.

A consideration of deposition of layers at different crystallization rates makes it possible to estimate whether or not a doping element is incorporated under conditions of equilibrium, *i.e.* in accordance with the phase diagram. To realize the equilibrium condition very slow cooling rates are frequently required. Two examples may illustrate this problem [90]. During the deposition of zinc-doped GaAs layers on (100) GaAs substrates the distribution coefficient for different cooling rates attains different values:

$$k_{Zn} = 1 \cdot 10^{-2} \text{ for } 5° \text{ min}^{-1},$$

$$k_{Zn} = 1.5 \cdot 10^{-2} \text{ for } 1.67° \text{ min}^{-1},$$

Obviously, in this case equilibrium conditions cannot be assumed. Epitaxy experiments on both (111) faces by the same authors resulted in two distribution coefficients, these being dependent on the substrate orientation but independent of the cooling rate:

(111) GaAs $k_{Zn} = 2 \cdot 10^{-2}$ for $1.67°$ min^{-1} and $5°$ min^{-1}

($\overline{111}$) GaAs $k_{Zn} = 1 \cdot 10^{-2}$ for $1.67°$ min^{-1} and $5°$ min^{-1}.

Since essentially the same values are valid for both cooling rates, equilibrium conditions can be assumed for these cases. The variations of the zinc concentration are found to be less than 10% throughout the whole layer. The thickness of the epitaxial layers ranged from 55 to 75 μm. This result indicates a zinc distribution coefficient essentially independent of the temperature in the cooling range from 650 to 800°C. As a consequence of the variation in the rate of incorporation and, hence, of the variation in the distribution coefficient during the controlled cooling procedure, doping concentration gradients of the materials arise in the direction of layer growth. These facts had been thoroughly investigated by *Grobe* and *Salow* [75] for pure GaAs (for the apparatus used see Fig. 14; cooling rate: 20 to 30 degrees/hour). Figure 27 shows the density of charge carriers, $(N_D - N_A)$, and their mobility as a function of the layer thickness for various growth conditions. It is of particular interest that the purity of layers can be considerably enhanced by depositing a small amount of GaAs onto the surface of the melt before the latter is brought into contact with the substrate (Sample E 364 in Fig. 27). Obviously, the fine crystalline GaAs causes an additional getter effect due to its large specific surface and, therefore, the major part of impurities with $k > 1$ is incorporated at these crystallites. This result has been experimentally substantiated by comparing the dependence of N_A and N_D on the temperature of growth (Fig. 28). Whereas in the case of the "end layers" all the solved GaAs had been deposited onto the substrate by a cooling procedure to room temperature, the melt had been prematurely removed at a certain tilting temperature in the case of the "terminal layers". Both layer types were grown under the same con-

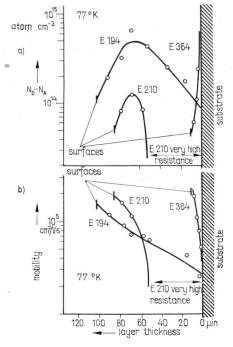

Fig. 27. Profile of charge carrier densities $N_D - N_A$ (a) and carrier mobilities (b) in GaAs layers. Tilting temperatures for E 210: 845°C, and for E 194: 805°C. E 364 represents a layer with premature deposition at a tilting temperature of 650°C (see text) [75]

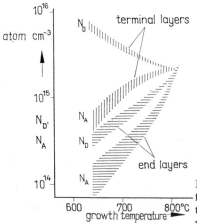

Fig. 28. Donor and acceptor densities of layers grown at different temperatures in a typical final layer, compared with the values present in a series of terminal layers with corresponding tilting temperatures [75]

ditions of deposition. It seems correct to assume that the end layers are composed of terminated layers. In the latter the densities of acceptors N_A and donors N_D are measured for the same layer thickness. The different rates of incorporation of donors and acceptors are shown in Fig. 28. In the case of terminal layers, the growth temperature T_k, on the abscissa, represents the tipping temperature. In the case of end layers, T_k means the temperature of formation of the different terminal layers composing these end layers. Growth temperatures of $T_k > 810°C$ for both layer types result in a high resistivity, highly compensated material with the same density of donors and acceptors. Towards lower growth temperatures or with progressive cooling N_A and N_D in the end layers decrease rapidly, whereas N_D increases in the terminated layers at the same tipping temperature and N_A decreases more slowly than in the end layers.

The growth temperature is especially significant for the amphoteric doping elements of the 4th main group of the periodic system of elements since the type of conduction can change both as a function of temperature and of the doping concentration. In the case of GaAs, at higher temperatures Si is preferably incorporated at Ga sites as a donor and at low temperatures preferably at As sites as an acceptor (Fig. 29).

From the phase diagram (Fig. 30) we conclude that a layer which crystallizes from a melt of composition A is n-conductive at the beginning. With the change of composition and temperature according to B it becomes p-conductive. At a further cooling procedure to C the composition follows the eutectic line and finally the Ga corner is reached. Hence it is possible for the LPE method to deposit p–n transitions by using only one melt. Compared with GaAs the transition temperature of $p \rightarrow n$ decreases for $Ga_{1-x}Al_xAs$ with increasing Al/As ratio [92]. Isothermal growth processes (Fig. 20) are particularly suitable for an exact investigation of these problems since deposition processes can be carried out also at very low temperatures.

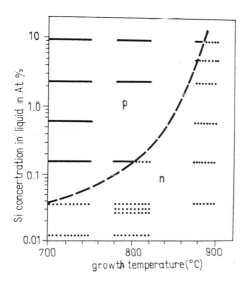

Fig. 29. Effect of growth temperature and Si concentration in the melt on the behavior of Si and GaAs crystals [72]. (———— p type and . . . . n type)

251

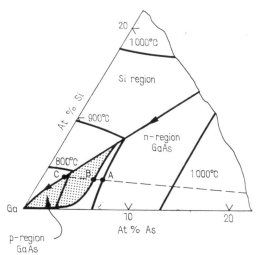

Fig. 30. Ga–As–Si phase diagram (see text) according to [91]

There is a wide range of further possibilities of influencing layer qualities by varying the crystal growth parameters. For instance, it was found for $Ga_{1-x}Al_xAs$ that partial dissolution of the GaAs substrates prior to the epitaxial process increases the efficiency of luminescence diodes since surface layers containing impurities and degeneracies are thus removed [93]. On depositiong n-GaP onto p-GaP : ZnO substrates it was found that the exterior quantum efficiency plotted as a function of the time of contact between the melt and the substrate passes through a maximum. This behavior reveals two competitive processes: in the case of shorter periods of contact the luminescence efficiency is increased due to the surface etching effect. As the etching rate decreases with increasing saturation of the solution the diffusion of Zn out of the substrate becomes more important. Thus, the ZnO pair concentration decreases in the active region and as a consequence the quantum efficiency decreases [94].

All these examples reveal that the layer properties can be influenced by a controlled process of crystal growth. The physico-chemical process of incorporating impurities is one of the basic problems of current research.

List of symbols

a	lattice parameter
E_g	band gap
$E_i (i = 0, 1, 2)$	transitions in the critical points of the optical spectrum
f_i	ionicity according to *Phillips*
$K_{i,j}$	equilibrium constant
N_i	mole fraction in the liquid phase
n_{ij}	number of i-j pairs
n_i	number of i atoms
R	gas constant
$T_{i,j}^F$	melting temperature
$V_{i,j}$	molar volume
X_{ij}	mole fraction in the solid phase
Z	coordination number

$\gamma_{i,j}$ activity coefficient
γ_i^0 activity coefficient of the component i in infinite dilution
ΔG^E free excessive enthalpy
$\Delta \bar{G}_{M,i}$ free partial molar enthalpy of mixing of the component i
ΔH_M enthalpy of mixing
$\Delta \bar{H}_{M,i}$ partial molar enthalpy of mixing of the component i
$\Delta H_{i,j}^f$ heat of formation
$\Delta H_{i,j}^F$ enthalpy of melting
$\Delta \bar{S}_{M,i}$ partial molar entropy of mixing of the component i
ΔS_M^E excessive entropy of mixing
$\Delta S_{i,j}^F$ melt entropy
$\delta_{i,j}$ solubility parameters according to *Hildebrand*
$\kappa_{i,j}$ electro-negativity defined by *Pauling*
$\Omega_{i,j}$ interaction parameter
$\mu_{i,j}$ chemical potential of the mixture
$\mu_{i,j}^0$ chemical potential of the pure components

4.1.4.   References

[1] *Casey, H. C., Trumbore, F. A.:* Mat. Sci. Eng. 6 (1970) 69.
[2] *Mataré, H. F.:* Intern. Electron. Rundschau 8 (1972) 177.
[3] *Panish, M. B.:* J. Phys. Chem. Solids 30 (1969) 1083.
[4] *Osamura, K., Inoue, J., Murakami, Y.:* J. Electrochem. Soc. 119 (1972) 103.
[5] *Köster, W., Thoma, B.:* Z. Metallkde. 46 (1955) 293.
[6] *Woolley, J. C., Smith, B. A.:* Proc. Phys. Soc. B 72 (1958) 214.
[7] *Wu, T. Y., Pearson, G. L.:* J. Phys. Chem. Solids 33 (1972) 409.
[8] *Hall, R. N.:* J. electrochem. Soc. 110 (1963) 385.
[9] *Rupprecht, S. A.:* Proc. Int. Symp. on GaAs 1966. Inst. Phys. Phys. Soc., p. 57.
[10] *Ilegems, M., Pearson, G. L.:* Proc. Int. Symp. on GaAs 1968. Inst. Phys. Phys. Soc., p. 3.
[11] *Panish, M. B., Arthur, J. R.:* J. Chem. Thermodynamics 2 (1970) 299.
[12] *Panish, M. B.:* J. Chem. Thermodynamics 2 (1970) 319.
[13] *Woolley, J. C., Smith, B. A., Lees, D. G.:* Proc. Phys. Soc. B 69 (1956) 1339.
[14] *Woolley, J. C., Smith, B. A.:* Proc. Phys. Soc. B 70 (1957) 153.
[15] *Foster, L. M., Scardefield, J. E.:* J. Electrochem. Soc. 117 (1970) 534.
[16] *Foster, L. M., Scardefield, J. E.:* J. Electrochem. Soc. 118 (1971) 495.
[17] *Panish, M. B., Sumski, S.:* J. Phys. Chem. Solids 30 (1969) 129.
[18] *Panish, M. B., Sumski, S., Lynch, R. T.:* Trans. Met. Soc. AIME 245 (1969) 559.
[19] *Panish, M. B.:* J. Electrochem. Soc. 117 (1970) 1202.
[20] *Kühne, G., Leonhardt, A.:* Kristall und Technik 7 K 57 (1972).
[21] *Guggenheim, E. A.:* Mixtures. Oxford University Press, London (1952).
[22] *Laugier, A.:* C. R. Acad. Sc. C 273 (1971) 404.
[23] *Vieland, L. J.:* Acta Met. 11 (1963) 137.
[24] *Stringfellow, G. B., Greene, P. E.:* J. Phys. Chem. Solids 30 (1969) 1779.
[25] *Hildebrand, J. H.:* J. Amer. chem. Soc. 51 (1929) 66.
[26] *Kajiyama, K.:* Jap. J. appl. Phys. 10 (1971) 561.
[27] *Thurmond, C. D.:* J. Phys. Chem. Solids 26 (1965) 785.
[28] *Arthur, J. R.:* J. Phys. Chem. Solids 28 (1963) 2257.
[29] *Panish, M. B., Ilegems, M.:* Proc. Int. Symp. on GaAs 1970. Inst. Phys. Phys. Soc., p. 7.
[30] *Blom, G. M.:* J. Electrochem. Soc. 118 (1971) 1834.
[31] *Blom, G. M., Plaskett, T. S.:* J. Electrochem. Soc. 118 (1971) 1831.
[32] *Darken, L. S.:* Trans. Met. Soc. AIME 239 (1967) 80.
[33] *Turkdogan, E. T., Darken, L. S.:* Trans. Met. Soc. AIME 242 (1968) 1997.
[34] *Turkdogan, E. T., Gruehan, R. J., Darken, L. S.:* Trans. Met. Soc. AIME 245 (1969) 1003.

[35] *Antypas, G. A., James, L. M.:* J. Appl. Phys. 41 (1970) 2165.
[36] *Antypas, G. A.:* J. Electrochem. Soc. 117 (1970) 700.
[37] *Antypas, G. A.:* J. Electrochem. Soc. 117 (1970) 1393.
[38] *Antypas, G. A., Yep, T. O.:* J. Appl. Phys. 42 (1971) 3201.
[39] *Stringfellow, G. B.:* J. Electrochem. Soc. 117 (1970) 1301.
[40] *Huber, D.:* Proc. Int. Conf. of Phys. and Semicond. Heterojunc. and Layer Struct. Vol. I, p. 195. Akadémiai Kiadó, Budapest (1971).
[41] *Huber, D.:* Private communication.
[42] *Van Vechten, J. A.:* 10th Int. Conf. on Phys. of Semiconductors, Cambridge, Mass., August 1970.
[43] *Foster, L. M., Woods, J. F.:* J. Electrochem. Soc. 118 (1971) 1175.
[44] *Richman, D.:* J. Phys. Chem. Solids 24 (1963) 1131.
[45] *Stringfellow, G. B.:* Mat. Res. Bull. 6 (1971) 371.
[46] *Hildebrand, J. H., Scott, R. L.:* The Solubility of Nonelectrolytes Dover Publ. Inc., New York (1964), p. 323.
[47] *Pauling, L.:* Natur der Chemischen Bindung, Verlag Chemie, Weinheim — Bergstr. (1964).
[48] *Hansen, M.:* Constitution of Binary Alloys. McGraw Hill Book Co., New York (1958).
[49] *Yazawa, A., Lee, Y. K.:* Trans. JIM 11 (1970) 411.
[50] *Phillips, J. C., Van Vechten, J. A.:* Chem. Phys. Letters 5 (1970) 159.
[51] *Phillips, J. C., Van Vechten, J. A.:* Phys. Rev. B 2 (1970) 6.
[52] *Phillips, J. C.:* Rev. Modern Physics 42 (1970) 317.
[53] *Sirota, N. M.:* Semiconductors and Semimetals. (Eds. R. W. Willardson and S. Z. Beer), Vol. 4. Academic Press, New York (1960), p. 86.
[54] *Kischio, W.:* Z. anorg. allg. Chem. 328 (1964) 187.
[55] *Lichter, B. D., Sommelet, D.:* Trans. Met. Soc. AIME 245 (1969) 1021; 245 (1969) 99.
[56] *Stringfellow, G. B.:* J. Phys. Chem. Solids 33 (1972) 665.
[57] *Steininger, J.:* J. Appl. Phys. 41 (1970) 2713.
[58] *Prince, A.:* Alloy Phase Equilibria. North Holland Publ., Amsterdam (1966).
[59] *Furukawa, Y., Thurmond, C. D.:* J. Phys. Chem. Solids 26 (1965) 1535.
[60] *Panish, M. B.:* J. Phys. Chem. Solids 27 (1966) 291.
[61] *Panish, M. B.:* J. Electrochem. Soc. 113 (1966) 224.
[62] *Panish, M. B.:* J. Less Common Metals 10 (1966) 416.
[63] *Jordan, A. S.:* Met. Trans. 2 (1971) 1959.
[64] *Jordan, A. S.:* Met. Trans. 2 (1971) 1965.
[65] *Ladany, I., Wang, C. C.:* J. Appl. Phys. 443 (1972) 236.
[66] *Hicks, H. G. B., Greene, P. D.:* Proc. 3rd. Int. Symp. on GaAs Inst. Phys. Soc. Conf. Ser. No. 9 (1970) 3.
[67] *Solomon, R.:* Proc. 3rd. Int. Symp. on GaAs, Inst. Phys. Soc. Conf. Ser. No. 9 (1970) 3.
[68] *Nelson, H.:* R. C. A. Review 24 (1963) 603.
[69] *Trumbore, F. A., Kawalchik, M., White, H. G.:* J. Appl. Phys. 38 (1967) 1987.
[70] *Saul, R. H., Armstrong, J., Mackett, W. H.:* J. Appl. Phys. Letters 15 (1969) 229.
[71] *Minden, H. T.:* J. Cryst. Growth 6 (1970) 288.
[72] *Rosztoczy, F. E.:* Electroch. Div. Abstr. of the Electrochem. Soc. 17 (1968) 516.
[73] *Donahue, J. A., Minden, H. T.:* J. Cryst. Growth 7 (1970) 221.
[74] *Maruyama, S.:* Jap. J. appl. Phys. 11 (1972) 424.
[75] *Grobe, E., Salow, H.:* Z. Angew. Physik 32 (1972) 381.
[76] *Beneking, H., Mischel, B., Schul, G.:* Electronics Letters 8 (1972) 16.
[77] *Panish, M. B., Hayashi, J., Sumski, S.:* Appl. Phys. Letters 16 (1970) 326.
[78] *Blum, J. M., Shih, K. K.:* J. Appl. Phys. 43 (1972) 1394; 43 (1972) 3094.
[79] *Blum, J. M., Shih, K. K.:* Proc. of the IEEE 59 (1971) 1498.
[80] *Miller, B. I., Pinkas, E., Hayashi, I., Capik, R. J.:* J. Appl. Phys. 43 (1972) 2817.
[81] *Woodall, J. M.:* J. Cryst. Growth 12 (1972) 32.
[82] *Woodall, J. M., Rupprecht, H., Reuter, W.:* J. electrochem. Soc. 116 (1969) 899.
[83] *Peaker, A. R., Sudlow, P. D., Mottram, A.:* J. Cryst. Growth 13/14 (1972) 651.
[84] *Deitch, R. H.:* J. Cryst. Growth 7 (1970) 69.
[85] *Tietze, H. J., Andrä, R., Butter, E.:* Kristall und Technik 6 (1971) 747.

[86] *Fischer, P., Kühn, G., Bindemann, R., Rheinländer, B., Hörig, W.:* Kristall und Technik 8 (1973) 167.

[87] *Woodall, J. M.:* J. electrochem. Soc. 118 (1971) 151.

[88] *Potemski, R. M., Woodall, J. M.:* J. Electrochem. Soc. 119 (1972) 278.

[89] *Takahashi, K., Moriizumi, T., Shirose, S.:* J. electrochem. Soc. 118 (1971) 1639.

[90] *Keller, K., v.Münch, W.:* Solid State Electronics 14 (1971) 526.

[91] *Panish, M. B.:* J. electrochem. Soc. 113 (1966) 1226.

[92] *Rado, W. G., Johnson, W. J., Crawley, R. L.:* J. Appl. Phys. 43 (1972) 2763.

[93] *Shih, K. K., Blum, J. M.:* J. electrochem. Soc. 118 (1971) 1633.

[94] *Kasami, A., Naito, M., Toyama, M.:* Jap. J. Appl. Phys. 10 (1971) 117.

[95] *Hockings, E. F., Kudman, J., Seidel, T. E., Schülz, C. M., Steigmeier, E.:* J. Appl. Phys. 37 (1966) 2879.

[96] *Van Hook, H. J., Lenker, E. S.:* Trans. Metall. Soc. AIME 227 (1963) 220.

A KINETIC PHASE TRANSITION IN A BINARY CRYSTAL GROWING FROM THE 50% ALLOYS MELT INITIATED BY THE MOVEMENT OF ROUGH STEPS

Yu. D. Čistyakov and Yu. A. Baikov

The Moscow Institute of Electronic Engineering, Moscow, USSR

4.2.1. INTRODUCTION

This section is dedicated to an investigation of kinetic phase transition in a binary crystal growing from the 50% alloys melt. This kinetic phase transition is a transition of growing crystals from an ordered to a disordered phase when the thermodynamic state of the melt/crystal system is exposed to some specific conditions. Kinetic phase transitions were first described in a well known publication [1] as an effect of irregulating, observed when a binary crystal grows from the vapor phase. In this article, the authors suggested that a growing crystal surface moved as a result of a lateral movement of some rough step which was on the growing crystal surface. The concept of a rough interface separating the melt from the growing crystalline phase [2], [3] is confirmed by the theory of growing crystals. The roughness is connected with the rough steps observed on the surface of crystalline phases. There is a critical transition temperature, T_{cr}, above which the interface between the melt and the crystal becomes noticeably rough. It should be noted, however, that the existence of T_{cr} does not have any influence on the kinetics of growing crystalline phases. On the other hand, in real crystals these steps always exist as a consequence of surface imperfections (*e.g.* dislocations); these steps are necessary for an explanation of crystal growth when the interface temperature T_m is lower than the critical transition temperature T_{cr}. Moreover, even when the critical transition temperature is below the melting point of the crystal under consideration, the difference between the crystallization hates corresponding to temperatures lower and higher than T_{cr} must vanish.

tiois section considers two aspects. The first aspect to be discussed refers to some conditions which have to be fulfilled according to the thermodynamics of the melt/crystal system, so that the crystal phase grows at certain interfaces due to lateral movement of the rough steps. The second aspect refers to some approximations for a complex interaction problem of particles on the crystal surface. This complex interaction of particles is responsible for the realization of kinetic phase transition.

4.2.2. A MODEL OF A CRYSTAL SURFACE WITH ONE STEP WHICH HAS MANY FRACTURES

The present model of crystallization suggests that on the (001) surface of a simple cubic crystal there is one step with the mean direction [010] which has many fractures of various heights and signs (Fig. 1). It is assumed that the entire melt/crystal system

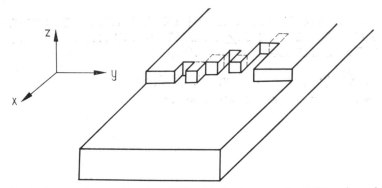

Fig. 1. A step of mean direction [010] with fractures on the (001) surface of a simple cubic crystal. The dashed areas represent single lattice sites

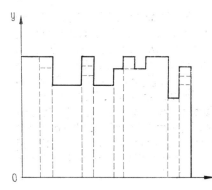

Fig. 2. A projection of a step into the (XOY) plane. The separate columns are dashed

consists of two different kinds of atoms, A and B, respectively; the concentration of the A-atoms being equal to the concentration of the B-atoms. The lattice of the simple cubic crystal consists of two sublattices i and k with alternating atomic centers. Processes of diffusion of atoms in the melt were not taken into account. The melt is supposed to be a homogeneous substance from which the A- and B-atoms adjacent to the columns which form the fractures (Fig. 2) participate in the exchange processes with the particles from the crystalline phase. The problem of particle exchange between the crystal and the melt has been solved approximately here. The main approximation is the choice of the so-called distribution functions of columns at the step:

$$X_\alpha^{(i)}(k_1, k_2), \qquad X_{\alpha\beta}^{(i)}(k_1, k_2), \qquad X_{\alpha\beta\gamma}^{(i)}(k_1, k_2).$$

These distribution functions are the probabilities of finding a column which ends with the particles α; α and β; α, β, and γ, where α belongs to the i-sublattice; such a column is situated between two neighboring fractures with heights k_1, on the left, and k_2, on the right side, respectively. Further, let $X_{\alpha\beta}^{(i)}(k_1)$ be the probability to find a column which ends with two particles α and β (particle α belongs to the i-sublattice) at a fracture of height k_1.

An approximate kinetic equation written for the function $X_{\alpha\beta}^{(i)}(k_1)$ takes into account only the acts of incorporation and of separation of particles at the columns. The appropriate distribution functions of the columns at the step, $X_\alpha^{(i)}(k_1, k_2)$, $X_{\alpha\beta}^{(i)}(k_1, k_2)$, and $X_{\alpha\beta\gamma}^{(i)}(k_1, k_2)$ may be written for the case of a stationary distribution as

$$
\frac{dX_{\alpha\beta}^{(i)}(k_1)}{dt} = \sum_{k_2=-\infty}^{+\infty} \omega_+ \left[X_\alpha^{(ki)}(k_1 - 1, k_2) + X_\alpha^{(i)}(k_2, k_1 + 1) \right] +
$$

$$
+ \sum_{\gamma, k_2=-\infty}^{+\infty} \left[\omega_{-\beta\gamma}^{(ki)}(k_1 + 1, k_2) X_{\alpha\beta\gamma}^{(i)}(k_1 + 1, k_2) + \omega_{-\beta\gamma}^{(ki)}(k_2, k_1 - 1) X_{\alpha\beta\gamma}^{(i)}(k_2, k_1 - 1) \right] -
$$

$$
- \sum_{\gamma, k_2=-\infty}^{+\infty} \omega_+ \left[X_{\alpha\beta}^{(i)}(k_1, k_2) + X_{\alpha\beta}^{(i)}(k_2, k_1) \right] - \sum_{k_2=-\infty}^{+\infty} \left[\omega_{-\alpha\beta}^{(ik)}(k_1, k_2) X_{\alpha\beta}^{(i)}(k_1, k_2) + \right.
$$

$$
+ \omega_{-\alpha\beta}^{(ik)}(k_2, k_1) X_{\alpha\beta}^{(i)}(k_2, k_1) = 0. \tag{1}
$$

It is supposed that the frequencies of incorporation of particles into columns, ω_+, are constant, i.e. they depend neither on the local surface configuration, where a particle is added to a column, nor on the kind of interacting particles. The term $\omega_{-\alpha\beta}^{(ik)}(k_1, k_2)$ is the frequency of separation of a particle β of the k-sublattice from a particle α of the i-sublattice at a column which is situated between two adjacent fractures with heights k_1, on the left, and k_2, on the right side, respectively. The frequencies $\omega_{-\alpha\beta}^{(ik)}(k_1, k_2)$ were averaged over the interaction energies of particles β in respect to other particles apart from α which were nearest neighbors of the particles β. Usually the number of such nearest neighbors varies from 2 to 4 (including the particles α) and depends on the signs of k_1 and k_2. Furthermore, it is supposed that the interaction energies

$$
E_{\alpha\beta}(\alpha, \beta = A, B)
$$

of particles β and α which are taken with inverse signs are interrelated by the following equations

$$
E_{AA} = E_{BB}, \quad E_{AB} = E_{BA} = mE_{AA} \ (m = 2). \tag{2}
$$

An interaction energy of the melt in respect of the crystal has not been taken into account here because $\omega_+ = \text{const}$. The mean growth rate of the step at the interface separating the melt from the crystal may be written as:

$$
r = 2\omega_+ - \sum_{k_1, k_2=-\infty}^{+\infty} \sum_{\alpha, \beta} \omega_{-\alpha\beta}^{(ik)}(k_1, k_2) X_{\alpha\beta}^{(i)}(k_1, k_2). \tag{3}
$$

It is supposed that the functions $X_{\alpha\beta}^{(i)}(k_1, k_2)$, $X_{\alpha\beta\gamma}^{(i)}(k_1)$ etc. must satisfy the following relations:

$$
X_{\alpha\beta}^{(i)}(k_1, k_2) = -\frac{X_{\alpha\beta}^{(i)}(k_1) \, X_{\alpha\beta}^{(i)}(k_2)}{X_{\alpha\beta}^{(i)}} \tag{4a}
$$

$$
X_{\alpha\beta\gamma}^{(i)}(k_1) = \frac{X_{\alpha\beta}^{(i)}(k_1 - 1) \, X_{\beta\gamma}^{(k)}(k_1)}{X_\beta^{(k)}(k_1 - 1)} \tag{4b}
$$

$$X_{\alpha\beta}^{(i)}(k) = \begin{cases} X_{\alpha\beta}^{(i)}(0)\,\lambda^k, & \text{if } k > 0 \\ X_{\alpha\beta}^{(i)}(0)\,\lambda^{|k|}, & \text{if } k < 0. \end{cases} \tag{4c}$$

That these relations are satisfied is a necessary condition in order to obtain from the kinetic equation (1) a system of six independent equations with six unknown quantities by which the kinetic phase transition problem may be solved.

In eqs. (4) the abbreviation

$$X_{\alpha\beta}^{(i)} = \sum_{k_1, k_2}^{+\infty} X_{\alpha\beta}^{(i)}(k_1, k_2)$$

has been used; $X_{\alpha\beta}^{(i)}(0)$ represents the probability of finding a column which ends with two particles α and β, where α belongs to the i-sublattice at a fracture of height zero, $(k = 0)$; and λ is the so-called step's roughness which may attain any value between zero and one, $0 < \lambda < 1$.

A method has been described elsewhere [1] of the subsequent analysis of the system of six equations which was obtained from the kinetic equation (1) and which contains the six unknown quantities

$$X_{\alpha\beta}^{(i)}(0)\,[\eta, q] \quad (\alpha, \beta = \text{A, B}), \quad \lambda[\eta, q]$$

and η, the latter being a far-order parameter which characterizes irregularities of the growing crystal, and where q is defined by

$$q = \exp\left[-\frac{E_{\text{AA}}}{T}\right].$$

A point of equilibrium, $q_e = \exp\left[-\dfrac{E_{\text{AA}}}{T_E}\right]$, was determined from the equation $v = 0$. The physical properties of the melt are characterized by ascribing to the parameter $R \approx \dfrac{1}{\omega_+}$ the value 10^2. This value for R takes into account the fact that the density of the melt is higher than the density of vapor (in Ref. [1] a case of growing crystals from the vapor has been considered where $R \approx \dfrac{1}{\omega_+} \sim 10^6$). In consequence of this, the frequency ω_+ for the melt could be approximated by the corresponding value for the case of growing crystals from the vapor. For the case considered for $T_E < T_C$, where T_C is the regulating temperature of a binary crystal, the calculation of

$$q_e = \exp\left[-\frac{E_{\text{AA}}}{T_E}\right] \quad \text{and} \quad q_k = \exp\left[-\frac{E_{\text{AA}}}{T_K}\right]$$ yields $q_e \approx 0.46$ and $q_k \approx 0.43$. Kinetic phase transition is characterized by the usual properties: a far-order parameter η tends to approach zero like the function $\sqrt{q - q_k}$ when $q \to q_k$ and the derivative $\dfrac{\partial v}{\partial q}$ has at $q = q_k$, a discontinuity step which is proportional to some constant A. This yields a mean growth rate of a step, v:

$$v = \begin{cases} k(q) & \text{when } q \gtrless q_k \\ k(q) - \eta^2 \bar{A} & \text{when } q \lessgtr q_k. \end{cases} \tag{5}$$

260

4.2.3. SOME CONDITIONS FOR THE REALIZATION OF KINETIC PHASE TRANSITION IN A CRYSTAL GROWING FROM THE 50% ALLOYS MELT

Now we shall consider some conditions which are necessary for the realization of kinetic phase transition in the crystalline phase of the melt/crystal system. From general ideas about this effect, it is evident that the kinetic phase transition temperature, T_k, and the observed minimum temperature of the supercooled melt, T_m^{min}, must satisfy the conditions:

$$T_m^{min} < T_k < T_c. \tag{6}$$

First of all, it is necessary to ascertain the conditions for the formation of rough steps on the surface of perfect crystals and for a negligible influence of accessory sources of the steps. This formation of rough steps is connected with a critical interface temperature, T_{cr}.

For the case of the melt/crystal system a condition determining whether a growth surface of a growing crystalline is rough or smooth has been established by *Jackson* [3]. For all crystalline growth surfaces, *Jackson* defined a function $\alpha(T_E) = \dfrac{const}{T_E}$ where T_E is the equilibrium temperature of the melt/crystal system and the value of the constant depends on the physical properties of the melt and the orientation of the respective growth surface. *Jackson* assumes that all those surfaces for which $\alpha(T_E) < 2$ are rough and the crystal surfaces with $\alpha(T_E) > 2$ are smooth. The value $\alpha(T_E) = 2$ corresponds to the transition of a crystal surface from a rough to a smooth surface. According to *Jackson*, the supercooling of the melt results in a quantitative change in the interface roughness. For the growth of a surface from the 50% alloys melt characterized by a function $\alpha(T_m) = \dfrac{const}{T_m}$, where T_m is the interface temperature and the constant is connected with the orientation of this surface, there is a critical temperature, T_{cr}, for which a surface becomes rather rough and the function $\alpha(T_m)$ reaches its critical value $\alpha(T_{cr}) = \dfrac{const}{T_{cr}} = \delta$. Thus, this critical temperature T_{cr} is analogous to the above-mentioned critical transition temperature. Here, it is supposed that for a perfect crystal with perfect surfaces no rough steps appear on its surface for $T_m < T_{cr}$. We suppose that $T_{cr} < T_E$, *i.e.* $\alpha(T_{cr}) > \alpha(T_E) = \gamma$ where γ, as well as δ, depend on the type of the melt and on the orientation of the respective growth surface. The melt temperature range at which crystal surfaces growing from the 50% alloys melt will have rough steps is:

$$T_{cr} = \frac{\gamma}{\delta} \cdot T_E < T_m < T_E, \text{ where } \gamma < \delta. \tag{7}$$

Thus, if kinetic phase transition occurs in a binary crystal, the above-mentioned temperature T_k must be within this interval. Hence, it is easy to establish an interrelation between the calculated magnitudes.

261

$$q_\text{e} = \exp\left[-\frac{E_\text{AA}}{T_\text{E}}\right] \text{ and } q_\text{cr} = \exp\left[-\frac{E_\text{AA}}{T_\text{cr}}\right], \text{ i.e.}$$

$$q_\text{cr} = \exp\left[-\frac{E_\text{AA}}{T_\text{cr}}\right] = \exp\left[-\frac{\delta}{\gamma}\frac{E_\text{AA}}{T_\text{E}}\right] = q_\text{e}^{\delta/\gamma}.$$

If $\delta = 2$, then for the following values of γ, 1.2, 1.8, and 1.9, the following values of q_cr, 0.25, 0.42, and 0.43, are obtained. Thus, for the binary system — the 50% alloys melt/crystal system — in the temperature range of the supercooled melt $T_\text{cr} \leq T_\text{m} \leq$ $\leq T_\text{E}$ there occurs growth of a crystal surface with rough steps, while in the tempera-ture range $T_\text{k} < T_\text{m} \leq T_\text{E}$ it is supposed that there will grow a regular crystal. In the temperature range $T_\text{cr} \leq T_\text{m} \leq T_\text{k}$, completely irregular crystals will be formed due to the movement of rough steps on the interface of the melt/crystal phases. It should be noted that the regulating temperature for a crystal T_c, must be either higher than T_E or at least close to it when $T_\text{c} < T_\text{E}$, i.e. condition (6) remains valid. Otherwise a realization of kinetic phase transition is impossible.

4.2.4. CONCLUSIONS

A consideration of kinetic phase transition of binary crystals growing from the 50% alloys melt lead to the following results.

(a) Realization of kinetic phase transition is possible when the interface temperature T_m approaches a so-called kinetic phase transition temperature T_k.
(b) The value of this temperature T_k is determined on the one hand by specific prop-erties of the melt and on the other hand by the crystallization mechanism of the crystalline phase (model of a rough step).
(c) It seems that an irregularity of a binary crystal occurs when T_m appreciably de-viates from the equilibrium temperature T_E.
(d) When a critical transition temperature T_cr is observed it seems that there are two temperature ranges between T_cr and T_E in which either regular or irregular crys-talline phases will grow.

Finally, it should be noted that the consideration of the problem of irregularity of binary crystals growing from the 50% alloys melt had too approximate a character. Particularly, this applies to both the methods by means of which the values of the magnitudes q_e, q_k were obtained and the determination of the supercooled melt tem-perature ranges where, due to the movement of rough steps a crystalline phase grows.

4.2.5. APPENDIX

In specific calculations the authors have employed for the separation frequencies $\omega_{-\alpha\beta}^{(ik)}(k_1, k_2)$ the following *Boltzmann* expression

$$\omega_{-\alpha\beta}^{(ik)}(k_1, k_2) = \nu \exp\left[-\frac{E_{\alpha\beta}^{(ik)}(k_1, k_2)}{T}\right].$$

262

Here, v is a frequency factor which determines the thermal oscillation of particles in the crystalline phase, T is the surface temperature of the crystal, and $E_{\alpha\beta}^{(ik)}(k_1, k_2)$ is the interaction energy of particles α and β in respect to other particles than α which are nearest neighbors of the particles β in all three directions:

$$E_{\alpha\beta}^{(ik)}(k_1, k_2) = \begin{cases} E_{\alpha\beta}^{(ik)} + \sum_\delta E_{\beta\delta}^{(ki)} X_\delta^{(i)} & *) \\[2mm] E_{\alpha\beta}^{(ik)} - 2\sum_\delta E_{\beta\delta}^{(ki)} X_\delta^{(i)} & **) \\[2mm] E_{\alpha\beta}^{(ik)} + 3\sum_\delta E_{\beta\delta}^{(ki)} X_\delta^{(i)} & ***) \end{cases}$$

  * when $k_1 > 0, \ k_2 < 0$

  ** when $k_1 > 0, \ k_2 \leqq 0$ or $k_1 \leqq 0, \ k_2 < 0$

*** when $k_1 \leqq 0, \ k_2 \geqq 0$.

Here $E_{\alpha\beta}^{(ik)}$ etc. are the interaction energies of particles α and β belonging to the i- and the k-sublattices; these interaction energies are taken with inverse signs, $X_\delta^{(i)}$ is the probability of finding a particle δ ($\delta = A, B$) at a lattice point belonging to the i-sublattice ($i = I, II$).

4.2.6.   REFERENCES

[1] *Černov, A. A.:* ZHETF 58 (1967) 2090.
[2] *Burton, W., Cabrera, N., Frank, F.:* The elementary processes of crystal growth. Akademisdat, Moscow (1959).
[3] *Jackson, K. A., Ulman, I., Hunt, J. D.:* J. of Crystal Growth 1 (1967) 1.
[4] *Onsager, L.:* Phys. Rev. 65 (1944) 117.
[5] *Wannier, G. H.:* Rev. Mod. Phys. 17 (1945) 50.

5. CHEMOEPITAXIAL AND CHEMOENDOTAXIAL LAYER GROWTH ON METALS

5.1. NUCLEATION AND GROWTH DURING GAS-METAL REACTIONS

J. Oudar

École Nationale Supérieure de Chimie de Paris, France

5.1.1. INTRODUCTION

When a metallic surface comes into contact with a gas of appreciable chemical re-activity, the reaction occurring on the surface may be described in terms of several distinct steps. The first step is that of adsorption, which involves very strong chemical bonds in highly reactive systems. The termination of this step occurs when the surface becomes entirely covered by a mono-layer of atoms or molecules. If the pressure of the reactive gas is higher than the dissociation pressure of the product formed — *e.g.* oxide or sulphide in the cases of the reactant gases oxygen or sulphur, respectively — then the surface may be uniformly covered by the product, provided that it is not volatile. It was shown twenty years ago that the formation of a continuous layer involves a transitional process of nucleation and growth.

It is the objective of this section to illustrate the influence of the crystalline orientation of the metal substrate on each of the reaction steps in the light of most recent experimental results. In particular, it will be shown that epitaxy governing the relationship between the lattices of the metal and of the reaction product, may be just as important during the adsorption stage as during the formation of the product, provided that the pressure of the reactive gas is not too high, and that the temperature is not too low. Until recently, precise results were only available for the epitaxy of the bulk compound on a metal substrate. Most of the reported results refer to the case of oxidation. During the last ten years research has been concentrated on the initial adsorption step, and some results have also been obtained for the transition between the adsorption layer and the bulk product. Numerous investigations concerning the influence of the orientation of the metal will also be considered.

5.1.2. THE EPITAXY OF ADSORPTION LAYERS

The most important results concerning structure and epitaxy of adsorption layers have been obtained by Low Energy Electron Diffraction techniques (LEED). Significant results have also been obtained by High Energy Electron Diffraction methods (HEED) at grazing incidence.

5.1.2.1.　Sulphur on metals

The most extensive studies concern the adsorption of sulphur on different metals, as have been reported for copper by *Domange* [1], [2], for nickel by *Perdereau* [3], for iron by *Margot* [4], [5] and for gold by *Kostelitz* [6]. The investigators employed LEED techniques and radiochemical analysis of the weight of adsorbed sulphur. The latter information is often very important for the interpretation of LEED patterns.

The following scheme has been proposed for a metal surface exposed to hydrogen sulphide [2]. A relatively straightforward precursory state, resulting from the quasi-instantaneous dissociation of the hydrogen sulphide in contact with the metal surface is initially observed, in which the adsorbed atoms are localized on the sites of maximum coordination: coordination number 3 on the (111) face of face-centered cubic (fcc) metals, coordination number 4 on the (001) face of body centered cubic (bcc) and fcc metals, coordination number 5 on the (110) face of bcc metals.

In this first stage of reaction the minimum distances between occupied sites are always compatible with the size of the S^{2-} ion. This suggests that the ions are in this state of ionisation, and that the existence of repulsive forces between the adsorbed atoms, together with a certain surface mobility, assures their regular distribution over the surface. The surface topography of the substrate is a determining factor for the lattice structures of these adsorption states. As the reaction proceeds, the surface phase becomes unstable and changes to a true two-dimensional metal/sulphur compound by means of a *reconstruction* of the metal surface. Although there is a strict epitaxial relationship between the surface compound and the underlying metal, the adsorbed sulphur atoms do not necessarily occupy positions corresponding to sites of a maximum coordination number. In this surface phase, the lateral binding forces between sulphur atoms, transmitted via the metal atoms, seem to prevail over the binding forces with the substrate. When the surface is entirely covered by the two-dimensional adsorbed compound, the reaction may be considered to be complete, and this state of saturation exhibits a remarkable chemical inactivity, even for conditions under which sulphuration traditionally takes place.

Although the initial stage conforms to the traditional picture of a metal surface being composed of a number of reaction sites, the idea of a surface compound arose only recently as a result of a number of LEED experiments, particularly those of *Farnsworth* (1961), *Germer* (1962) and their co-workers.

The term *corrosive chemisorption* has recently been introduced in order to describe a phenomenon which seems to be quite general for highly reactive systems, and the formation of a reconstructed layer has been explained in terms of a number of hypotheses, based upon the extraction of metal atoms either from the edges of steps, or from terraces [7].

The adsorption layers formed at saturation on the metal faces of highest atomic density for copper, nickel, iron, and gold are depicted in Figs. 1, 2, 3, and 4, respectively. It should be noted that the results for gold were not obtained by means of LEED, but by diffraction of fast electrons at a metal surface previously treated by exposure to an atmosphere of a H_2S/H_2 mixture of appropriate composition [6]. Although only one domain is shown for each face, the symmetry requirements of the substrate

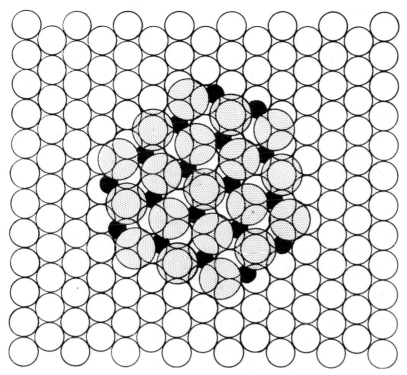

Fig. 1. Structure of a two-dimensional sulphide deposit formed by adsorbed sulphur atoms on a (111) surface of copper [1]. (The sulphur atoms are grey, the metal atoms black)

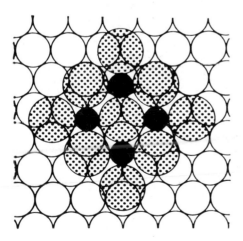

Fig. 2. Structure of a two-dimensional sulphide deposit formed by adsorbed sulphur atoms on a (111) surface of nickel [38]

necessitate to consider several domains generated from one another by rotation about an axis perpendicular to the surface.

Certain reservations should be considered concerning the proposed model, particularly with regard to the internal stoichiometry of the surface compound. Even for those

267

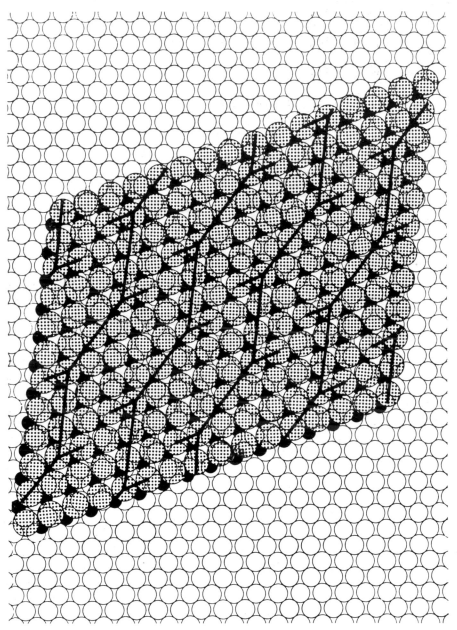

Fig. 3. Structure of a two-dimensional sulphide deposit formed by adsorbed sulphur atoms on a (110) surface of iron [4], [5]. The sulphur atoms in special positions in relation to the substrate (binary and ternary sites) are connected by solid lines

268

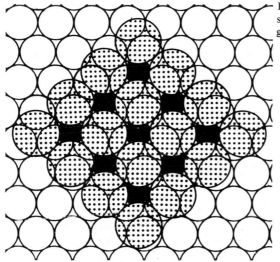

Fig. 4. Structure of a two-dimensional sulphide deposit on a (111) surface of gold [6]

cases for which the intensities of the diffracted beams have been carefully analysed, or where spectroscopy analyses have been carried out, it has been proved to be impossible to obtain precise information about the number and positions of metal atoms, and this problem still remains a difficult one. Consequently, the metal cations have been distributed rather arbitrarily, according to the assumptions about the stoichiometry of the surface compound. Other possibilities hit the experimental data. Then, for the compound formed on the (111) face, it is possible to consider a neutral layer of Cu_2S. For the (111) face this would be equivalent to putting one cation in each of the ternary sites of the sulphur lattice, rather than one cation in one site out of two. Taking the size of the Cu^+ ions into account, it is possible for all the cations to be in essentially the same plane on this face, where the atomic density is high. Similarly, one might have two layers on the (111) face, one of Ni^+ ions in contact with the metal, and one of S^{2-} ions, giving the stoichiometry of Ni_2S. *Auger spectroscopy* studies of the adsorption layers and the sulphides formed on copper and nickel seem to favor this last hypothesis [8].

It is possible to overcome the uncertainty concerning the stoichiometry of the surface compound if one only considers the sulphur lattice and its orientation in respect to the substrate. It is reasonable to assume that the metal atoms of the compound are in the interstices of the anionic lattice, as is generally supposed for bulk sulphides. According to the results obtained not only for the densest faces but also for other high density faces of copper, nickel, iron, and gold, the different two-dimensional metal–sulphur compounds in a very schematic way could be divided into two categories.

In the first stage the sulphur atoms tend to get arranged according to the symmetry of the substrate, or to assume a close-packed hexagonal arrangement (iron–sulphur and copper–sulphur compounds). In this case the metal–sulphur bond may be considered to possess a pronounced ionic character.

269

In the second stage the surface topography of the substrate does not exert a determining influence on the arrangement of sulphur atoms, and a pronounced tendency towards a square arrangement is observed. This applies even for such cases for which the symmetry of the face concerned differs markedly from the square arrangement (gold–sulphur and nickel–sulphur systems). It may be assumed in this last case that there are metal–sulphur bonds which are considerably stronger and more directed, providing the bonds with some covalent character.

These differences are in agreement with the accepted ionic character of Cu_2S and FeS on the one hand, and with the accepted covalent character of the Ni–S bonds in Ni_2S_2 on the other hand. Little is known about Au_2S.

This analogy between the two and three-dimensional sulphides is particularly clear in the case of copper. In this case the arrangement of sulphur atoms on the (111) face is identical with that of sulphur atoms in the densest plane of Cu_2S, and the S–S distances differ only slightly (3.90 Å and 3.96 Å, respectively). The same applies for the iron–sulphur system, where the two-dimensional sulphide corresponds to the basal plane (0001) of the hexagonal sulphide FeS. However, there are more pronounced deformations due to epitaxial stresses for the surface compound formed on iron. These stresses may be explained by the displacement of sulphur atoms in order to occupy certain favorable positions (binary or ternary sites). This case of deformation seems to be characteristic for two-dimensional iron–sulphur compounds (see below). We shall now consider more closely certain factors which may be important for the epitaxy of adsorption layers, emphasizing the copper–sulphur and iron–sulphur systems. It is important to realize that the conditions under which adsorption occurs in a low energy electron diffractometer may be far from those for equilibrium, particularly for a very reactive system in which the metal is in contact with a highly electronegative element such as oxygen or sulphur. Therefore it is necessary to consider whether the observed epitaxial relations really correspond to the minimum free energy state for the system.

This problem has been considered by *Werlen-Ruze* and *Oudar* [9] who studied adsorption layers formed under quasi-reversible conditions. In these experiments adsorption layers were formed between 500°C and 650°C on low index planes of copper exposed to H_2S/H_2 mixtures whose partial pressures were slightly lower than those at which the sulphide Cu_2S is formed. It is known that the surface is saturated with adsorbed sulphur under such conditions. As has been shown by radioactivity measurements, within the limits of experimental error (of the order of 2–3%) this state corresponds to a surface density of sulphur reached at saturation by irreversible incorporation of sulphur atoms. The experiment was carried out at room temperature using HEED techniques. Comparison of the results obtained for the arrangement of adsorbed sulphur atoms on the (111), (001), and (210) faces of copper, for the two different methods of formation, shows that the epitaxial relations are different for the two low index planes (111) and (001), but they are identical for the faces (110) and (210) which are "rougher" at the atomic scale. In Fig. 5 the unit lattice cell containing one sulphur atom is depicted for epitaxy on the substrate. Within low index planes the directions of high atomic density in the substrate are aligned with those in the sulphur compound if the latter has been formed under reversible conditions, but not otherwise.

270

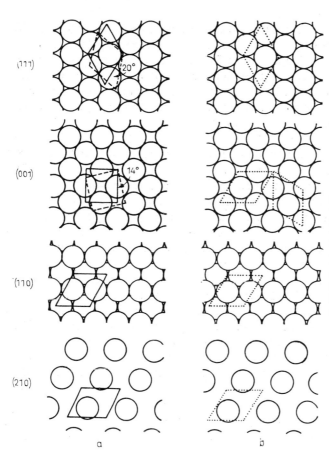

Fig. 5a. Unit lattice cell of sulphur in a two-dimensional sulphide deposit formed by adsorbed sulphur atoms on (111), (001), (110), and (210) surfaces of copper [9].

Solid lines characterize the atomic sites within the two-dimensional sulphide deposit formed under the influence of the H_2/H_2S mixture (reversible conditions).

Dashed lines characterize the atomic sites within the two-dimensional sulphide deposit formed under the influence of pure H_2S (irreversible conditions)

Fig. 5b. The unit lattice mesh of sulphur in a three-dimensional Cu_2S deposit at the interface on (111), (001), (110) and (210) faces of copper [9]

(111)

(001)

(110)

(210)

a   b

It is difficult to say why such differences arise. When the process proceeds under quasi-reversible conditions, it is reasonable to assume that there is a greater tendency towards the formation of most stable configurations for the adsorption layer, since an exchange with the gas phase allows relaxation to occur. Theoretical considerations show that the alignment of the directions of high density in the deposit and in the substrate generally leads to a minimization of interfacial energy [10]. When adsorption takes place irreversibly, the problem is to know whether those defects which are formed during the atomic rearrangement of the surface can be eliminated later. In particular, it is possible that the excess metal atoms may remain isolated as interstitial atoms, which may according to the circumstances migrate towards mono-atomic discontinuities. These defects induced by adsorption may play an important part in determining the final orientation of the surface compound. It has also been shown that the existence of mono-atomic steps, present in even the best possibly defined surfaces, may have a determining influence on the epitaxy of the surface compound. Thus, an increase in the population of steps, whose orientation is fixed by a displacement of 1 or 2 with respect to the high density planes, favors the preferential formation of one of the domains of the surface compound which had been observed [1]. A structural study of the adsorption of sulphur on the (001) plane of iron has been carried out by means of the LEED technique [4], [5]. As the surface concentration of sulphur rises, a succession of structures, characterized by coincidence of their lattice meshes (Fig. 6), is observed. The numbers of sulphur atoms in the centered rectangular coincidence mesh as determined by radio-active tracer are 12, 10, 8, and 6 for the C (22×2) R $45°$, C (18×2) R $45°$, C (14×2) R $45°$, and C (10×2) R $45°$ structures, respectively.

The coincidence lattices of the different structures have one characteristic in common: the parameter in the $[1\bar{1}]$ direction is identical with that of the metal substrate. This may be explained by the assumption that the crystallization of the surface compound starts at the mono-atomic steps on the metal surface. The model proposed for the C (10×2) R $45°$ structure, shown in Fig. 7, assumes that the smallest coincidence mesh is the oblique mesh. In order to conform to the symmetry of the diffraction pattern, which contains a 2^+ fold axis, another model has been proposed for the coincidence mesh, the centered rectangular lattice.

This model may be deduced from the preceding one by displacing those sulphur atoms which are free to move in the $[\bar{1}1]$ and $[1\bar{1}]$ directions so that they are as close as possible to a quaternary site in the substrate. Thus, a new structure results for the C (10×2) R $45°$ structure, as shown in Fig. 8a. It may be described in terms of two elementary unit patterns, a rectangle and pseudo-equilateral triangle (Fig. 8b).

In order to preserve the stoichiometry of FeS, one Fe^{2+} ion is introduced into every second triangle formed by the positions of sulphur atoms. This is equivalent to introducing a periodic displacement of the parallel rows by $\frac{1}{2}$ of a lattice parameter in the $[11]$ direction in the arrangement of S-atoms. This displacement, or shearing, may be considered to minimize the interfacial energy between the surface compound and the substrate by moving the sulphur atoms involved closer to the quaternary sites of the metal. There may also be a certain relaxation of the first surface layers of the metal, which are not of maximum compactness.

Now the change of the epitaxial relations of the surface compound due to an increasement of the surface density of sulphur atoms should be considered. For every two-dimensional nucleus with a length equal to the lattice parameter of the coincidence mesh, it is necessary according to the second model proposed to introduce two displacements, each of which gives rise to a row of triangular patterns. Since the coincidence lattice becomes progressively smaller passing from the C (22×2) R $45°$ structure through intermediate structures, to the C (10×2) R $45°$ structure, the proportion of triangular patterns to rectangular patterns increases accordingly. This leads to the conclusion that the observed epitaxial relations result from a compromise among three tendencies: the tendency of the substrate to impose its own square symmetry; the influence of the steps, which deform the square arrangement, producing a rectangular arrangement by imposing a constant parameter in one direction; and finally, the tendency of the surface compound to form the most close-packed arrangement for the sulphur atoms, i.e., a hexagonal arrangement. The latter tendency increases as the amount of adsorbed sulphur, that is, as the number of lateral bonds per unit surface area increases. This leads to an increasing proportion of pseudo-

272

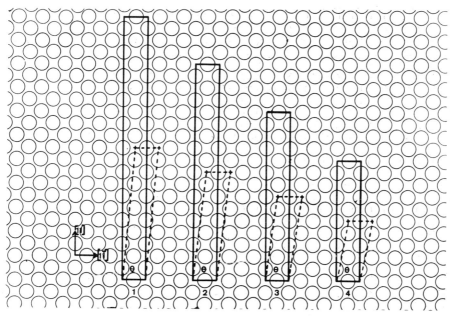

Fig. 6 Coincidence lattice meshes of different structures formed during adsorption of sulphur on a (001) surface of iron [4], [5]

The structures are:
1. C (22×2)R $45°$
2. C (18×2)R $45°$
3. C (14×2)R $45°$
4. C (10×2)R $45°$

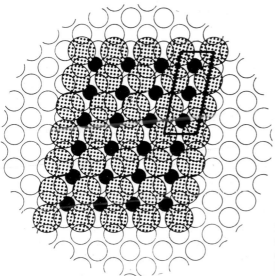

Fig. 7. Structure of a two-dimensional sulphide deposit formed by sulphur atoms adsorbed on the (001) face of iron [4], [5]. The structure is C (10×2)R $45°$

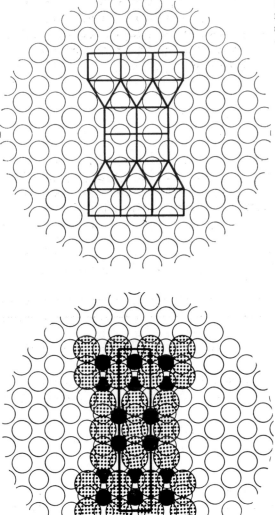

Fig. 8a. Relaxed structure of a two-dimensional sulphide deposit formed by sulphur atoms adsorbed on a (001) face of iron

Fig. 8b. Rectangular and triangular arrangements of the sulphur atoms in the previous structure

equilateral triangular patterns, or, in terms of the first model proposed, to a progressive rotation of the oblique coincidence lattice.

If the influence of the mono-atomic steps could be overcome, a different set of epitaxial relations would probably be found, with an arrangement of sulphur atoms which would most likely be still close to a hexagonal pattern. Uncertainty still exists in regard to the stoichiometry of the surface composition. It is not unreasonable to suppose that each structure may be characterized by its own stoichiometry. Such a hypothesis is supported by the analogy between the linear defects proposed for the adsorption layer and the planar defects proposed by *Wadsley* [11] to explain the non-stoichiometry of transition metal oxides. The remarkable influence of such structural defects on mono-atomic discontinuities may explain the different experimental results obtained under the same conditions of oxidation on the surface of whiskers of iron and on smooth surfaces prepared by traditional methods, which are apparently less perfect [12].

5.1.2.2.   Oxygen on metals

Several structural studies concerning the adsorption of oxygen on different metals, as Ni, [13]–[17], iron [12], [18], copper [19]–[22], and tungsten [23], [24] have been carried out by means of the LEED technique. Because of the inherent difficulties in the interpretation of diffraction intensities and because of the lack of precise data on the amount of adsorbed oxygen, the interpretation of the experimental results has led to a considerable controversy. Identical experimental results have beeen interpreted by different investigators in terms of two-dimensional and in terms of three-dimensional structures. An idea of the difficulties encountered in interpretation may be gained from the review by *Bauer* [25].

One of the most significant results is that of *May* and *Germer* [15] concerning the (110) face of nickel, where they have identified a surface compound with a lattice parameter very close to those of the oxide NiO (Fig. 9).

This compound, which has been shown to be only one atomic layer thick, has been termed a pseudo-oxide.

In other systems, complex diffraction patterns have sometimes been attributed to the formation of arranged adsorption layers. The case of iron in the presence of oxygen will be considered below.

5.1.2.3.   Carbon on metals

The hot carburation of tungsten in the presence of methane has been investigated by *Boudart* and *Ollis* [26]. These authors have detected the existence of two structures, (110) R (15×3) and (100) — (5×1), on the (110) and (100) faces, which might be interpreted as being due to the formation of a two-dimensional tungsten carbide W_2C (Figs. 10 and 11).

According to theoretical considerations by *Damai–Imelik* and *Alii* [27] a two-dimensional nickel carbide is formed by the decomposition of ethylene on the (100) and (111) faces of nickel.

5.1.3.   NUCLEATION AND GROWTH IN THREE-DIMENSIONAL COMPOUNDS

The important role of nucleation and growth during the initial stages of the formation of oxide layers is now well recognized.

After the first observations of oxidation processes on iron [28] and on copper [29],

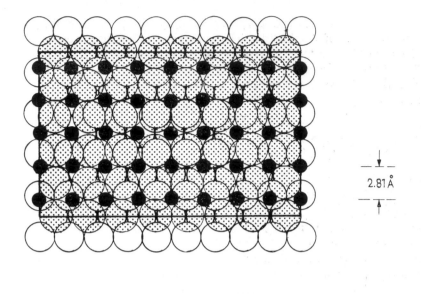

Fig. 9. Structure of the two-dimensional oxide (pseudo-oxide) formed by adsorption of oxygen on a (110) surface of nickel [16]. (The oxygen atoms are grey. The nickel atoms in the two-dimensional oxide are black)

Fig. 10. Structure of the two-dimensional carbide formed on a (110) surface of tungsten [26]. (Only the tungsten atoms of the carbide are visible)

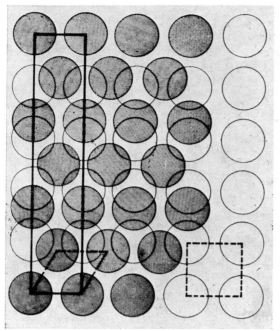

Fig. 11. Structure of the two-dimensional carbide formed on a (100) surface of tungsten [26]. (Only the tungsten atoms of the carbide are visible)

[30] at low oxygen pressures by *Benard* and *Bardolle*, this phenomenon has now been observed under very different conditions of oxidation. Thus, the presence of individual crystallites of NiO, with a mean diameter of the order of 30 Å, has been revealed by means of LEED techniques during the oxidation of nickel. The results which have been obtained for the growth rate of the oxide at room temperature on barium and magnesium may also be explained in terms of a mechanism of nucleation and growth. The nuclei of oxide are assumed to be not thicker than the equivalent of a few unit cells. The same phenomenon has also been detected at other gas–metal reactions carried out under controlled conditions, notably for sulphuration and halogenation.

The most complete systematic study is that of *Grønlund* [31] on copper exposed to low oxygen pressures. Numerous elements of this phenomenon have been enunciated by *Benard* in his book on oxidation [32] and at a recent colloquium [33]. At a given temperature, and under a known pressure of oxygen, the evolution of the process may be summarized as follows. There are three successive stages:

a) an incubation period, during which the surface remains specular as observed by means of optical microscope. This phase ends abruptly when a definite number of nuclei rapidly appears on the surface;

b) a period of lateral growth of the nuclei, which ends when the surface is covered by nuclei;

c) a period of uniform growth of the oxide layer on the entire surface.

Figure 12 shows the relationship between the incubation period, the period of nuclei growth, and the period of growth of the continuous film for copper at 550°C.

277

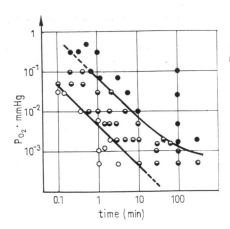

Fig. 12. Nucleation range of oxidation of copper at 550°C [31]. White points represent the incubation period, white-black points the nucleation period, and black points the growth of a continuous film

The influence of the metal structure, the temperature, and the pressure on the number of nuclei have been studied; in general, the length of the incubation period and the surface density of nuclei formed under specified conditions of temperature and pressure depend closely on the crystalline orientation of the metal surface. For a given orientation of the surface and a given pressure, the surface density of nuclei diminishes as the temperature increases. At a fixed temperature, an increase of the pressure leads to a corresponding increase of the surface density of nuclei. Thus, at low temperatures and high pressures, the number of nuclei becomes considerable, and the almost instantaneous formation of a continuous layer is observed. To a certain extent, this explains why the phenomenon has remained poorly understood for a long time. All these characteristics are found in numerous gas–metal systems.

Quite precise data are now available concerning the incubation period. As has been noted above, the initial stage of the reaction is the formation of an adsorption monolayer. During this period certain investigators have also noted the formation of a micro-crystalline film which is invisible under the optical microscope and which subsists during the growth of the nuclei. The formation of such a film is by no means a general phenomenon and may result from the oxidation of surface near layers of the cold-drawn metal as a consequence of poor preparation or from recrystallization of an initially present film. In this case nucleation would be a secondary effect. We consider the formation of such films to be accidental.

Under the influence of adsorption the micro-topography of the surface may undergo important modifications leading to the appearance of facetting.

In most cases the high index surface planes decompose into planes of high atomic density and complex recovery planes. It has been shown that the new profile is most often formed by surface diffusion. One of the systems that has been extensively studied is silver in the presence of oxygen [34]. The LEED method is quite useful for the investigation of facet formation. Particularly tungsten in the presence of oxygen has been studied from this point of view [35]–[37]. It may happen that a surface plane of high atomic density is unstable; thus a (111) surface of nickel decomposes in the presence of sulphur into a system of complex facets [3]. It is likely that this is due to the poor adaptability of the square symmetry of the structure of the surface compound to the hexagonal symmetry of the substrate. These modifications of the original

278

surface profile, which tend to occur during the incubation period may retard the formation of the oxide.

It is also necessary to take into account the solution of oxygen in the interior of the metal, because the formation of the oxide generally occurs when the metal is saturated with oxygen. The diffusion of oxygen proceeds by an interstitial mechanism of low activation energy; *e.g.* for copper diffusion should become appreciable, even at quite moderate temperatures of about 300°C. In some metals, such as titanium and zirconium, which are very avid for oxygen, the amount dissolved is very large and may appreciably modify the crystalline lattice of the metal by causing an expansion of the lattice. When the reaction takes place at low pressures, the solution of oxygen may considerably retard the oxide formation. In the case of sulphuration, however, the diffusion mechanism is different and only becomes effective when the mobility of vacancies in the metal is appreciable, *i.e.*, at temperatures of the order of $(2/3)T_f$, where T_f is the temperature of fusion of the metal [38].

Nevertheless, the solution of oxygen in the metal does not explain the very pronounced influence of the crystalline orientation of the surface upon the length of the incubation period. This applies particularly for metals with cubic lattices, where diffusion through the lattice may be considered essentially isotropic.

The differing lengths of the incubation period may partly be due to differences in the speed of formation of the adsorption layer according to the crystalline orientation. These differences may easily be demonstrated by means of LEED. However, experience reveals that when the saturation of adsorption is essentially complete the subsequent reactivity of the surface may nevertheless vary appreciably according to the arrangement of the planes constituting the surface. The problem of the very origin of the process of nucleation is discussed more extensively below (Section 5.1.5).

5.1.4.   EPITAXY OF THREE-DIMENSIONAL COMPOUNDS

The conditions of high temperature and low supersaturation required for the observation of nucleation and growth phenomena of surface compounds during gas–metal reactions are precisely those which favor epitaxy of these compounds. This particular aspect of the problem has been considered by numerous investigators. Among the more complete reviews which consider both, continuous and discontinuous films, are those of *Gwathmey* and *Lawless* [39], *Lawless* [40], *Bardolle* [41], and *Neuhaus* and *Gebhardt* [42].

Results have recently been obtained for the oxidation of nickel [17] and of tungsten [43], the sulphuration of nickel [44], copper [9], and of iron [45], [46], and the chlorination of copper [47], [48] and of nickel [49].

5.1.4.1.   General criteria

The very large number of experimental results enables one to formulate a number of criteria based on merely geometrical considerations concerning the two lattices in epitaxy. These criteria seem to be suitable for the majority of both, continuous and discontinuous films:

a) The lattice parameters of the deposit are characteristic of the product rather formed than modified in order to match the substrate lattice. The concept of pseudo-morphism seems to apply only in those rare cases where the two lattices have very similar structures at the interface.
b) The plane of the reaction product in the interface is a plane of high atomic density, although not necessarily that of highest density.
c) A parallelism of the directions of high atomic density in the metal and in the compound is observed.

The last criterion has initiated certain authors to consider the parametric disagreement along the directions, perpendicular to the directions of parallelism. The parallelism certainly plays a fundamental role in the epitaxial growth, as it enables the epitaxial growth to occur with a minimum displacement of the metal atoms.

It should be noted that the parallel directions do not necessarily correspond to the chains of highest atomic density, as has been suggested by certain authors. In some cases a less dense chain may take its part, particularly if there is only a small parametric disagreement along this chain. This has been clearly shown for the CuCl–Cu [47] and Cu_2S–Cu [9] systems.

It should be pointed out that an older criterion postulated by *Royer*, based on the existence of a simple or multiple mesh that was similar in form and dimension to the two lattices, is frequently not obeyed.

Figure 13 shows a representation of the epitaxy of FeO on a (001) surface of iron; it may be seen that there is a sort of continuity between the two lattices, in this case, the high density (110) plane of the iron becoming the high density (001) plane of the FeO.

The epitaxy of a compound may sometimes be described in terms of epitaxy on the surface planes of the metal facets. Thus, the epitaxy of the oxide WO_3 on the (100) face of tungsten may be explained if the facets of orientation [110], declined by 45° to the original surface, are considered as the interface planes. Furthermore, these

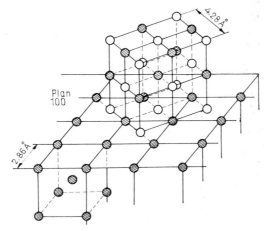

Fig. 13. Epitaxy of FeO on a (001) surface of iron [41]

facets appear when the oxide is evaporated by heating [43]. In the case of the sulphuration of copper, a disorientation of 1 or 2° in a high density plane is sufficient for the formation of facets following the high density plane on which preferential nucleation of the sulphur takes place.

5.1.4.2. The relationship between the morphology of the nuclei and their epitaxial character

Already since the first studies on nucleation there is no doubt about the strong influence of the orientation of the underlying substrate metal on the shape of the deposit crystals. Depending on the reaction conditions, the nuclei of the bulk compound may assume during growth either the shapes resulting from the competition between the

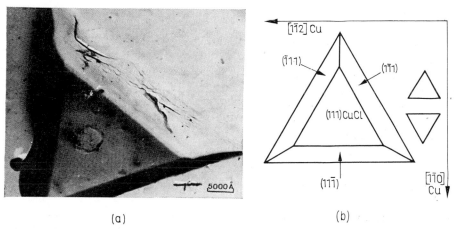

(a)   (b)

Fig. 14. Morphology of a CuCl nuclei formed on a (111) surface of copper [48]

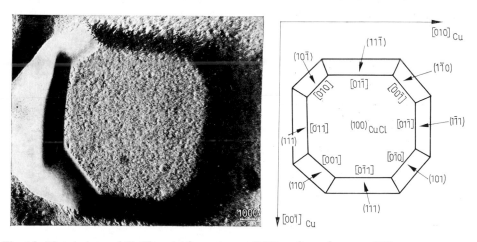

Fig. 15. Morphology of CuCl nuclei formed on a (100) surface of copper [48]

281

growth rates of the different faces, or the equilibrium forms. It seems that the equilibrium shape develops for the NiO nuclei with a diameter of a few tens of an Ångström during the oxidation of nickel under low pressure. The face of NiO which tends to form is the (100) face, which is characterized by the lowest number of missing neighbors, *i.e.*, the face of minimum free energy [14b].

For the CuCl nuclei formed on copper surfaces under low pressure of chlorine, the most developed facets also correspond to the planes of highest atomic density and the boundaries limiting these facets are the [110] directions of highest atomic density [47] (Figs. 14 and 15).

Interesting results concerning the interfacial energy may be obtained from an analysis of the morphology of the nuclei provided that equilibrium has really been reached. This particular aspect has been considered in some detail by *Sato* and *Shinozaki* [50].

5.1.4.3.   Models of the interface

One of the most elaborated models to describe the interface between two crystals of different lattice parameters has been established by *van der Merwe* [51]. This model is used for the special cases considered in this section, since it is not yet possible to apply the recently obtained results of *Woltersdorf* ([152] in Chapter 1 of this book) here as far as semi-choherent interfaces are concerned. According to this model, the misfit is reduced in quite large regions, providing relatively narrow bands of low agreement. The effect of this relaxation is the reduction of the energy of the bi-crystal system compared with that of the non-relaxed crystal. The difference in energy between the relaxed and the non-relaxed systems depends on:

a) The physical parameters of the systems, particularly the misfit between film and substrate;

b) The elastic properties of the two crystals. A further reduction in energy may be obtained by introducing into the film lateral compression or an extension which reduces the misfit between film and substrate. If the misfit is very small, the film may adopt the lattice parameters of the substrate and this is the case in pseudomorphism.

Consider two crystals A and B with lattice parameters a and b, respectively. It is assumed that these two crystals crystallize in similar cubic structures and that they are always oriented parallel to one another, the spacings a and b being different along one direction within the interface.

The misfit between the two crystals may be defined as

$$f = \frac{2(b-a)}{(b+a)} \tag{1}$$

which for small f reduces to

$$f = \frac{(b-a)}{a}. \tag{2}$$

282

For the dislocation model

$$P \cdot b = (P + 1)a \tag{3}$$

holds, where P is an integral and the number of dislocations per cm^2 and unit length is $(b - a)$ $(a \cdot b)$ (Fig. 16).

If δ_a and δ_b are the *Poisson coefficients* of A and B, respectively, μ_a and μ_b their shearing constants, and μ the shearing modulus at the interface of the two crystals, then c, β, and $\bar{\lambda}$ are defined as:

$$\frac{2}{c} = \frac{1}{a} + \frac{1}{b} \tag{4}$$

$$\frac{2}{\bar{\lambda}} = \frac{(1 - \delta_a)}{\mu} + \frac{(1 - \delta_b)}{\mu_b} \tag{5}$$

$$\beta = 2\mu \cdot \frac{(b - a)}{(b + a)\dfrac{\bar{\lambda}}{\mu}} \tag{6}$$

$\bar{\lambda}$ is thus the average of the elastic constants of the two crystals, and β, which expresses the difference $(b - a)$ between the lattice parameters of the two crystals, is proportional to the number of *interfacial dislocations. van der Merwe* has derived the law for the variation of *interfacial energy* α_i as a function of β. The curve obtained is depicted in Fig. 17.
There are two possible cases:

a) $\beta \gg \dfrac{1}{2}$.

The value of α_i is probably large (10^2 to 10^3 ergs/cm^{-2}) and α_i reduces to $\mu_c/4\pi^2$ and becomes independent of the number of interfacial dislocations. It may be shown that if the crystal is disoriented relative to the substrate, this is equivalent to the intro-

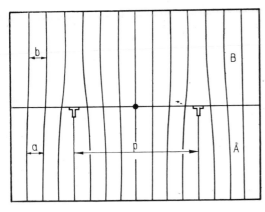

Fig. 16. The dislocation interface model [51]

283

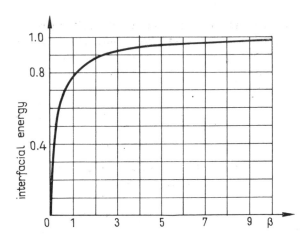

Fig. 17. The interfacial energy, α_i, as a function of β [51]

duction of additional interfacial dislocations. Since α_i is practically independent of the number of dislocations under the above conditions, all orientations yield practically the same value of α_i, and there is no epitaxy.

b) $\beta \ll \dfrac{1}{2}$.

Here α_i is strongly dependent upon the number of dislocations, a condition which favors epitaxy, since any disorientation introduces new interfacial dislocations. Thus, a necessary condition for epitaxy is:

$$\frac{2(b-a)}{(b+a)} \ll \frac{\mu}{2\pi\bar{\lambda}} \approx 0.1 \frac{\mu}{\bar{\lambda}}. \tag{7}$$

The higher the adherence between the two is (μ high) and the lower the elastic deformability of each of them is (μ_a and μ_b are assumed to be small), the better is condition (7) satisfied. Under these conditions the value of the misfit, $\dfrac{2(b-a)}{(b-a)}$, may be of the order of 0.2–0.5.

As has been noted by *Cabrera* [52], the above condition is necessary but not sufficient for epitaxial growth. It is also necessary that the growth rate is not too high, *i.e.*, that the supersaturating power of the atmosphere is not too high. According to classical nucleation theory the influence of this factor on the number N of nuclei being formed could be described explicitly:

$$N = N_0 \exp\left[\frac{-4\pi\alpha^2\Omega^2(\alpha - \alpha' + \alpha_i)}{kT(\Delta\mu)^2}\right], \tag{8}$$

where α is the surface tension (interfacial energy) between nucleus and atmosphere, α' is the surface tension between substrate and atmosphere, α_i is the surface tension between substrate and nucleus, Ω is the molecular volume in the nucleus, $\Delta\mu$ is the difference between the chemical potentials of the vapor and the nucleus.

284

This expression illustrates the well-known fact that the number of nuclei increases very rapidly as $\Delta\mu$ increases. It should be noted that the validity of such an expression has been questioned for the case of nuclei of very small dimensions. If condition (7) is satisfied, then the influence of a disorientation Θ on the interfacial energy may be written as

$$\alpha_i = \alpha + \left(\frac{\mu}{\pi}\right) c \,|\, \Theta \,| \tag{9}$$

where c is given by eq. (4) and α by (8), provided that Θ is small. The number $N(\Theta)$ of nuclei may be expressed in terms of Θ by substituting (9) into (8):

$$\frac{N(\Theta)}{N(\Theta_0)} = \exp\left[-\frac{|\Theta|}{|\Theta_0|}\right] \tag{10}$$

where

$$\Theta_0 = \frac{kT(\Delta\mu^2)}{4\alpha^2 \Omega^2 \mu c}. \tag{11}$$

Taking the usual values for the constants and expressing $\Delta\mu$ in [eV] yields $\Theta_0 \cong 0.1$ $(\Delta\mu)^2$ for $\Delta\mu^2 = 0.3\,\mathrm{eV}$ and $\Theta_0 \cong 10^{-2}$.

Two limiting cases may be considered:

a) $\Delta\mu$ is large, growth is rapid. There is initially a large number of nuclei of arbitrary disorientation and Θ_0 is large.

b) $\Delta\mu$ is small, growth is slow. There is initially a small number of nuclei, most of which are well oriented and Θ_0 is small.

Numerous experiments confirm that the supersaturating power of the atmosphere is important in deciding whether epitaxial nuclei are found or not.

5.1.4.4. Deformations in nuclei and thin films

The above theoretical considerations are based on the assumption that the lattice parameters of the two lattices remain unaltered. Deformations may occur, diminishing the number of misfit dislocations. For such a case Cabrera [52] proposed to apply the above model. If a_0 and b_0 are the lattice parameters of the two adjacent crystals without deformation the true lattice parameters may be expressed as

$$a = a_0(1 + \varepsilon_a) \tag{12a}$$

and

$$b = b_0(1 + \varepsilon_b). \tag{12b}$$

It can be shown that when condition (7) is fulfilled, the interfacial energy diminishes linearly with ε_a and ε_b according to the expression

$$\alpha_i = \alpha_{i_0} - \frac{\lambda}{\mu}(A\varepsilon_a + B\varepsilon_b)\ldots \tag{13}$$

where α_{i_0} is the value of α_i when there are no deformations and with the abbreviations

$$A = \frac{4a_0 b_0^2 (3a_0 - b_0)}{(a_0 + b_0)^3} \tag{14a}$$

and

$$B = \frac{4a_0^2 b_0 (3b_0 - a_0)}{(a_0 + b_0)^3}. \tag{14b}$$

These deformations give rise to a contraction of the nucleus and an expansion of the substrate in the interfacial plane and a tension in the substrate area around the nucleus. These changes are associated with energies that may be calculated on the assumption that the deformations are elastic.

The deformations corresponding to the minimum total energy may be expressed in terms of the radius R of the nucleus, which is assumed to be hemispherical. This leads to the following results:

$$\varepsilon_a = \frac{3A\lambda}{16R\lambda_a} \tag{15a}$$

and

$$\varepsilon_a = \frac{3B\lambda}{4R\lambda_b}, \tag{15b}$$

where $\lambda_a = \frac{\mu_a}{(1 - \delta_a)}$ $\qquad$ (16a)

and $\lambda_b = \frac{\mu_b}{(1 - \delta_b)}.$ $\qquad$ (16b)

Thus ε_a and ε_b depend upon the size of the nucleus. When R increases, ε_a and ε_b decrease, and the number of interfacial dislocations rises until the film is uniform and continuous corresponding to $\varepsilon_b = 0$.

It may be shown that under these conditions ε_b is given by

$$\varepsilon_b = \frac{1}{4(1 + \delta_b)} \frac{B}{\mu} \frac{\lambda}{\lambda_b}. \tag{17}$$

The contraction within the interfacial plane is compensated by an expansion along the normal to the surface which keeps the density of the oxide essentially constant.

286

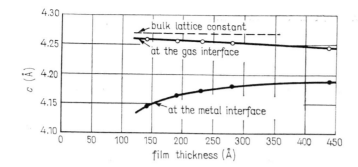

Fig. 18. Variation of the lattice parameter of Cu_2O as a function of the thickness of a film formed on a [110] surface of copper [54]. [Å] is the lattice parameter of the oxide in the direction parallel to the [110] direction of the copper substrate

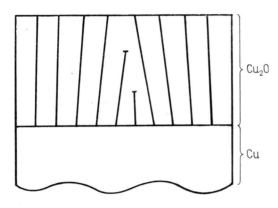

Fig. 19. Mosaic structure of a Cu_2O deposit formed on a (110) surface of copper [54]

It has been shown above that the adsorption layers may suffer considerable deformations as a result of their high plasticity. The substrate is expected to have less influence upon the lattice parameters of thicker deposits.

Observations by means of electron microscopy of *Hart* and *Maurin* [53] on the oxidation of aluminium are in agreement with the above theoretical predictions. These authors have shown that Al_2O_3 nuclei with a diameter of less than about 200 Å are entirely free from dislocations. This total accommodation of the substrate requires deformations of the order of 2%. For larger diameters, the existence of interfacial dislocations is evidenced by the presence of *Moiré fringes*.

The most significant results for thin films have been obtained by *Cathcart et al.* [54], who investigated the development of the lattice parameter from matching the oxide–metal interface to matching the oxide–gas interface by means of X-rays for the case of Cu_2O formed on the (110) surface of copper. The oxide film, between 150 Å and 500 Å thick, is laterally compressed at the metal–oxide interface, and retains its normal lattice parameter at the oxide–gas interface (Fig. 18).

Compression in a direction parallel to the interfacial plane is accompanied by an expansion in a direction perpendicular to it so that the lattice volume remains constant. In order to take into account the deformations a model based on the existence of a lattice with edge dislocations in the oxide has been proposed (Fig. 19). This results in an oxide which, although exhibiting an almost unique orientation, is not mono-crystalline, but has a mosaic structure. The influence of boundaries between

287

crystallites with low orientation differences on the rate of growth of the oxide will be discussed below.

In the case of Fe_3Si formed on iron, considerable variations of the lattice parameters have been observed [55]. These are due partly to variable departures from the stoichiometry going from one interface to the other, and partly to tensions parallel to the surface accompanied by a compression perpendicular to the surface. The deformations resulting from these stresses initiate a reduction of the misfit in the interfacial region, and may be understood if it is realized that the lattice parameter of cubic Fe_3Si (5.658 Å) is slightly less than twice that of iron ($2.866 \cdot 2 = 5.733$ Å). As in the preceding cases the stresses decrease with an increasing distance from the iron–silicon interface.

5.1.5. THE TRANSITION FROM TWO-DIMENSIONAL COMPOUNDS TO THREE-DIMENSIONAL COMPOUNDS

Various explanations have been developed in order to explain the origin of the phenomena of nucleation of bulk compounds at the end of the adsorption step.

In particular, it has been suggested that the localization of nuclei is controlled by the presence of structural defects or impurities which act as nucleation sites. Such a hypothesis has been proposed for the oxidation at room temperature of metal films of barium or magnesium deposited under vacuum. Heterogeneous nucleation along the polygonization walls is also observed when copper containing a large number of defects or an appreciable amount of impurities is sulphurated under controlled conditions. Nevertheless, when a reaction occurs at high temperature on pure, perfectly crystalline metals, this effect of the defects is less marked. In this case the surface density of nuclei is only a function of temperature and of the oxidizing power of the atmosphere. Indeed, when there is a considerable supersaturation, it is possible to obtain many more nuclei than dislocations. It is therefore not conceivable that the localization of nuclei is predetermined, and the nucleation is evidently homogeneous [32]. Such a process may be compared with the model proposed by *Stranski* and *Krastanov* in 1938 [56], which is based on thermodynamic considerations of the surface energies and the energy of the metal–compound interface [57].

At temperatures at which the solubility is appreciable, the formation of a stable three-dimensional nucleus is essentially not different from the precipitation of a new phase in the interior of the metal lattice. As soon as the nucleus forms the adsorption layer becomes unstable with respect to the oxide, and this results in a continuous transport of matter towards this growth center. This results in a lack of oxygen around each nucleus preventing the formation of new growth centers, which explains the fact that the number of such centers remains approximately constant during their growth, taking place essentially by surface diffusion. The nature of the diffusing species is still speculative. Nevertheless, according to the mixed nature of the adsorption layer, which in a sense acts as a nutritive medium, it is possible to regard the diffusing species as active chemical compounds of the two reactants. Recent observations by emission microscopy have revealed the existence of such compounds, molecules formed by a nucleus of three or four metal atoms around one oxygen atom [58].

The influence of orientation upon the number of growth centers is still poorly understood. This influence is illustrated by the results of *Bardolle* for iron-oxide at 850°C under low oxygen pressures (Fig. 20). A relationship between the density of nuclei and the diffusion rate of the diffusing species, itself influenced by the crystalline orientation of the substrate, has been postulated [59]. It is however not possible to separate from this relationship a contingent correlation between the metal-oxide interfacial energy and the surface density of nuclei since such a correlation is predicted by classical heterogeneous nucleation theory on the basis of thermodynamic considerations (see above).

A close relationship between the structure of the adsorption layer and the density of nuclei has been observed for the copper–sulphur system: With an increasing deformation of the adsorption layer due to epitaxial stresses the incubation period decreases and the density of nuclei increases [9]. It has also been noted that if the metal has been kept for some time in an atmosphere which is mildly reducing in respect to the formation of Cu_2S, a better crystallization of the adsorption layer is achieved and time is given for facets to form. In this case the incubation period increases considerably when the metal is transferred to a sulphurating atmosphere at the same temperature.

Similarly, if a (100) surface of tungsten is preheated to 1 200 K during an exposure to a vacuum of 10^{-4} torr or better, which causes facets to form, then the exposure necessary for the formation of WO_3 nuclei at 990 K increases considerably [43]. These examples underline the importance of the effects at the beginning of an experiment on the nucleation process.

Another important aspect with regard to epitaxy is the way in which the transition between the lattice of the adsorption compound and the bulk compound takes place. One of the most important studies of this aspect concerns the copper–sulphur system, for which the arrangements of sulphur atoms within the chemisorbed layer and of sulphide in the interface have been compared for the principal low index planes [9]. This comparison has been restricted to the adsorption structure formed under conditions similar to those for the formation of the sulphide Cu_2S. It is moreover this structure which subsists between the nuclei of Cu_2S during their growth. At the (111) surface of copper the lattice of the reversibly chemisorbed sulphur has the same hexa-

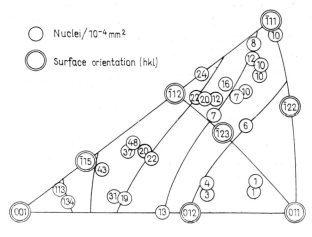

Fig. 20. Anisotropy of the density of nuclei on iron [32]. (850°C, $P_{O_2} = 10^{-3}$ torr)

gonal symmetry and the same orientation as the cuprous sulphide lattice. The small difference between the lattice parameters only causes a variation of the surface area of the elementary lattice mesh of -1.3%. When the chemisorbed state passes to the sulphurated state on the (001) surface of copper, a change of symmetry occurs. The chemisorption lattice is square, whilst that of Cu_2S is hexagonal. However, one of the two directions of high sulphur ion density in the square chemisorption lattice retains the orientation of the Cu_2S lattice. The two equivalent positions resulting from the fourfold symmetry of the (001) face of copper are depicted in Fig. 4. The difference in surface area of the elementary mesh is of the order of -3.4%. On the (110) surface of copper, the chemisorption lattice is pseudo-hexagonal. The direction corresponding to the [110] direction in copper (the dense row) is conserved in the orientation of the hexagonal lattice of Cu_2S. The small differences of the lattice parameters lead to a change in surface area of the unit cell of the lattices of $+3\%$. On the (210) surface of copper the chemisorption lattice is pseudo-hexagonal. The direction parallel to the [121] direction of copper is conserved in the orientation of the hexagonal lattice of Cu_2S. The differences of the lattice parameters are a little larger in this last case, but, nevertheless, give rise to a variation of the surface area of the elementary lattice mesh of only 8%. In conclusion, it may be stated that, except for the (001) face, the three-dimensional sulphur lattice is simply the continuation of the two-dimensional sulphur lattice without disorientation or substantial deformation. The same conclusion may be drawn for the (001) face, provided it is assumed that the hexagonal close-packed arrangement is transformed to a square arrangement, which conserves a direction of high sulphide ion density during cooling.

This continuity between the two-dimensional and the three-dimensional lattices is also observed on the (110) surface of nickel in the presence of oxygen [16]. The two-dimensional oxide has lattice parameters almost equal to those of the bulk NiO, and an identical positioning (Fig. 9). Similarly, on the (001) surface of iron, the rearranged $C(2 \times 2)$ structure identified by means of LEED techniques may be considered as an initial epitaxial layer of FeO compressed by 6% along the two principal directions. The copper–oxygen system has been investigated with some thoroughness by *Lawless* [40] and *Rhead* [59] using the LEED method. These authors observe the formation of a continuous epitaxial film about 10 Å thick on the (111) and (001) faces when oxidized at high temperatures and low pressures. By analogy with the copper–sulphur system, it is possible to conclude that this layer formed during the incubation period in fact corresponds to the rearranged adsorption layer. The orientation of this film is identical with that of the epitaxial Cu_2O nuclei, although there is a slight compression (Table 1):

Table 1

Surface planes	Determined direction	Observed value of lattice parameter	Theoretical value of lattice parameter	% compression
(111)	$2\overline{2}0$ Cu_2O	1.491 ± 0.004	1.510	$1.3\% \pm 0.2\%$
	$\overline{2}24$ Cu_2O	0.863 ± 0.001	0.8715	$1.1\% \pm 0.1\%$
(100)	022 Cu_2O	1.504 ± 0.004	1.510	$0.4\% \pm 0.3\%$

The study of the interaction of oxygen with the (110) face of a whisker of iron, carried out by *Molière* and *Portele* [12] utilizing LEED techniques has revealed a new phenomenon characterized by the existence of a transition layer between the lattice of the metal and that of the oxide FeO. This phenomenon may be observed when a thin layer of epitaxial FeO is heated to 270°C. A complex diffraction pattern is thus obtained which may be explained by the superposition of three lattices (the iron substrate, a layer of FeO adapted to the iron lattice, and a layer of normal FeO) (Fig. 21). It is conceivable that the adjustment of the deformed FeO lattice to the iron lattice is possible because the difference between the two lattice parameters is very small in the [001] direction in iron. A subsequent heating at 300°C, followed by a short annealing period at 450°C causes a further modification: the intermediate FeO layer becomes partially adapted to the substrate by a shearing deformation, so that its lattice parameter coincides with that of iron in the [001] direction (Fig. 22). The regular FeO structure appears in several domains rotated by 5.3° with respect to the substrate, which corresponds to a rotation by the angle of shear. One axis of the deformed structure is parallel to the [001] direction in iron, whilst the other is parallel to the [110] direction of the regular FeO layer. In this way the transition layer partially adapts to each of the lattices with which it is in contact. The existence of very clear multiple diffraction phenomena between the deformed and the undeformed layers of FeO indicates that the thickness of the layers is very small, of the

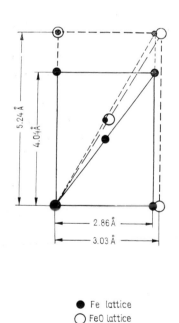

● Fe lattice
○ FeO lattice
● FeO$_{def}$ lattice

Fig. 21. Geometric relations between the Fe and the FeO lattices [12]

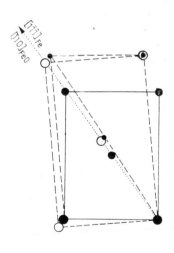

● Fe lattice
○ FeO lattice
● FeO$_{def}$ lattice

Fig. 22. Geometric relations between the Fe and the FeO lattices [12]

order of a few atomic layers, or even just one. The existence of transition layers during the formation of epitaxial deposits has been observed in other systems. This particular aspect of epitaxy has been reviewed very recently by *Mayer* [60].

5.1.6.   INFLUENCE OF THE CRYSTALLINE ORIENTATION
           ON THE RATE OF OXIDE GROWTH

5.1.6.1.   **Growth of the oxide nuclei**

Several theoretical studies have been accomplished concerning the rate of oxide nuclei growth.

One of the most comprehensive ones is that of *Orr* [61] which applies to the case of oxide nuclei whose thickness remains constant during growth. Three surface phenomena have been considered which control the growth rate of the oxide nuclei:

a) adsorption of the gas atoms;
b) surface diffusion;
c) capture of the diffusing particles by the nuclei.

On the basis of this model the sticking coefficient S may be written as:

$$S = A\Theta^n, \tag{18}$$

where Θ is the fraction of surface covered, n is a constant less than unity, determined by the geometry of the nuclei of the oxide, and A is a constant determined by the distance between the nuclei and the boundary of the sink zone surrounding each nucleus within which every particle that hits the surface is incorporated in the oxide. This model agrees quite well with the results obtained for the oxidation of evaporated films of magnesium between 86 K and 300 K, which follows a law of the form

$$S = A\Theta^{1/2}, \tag{19}$$

where S is independent of the pressure. *Orr* concluded from the value of A obtained experimentally that the process of capture was determinant, and that surface diffusion could not be a rate controlling process. It should be noted that in these experiments the number of nuclei was particularly high, namely of the order of $10^{11}\,\mathrm{cm}^{-2}$. *Grønlund* and *Benard* [32] were the first to note the importance of surface diffusion in the case of high temperature and low oxygen pressure; under these conditions a considerably smaller number of nuclei but with a greater thickness is formed. *Rhead* [59] has proposed a growth model with surface diffusion as the rate controlling process for nuclei of constant thickness and circular form. This results in a concentration gradient of diffusing atoms around each particle. If the concentration of atoms is C within the particle, C_s close to the particle, and C_∞ at a large distance from the particle, and if $C - C_s \gg C_\infty - C_s$ then the radius R of the particle may be expressed as

$$R = 2\lambda (D_s t)^{1/2} \tag{20}$$

292

and if A is the area of the particle:

$$\frac{dA}{dt} = 4\mu \lambda^2 D_s,\tag{21}$$

where D_s is the coefficient of surface diffusion, t is the time, and λ is given by:

$$\lambda^2 \ln(e^{C'}\lambda^2) + (C_\infty - C_s)(C - C_s) = 0\tag{22}$$

with *Euler's constants* $C' = 0.5772$.

This expression accounts qualitatively for the experimental results obtained during the oxidation of copper, for which the reaction rate is approximately constant. *Bouillon* and *Jardinier-Offergeld* [62] have proposed two hypotheses based upon the above model and using their experimental data obtained between 600°C and 800°C under a pressure of $5 \cdot 10^{-5}$ torr. According to the first hypothesis it is accepted that C_s is very small compared with C_∞, the latter corresponding to one monolayer. Under these conditions, if $C = 10^3$ atomic layers

$$\frac{(C_\infty - C_s)}{(C - C_s)} = 10^{-3}$$

holds and D_s may be calculated.

Following the second hypothesis the substitution of an experimentally determined value for D_s allows the calculation of C_s. The values thus obtained for C_s are very close to unity. This second interpretation seems to be the more reasonable one. Similarly, the value obtained for the activation energy for the growth of a particle (13 to 14 kcal mole^{-1} on the (001) face) is in satisfactory agreement with that obtained for surface diffusion (18 kcal mole^{-1}).

5.1.6.2. Growth of thin films

Recent studies have revealed the existence of numerous defects, related to the kind of epitaxial relations in thin films. These defects may appreciably affect the growth rate of the film, because they represent preferential paths for diffusion. The most significant results are those of *Cathcart* and his co-workers on the oxidation of copper and nickel [54], [63].

The epitaxial relations between copper and cuprite, Cu_2O, which both crystallize according to the cubic system, are given in Table 2.

At (001) and (111) surfaces of copper, there are several directions in which the oxide may preserve the symmetry of the plane concerned. Thus there are four orientations on the (001) face which are 2 to 2 in the twinned position, and which may be generated by rotations of 90° about an axis perpendicular to the surface; on the (111) face, there are two orientations in the twinned position. There is however one orientation for the oxide on the (110) and (113) faces.

Examination of the results for oxidation at 250°C and at an oxygen pressure of one

Table 2. Epitaxial relations between copper and cuprite, Cu_2O

Parallel planes		Parallel axes	
Metal	Oxide	Metal	Oxide
(001)	(111)	[1$\bar{1}$0]	[1$\bar{1}$0]
		[1$\bar{1}$0]	[$\bar{1}$10]
		[110]	[1$\bar{1}$0]
		[110]	[$\bar{1}$10]
(110)	(110)	[$\bar{1}$10]	[1$\bar{1}$0]
(111)	(111)	[1$\bar{1}$0]	[1$\bar{1}$0]
		[1$\bar{1}$0]	[$\bar{1}$10]
(113)	(110)	[1$\bar{1}$0]	[1$\bar{1}$0]

atmosphere shows that the two planes which are most rapidly oxidized are those for which the possibility of several orientations of the oxide exists, and that the (001) face is oxidized more rapidly than the (111) face (Fig. 23) [64]. From this it may be concluded that there is a close correlation between a high rate of oxidation and the presence of paths favoring rapid diffusion, such as the boundaries of incoherent twins along the (001) and (111) faces. These boundaries are the direct result of a multiple positioning of the oxide nuclei which precedes the formation of the continuous film. The defects present in the film at the (110) and (311) faces are less favorable to diffusion being low angle boundaries related to the mosaic structure of the oxide revealed by X-ray studies (Fig. 19). Thus the lower rate of oxidation on the (311) face and a smaller disorientation between the blocks of the mosaic appear jointly. It is significant

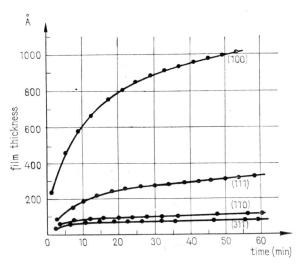

Fig. 23. Anisotropy of oxide growth on a copper surface at 250°C and $P_{O_2} = 1$ atm [64]

that when the dispersion of the mosaic is increased by the incorporation of CO_2 in the oxide, the rate of oxidation increases as well.

The same conclusions may be drawn concerning the oxidation of nickel, for which the structure of the thin NiO film formed has been studied in regard to the crystalline orientation of the metal. Except for the formation of a polycrystalline film on the (110) face, the NiO–Ni system exhibits the same epitaxial relations as Cu_2O–Cu. The result of this is that the NiO film formed on the (311) face is a pseudo-monocrystal characterized by a mosaic structure. The grain boundaries between the mosaic blocks are boundaries of small disorientation. The films formed on the (111) and (100) faces are composed of small crystallizes which exhibit two or four orientations, resulting in the existence of incoherent boundaries of twins. Finally, the oxide formed on the (110) face is essentially polycrystalline and this contains a high proportion of very distorted boundaries. Corresponding to a diminishing rate of oxidation, the crystallographic faces are (110), (100), (111) and (311). As in the case of the oxidation of copper, there is a strong correlation between the rate of oxidation and the existence of paths of rapid diffusion in the oxide. All these observations emphasize the determinant role of structural differences for growth rates of thin films during a chemical reaction. Very few theories take into account this important factor.

5.1.7.   REFERENCES

[1] *Domange, J. L., Oudar, J.:* Surface Science 11 (1968) 124.
[2] *Domange, J. L., Oudar, J., Benard, J.:* in: "Molecular Process on Solid Surface" (Eds.: E. Drauglis, R. D. Gretz, R. I. Jaffee). McGraw-Hill Book Co.
[3] *Perdereau, M., Oudar, J.:* Surface Science 20 (1970) 80.
[4] *Oudar, J., Margot, E.:* in: "Structure et propriétés des surfaces des solides", Colloque CNRS N° 187, Edition du Centre National de la Recherche Scientifique (1970) p. 123.
[5] *Oudar, J.:* Bull. Soc. fr. mineral. cristallog. 94 (1971) 225.
[6] *Kostelitz, M., Oudar, J.:* Surface Science 27 (1971) 176.
[7] *May, J. W.:* Surface Science 18 (1969) 431.
[8] *Perdereau, M.:* C. R. Acad. Sci. Paris (1972).
[9] *Werlen-Ruze, B., Oudar, J.:* Journal of Crystal Growth 9 (1971) 47.
[10] *Bettman, M.:* in: "Single-Crystal Films" (Eds.: M. H. Francombe and H. Sato), Pergamon Press (1964) 177.
[11] *Wadsley, A. D., Andersson, S.:* in: "Perspectives in structural chemistry" Vol. 1 (Eds.: J. D. Dunitz and J. A. Ibers) Wiley and Sons, New York—London.
[12] *Molière, K., Portele, F.:* in: "The structure and chemistry of solids surface" (Ed.: G. A. Somorjay) Wiley and Sons, New York (1968) 69—1.
[13] *Farnsworth, H. E., Madden, H. H. Jr.:* J. Appl. Phys. 32 (1963) 61.
[14] *Farnsworth, H. E.:* Appl. phys. letters 2 (1963) 199.
[15a] *MacRae, A. U.:* Surface Science 1 (1964) 319.
[15b] *MacRae, A. U.:* Science 139 (1963) 379.
[16] *May, J. W., Germer, L. H.:* Surface Science 11 (1968) 443.
[17] *Germon, L. B., Lawless, K. R.:* in: "Structure et propriété des surfaces des solides" Colloque CNRS N° 187, Edition du Centre National de la Recherche Scientifique (1970) p. 61.
[18a] *Pignocco, A. J., Pellissier, G. E.:* J. Electrochemical Soc. 112 (1964) 1188.
[18b] *Pignocco, A. J., Pellissier, G. E.:* Surface Science 7 (1967) 261.
[19] *Ertl, G.:* Surface Science 6 (1967) 208.
[20] *Trepte, L., Menzel-Kopp, C., Menzel, E.:* Surface Science 8 (1967) 223.
[21] *Simmons, G. W., Mitchell, D. F., Lawless, K. R.:* Surface Science 8 (1967) 130.

[22] *Grønlund, F., Hojlund-Nielsen, P. E.:* Surface Science (to be published).
[23] *Germer, L. H., May, J. W.:* Surface Science 4 (1966) 452.
[24] *Tracy, J. C., Blakely, J. M.:* Surface Science 15 (2969) 257.
[25] *Bauer, E.:* Bull. Soc. Fr. Miner. Cristallog. 94 (1971) 204.
[26] *Ollis, D. F., Boudart, M.:* Surface Science 23 (1970) 320.
[27] *Bertolini, J. C., Dalmai-Imelik, G.:* in: "Structure et propriétés des surfaces des solides", Colloque CNRS N° 187, Edition du Centre National de la Recherche Scientifique (1970) p. 135.
[28] *Bardolle, J., Bernard, J.:* C. R. Acad. Sci. 232 (1951) 231.
[29] *Menzel, E., Stossel, W.:* Naturwiss. 41 (1954) 302.
[30] *Grønlund, F., Benard, J.:* C. R. Acad. Sci. 240 (1955) 624.
[31] *Grønlund, F.:* J. Chem. phys. 53 (1956) 660.
[32] *Benard, J.:* "L'oxydation des métaux" Vol. 1. Gauthier Villars, Paris (1962/64).
[33] *N. N.,* "Processus de moléculation dans les réactions des gaz sur les métaux". Colloque CNRS N° 122, Paris (1963).
[34] *Moore, A. J. W.:* "Metal Surface". American Society for Metals, (1963) p. 155.
[35] *Taylor, N. J.:* Surface Science 2 (1964) 544.
[36] *Germer, L. H. et al.:* Surface Science 8 (1967) 115.
[37] *Tracy, J. C., Blakely, J. M.:* Surface Science 13 (1969) 313.
[38] *Moya, F., Moya-Gontier, G. E., Cabane-Brouty, F., Oudar, J.:* Acta Metallurgica 19 (1971) 1189.
[39] *Gwathmey, A. T., Lawless, K. R.:* in: "The surface chemistry of metals and semiconductors (Ed.: H. C. Gatos), Wiley and Sons, New York—London (1960) p. 483.
[40] *Lawless, K. R.:* "Oxidation of Metals (Energetics and Metallurgical Phenomena)". (Ed.: W. Mueller) Blackie and Son (1962) p. 347.
[41] *Bardolle, J.:* "L'oxydation des métaux" (Ed.: J. Benard), Gauthier Villars, Paris (1962) p. 134.
[42] *Neuhaus, A., Gebhardt, M.:*
[42a] Werkstoff und Korrosion 7 (1966) 567.
[42b] Nova Acta Leopoldina 1 (1968) 208.
[43] *Lee, A. E., Singer, K. E.:* Journal of Crystal Growth 9 (1971) 68.
[44] *Pernoux, E., Merle, P.:* C. R. Acad. Sci. 265 (1967) 960.
[45] *Plumensi, J. P., Thomas, B.:* C. R. Acad. Sci 268 (1969) 2069.
[46] *Plumensi, J. P.:* "Métaux Corrosion Industrie" (1970) 533, 534.
[47] *Dagoury, G., Vincent, L. M., Oudar, J.:* C. R. Journées Internationales sur l'oxydation des métaux, SERAI, Bruxelles (1965) p. 262.
[48] *Dagoury, G.:* Thésis, Orsay (1970).
[49] *Vigner, D.:* Rapport CEA. R. 3847, CEN. Saclay (1969).
[50] *Sato, H., Shinozaki, S.:* Surface Science 22, 229, (1970).
[51] *van der Merwe, J. H.:*
[51a] in: "Single crystal films" (Eds.: H. Francombe and H. Sato), Pergamon Press (1964) p.139;
[51b] *Journal of Applied Physics* 34 (1965) 123;
[51c] in: "Advances in Epitaxy and Endotaxy" (Eds.: H. G. Schneider and V. Ruth). VEB Deutscher Verlag für Grundstoffindustrie, Leipzig (1971) p. 129.
[52] *Cabrera, N.:* in: "Processus de nucléation dans les réactions des gaz sur les métaux", Colloque CNRS N° 122, Paris (1963), Ed. du CNRS (1965).
[53] *Hart, R. K., Maurin, J. K.:* Surfaca 20 (1970) 285.
[54] *Borie, B. S., Sparks, C. J., Cathcart, J. V.:* Acta Met. 10 (1962) 691.
[55] *Audisio, S., Mai, C., Schaeffer, C., Monnier, G., Riviere, R.:* Bull. Soc. Chim. Fr. 3 (1971) 742.
[56] *Stranski, I. N., Krastanov, L.:* Akad. Wiss. Wien, Math-Naturwiss. Kl. II b 146 (1938) 797.
[57] *Bauer, E.:* Zeitschrift für Kristallographie 110 (1958) 372.
[58] *Cranstoun, G. K. L., Anderson, J. S.:* Nature 219 (1968) 365.
[59] *Rhead, G. E.:* Trans. of the Faraday Society 61 (1965) 797.
[60] *Mayer, H.:* in: "Advances in Epitaxy and Endotaxy" (Eds.: H. G. Schneider and V. Ruth). VEB Deutscher Verlag für Grundstoffindustrie, Leipzig (1971) p. 63.

[61] *Rhodin, T. N., Orr, W. H., Walton, D.:* in: [33].

[62] *Jardinier-Offergeld, M., Bouillon, F.:* in: "International conference on solid surfaces", Boston. October 1971, Journal of Vacuum Science and Technology . . . . (1972).

[63a] *Cathcart, J. V., Petersen, G. F., Sparks, C. J. Jr.:* in: "Surface and Interface". Syracuse University Press, N. Y. (1967) p. 333.

[63b] *Cathcart, J. V., Petersen, G. F., Sparks, C. J. Jr.:* J. Electrochemical Soc. 116 (1969) 664.

[64] *Lawless, K. R., Gwathmey, A. T.:* Acta Met. 4 (1956) 153.

M. Huber and *J. Oudar*

In the previous article models have been proposed for surface compounds which are mainly based on the hypothesis that the absorbed atoms are arranged in a compact lattice ("compact model"). Model metal atoms are inserted in the sulphur atoms lattice (mixed layer formed by reformation of the metallic substrate).

Very recently several structures have been studied again assuming that all adsorbed atoms are localized on regular sites [1], [2]. In general, this new hypothesis yields models with a higher symmetry than previous "compact models". The concept of surface compound or adsorption compound applies to the layer of adsorbed atoms and metal atoms directly in contact with adsorbed atoms as well. These double layers have a well-defined stoichiometry which can be determined from the content of the coincidence mesh. In some cases, as a first approximation, the metallic substrate can be considered to be undisturbed.

In other cases it is necessary to assume some perturbation of the first layer of the substrate which modifies the nature and the distribution of sites occupied by adsorbed atoms.

We give here some examples concerning the adsorption of sulphur on copper and iron. Some of them were analyzed in the previous article whith the hypothesis of "compact model".

I. *On a non-disturbed substrate,* two cases can be distinguished according to the superficial concentration of sulphur atoms compared to the critical concentration of 1 sulphur atoms for 2 metal atoms.

I.-1 Structure $p(8 \times 8)$ S/Cu(111) (Fig. 1)

This structure has been observed in the case of adsorption of sulphur on (111) copper exposed to a H_2S/H_2 mixture. The mean concentration measured corresponds to 26 atoms for 64 copper atoms in the coincidence mesh $p(8 \times 8)$. The deficiency in sulphur atoms from the stoichiometry of 1 atom for 2 copper atoms is balanced by vacancies regularly distributed.

I.-2 Structures corresponding to $Fe_{2n-2}S_n$ on Fe(100) (Fig. 2)

On a (100) iron surface the first structure which appears corresponds to a $c(2 \times 2)$ structure where a half of the sites of the fourfold lattice may be occupied by sulphur atoms. When we increase the superficial concentration of sulphur atoms, the distortion of the coincidence mesh described in the previous article can be related to a series of structures with periodic shears of sulphur atom chains which gives rise to

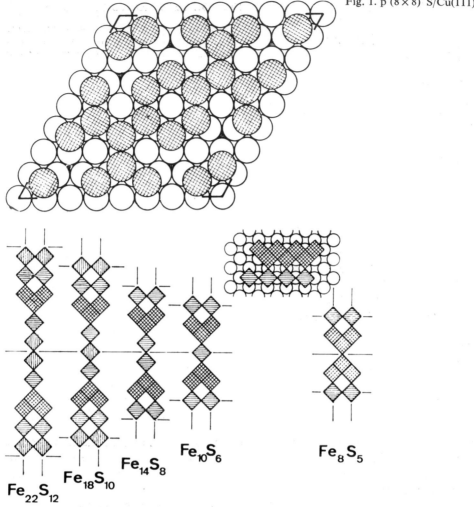

Fig. 1. p (8 × 8) S/Cu(111)

Fe$_{22}$S$_{12}$ Fe$_{18}$S$_{10}$ Fe$_{14}$S$_8$ Fe$_{10}$S$_6$ Fe$_8$S$_5$

Fig. 2. Series of surface compounds Fe$_{2n-2}$S$_n$. Upper part: On the right-side a single chain and a zig-zag chain on the substrate

zig-zag chains and single chains regularely disposed. The stoichiometry is given by the general Fe$_{2r-2}$S$_{2r}$ formula with $n = 10, 10, 8, 6, 5$.

II. Structure with reformation of the metallic substrate

The surface compound $\begin{bmatrix} 3\Gamma \\ 1\,2 \end{bmatrix}$ S/Cu(111) (Fig. 3) has a concentration corresponding to 3 sulphur atoms for 7 metallic atoms. If one turns a group of 3 metal atoms inside the coincidence mesh around the three-fold acxis (approximately 10°) one obtains 3 equivalent pseudosquare sites for the 3 adsorbed sulphur atoms. These

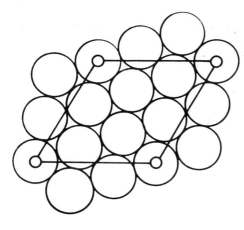

Fig. 3. Surface compound $\begin{bmatrix} 3 & \varGamma \\ 1 & 2 \end{bmatrix}$ S/Cu(111)
Upper part: Coincidence mesh on undisturbed substrate. Lower part: Substrate after reconstruction with pseudosquare sites (in black)

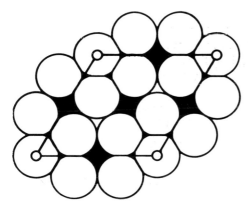

rotations take place cooperatively and correspond to a slight displacement of the metal atoms (of the order of a 1/4 of the atomic diameter of the metal). Structures having a coincidence mesh with a low symmetry seem to be particularly concerned by such a reformation.

These new proposed models have been discussed on the basis of chemical and structural considerations and seem generally to be more consistent than compact models. Some of them could be suitable for experimental check by LEED intensity mesurements, ion reflexion on neutral atoms, or photoelectron spectroscopy.

5.1.9.   REFERENCES

[1] *Huber, M.:* Paper presented at the "Séminaire interdisciplinaire adsorption sur les métaux de transition" Mont St ODILE (France) Mai 1974.
[2] *Huber, M., Oudar, J.:* in press

5.2.

NUCLEATION AND CRYSTAL GROWTH DURING THE FORMATION OF TARNISH LAYERS ON METALS

Chr. Weissmantel, G. Hecht and *J. Herberger*

Department of Physics — Electronic Components
Technische Hochschule Karl-Marx-Stadt, German Democratic Republic

5.2.1. INTRODUCTION

As fundamental processes of atmospheric corrosion, reactions between pure metals and gases leading to the formation of tarnish layers, have gained much theoretical and practical interest. During the last few decades a considerable amount of work has been devoted to the investigation of the elementary steps and the kinetics of layer formation on the surface of metals in the course of interaction with corrosive gases under defined conditions of temperature and pressure. Early experimental work dealt mainly with measurements concerning the rate laws of the reactions, whereas in more recent investigations additionally the influence of nucleation and crystal growth has been studied by electron microscopy, and other methods. Correspondingly, the theoretical efforts to describe the kinetics of layer growth followed the basic concept of a uniform formation of the tarnish layer with rate-controlling diffusion via point defects. However, authors as *Evans* [1], *Bénard* [2], and *Kofstad* [3] have emphasized that inhomogeneous growth, caused by nucleation and recrystallization, may have a strong if not prevailing influence on the kinetics, especially in the region of thin tarnish films.

Summarizing the results of the present experimental work, it seems well established that many different elementary processes such as chemisorption, nucleation, lattice diffusion, crystal growth and recrystallization are occurring during the layer growth, simultaneously or consecutively. Therefore, the theoretical interpretation of observed rate laws or *reaction isotherms* may often be ambiguous and identical laws may be explained by different mechanism. Furthermore, as *Evans* [1] and *Kofstad* [3] have pointed out, in smaller intervals different expressions for the rate equations may be applied as sufficient approximations depending only on a suitable choice of the respective constants. These difficulties are pronounced in the stage of very thin layers, but with increasing temperatures and layer thickness the process is approaching, in most cases, the *square root law* [4]

$$a = c_1 \cdot t^{1/2} \tag{1}$$

where

a is the film thickness and
t is the time,

explained by *Wagner* [5] on the basis of a rate-controlling *ambipolar diffusion* of the components via point defects of a more or less uniform corrosive layer.

On the other hand, in the region of thin tarnish layers ($a \leq 10^2$ nm) formed at low or medium temperatures mainly the following rate laws have been observed and interpreted:

(a) The inverse logarithmic law

$$a = a_0 - c_2 \ln (k_1 t + 1)^{-1} \tag{2}$$

first derived by *Cabrera* and *Mott* [6] and assuming a rate-controlling transport mechanism neglecting the influence of space charges or nucleation.

(b) The direct logarithmic law

$$a = c^3 \ln (t - t_0) \tag{3}$$

has been frequently observed, but as *Hauffe* stated [7], the different efforts to interpret this law on the basis of rate-determining electron tunneling, chemisorption, or special space charge effects, are not very satisfying. It seems much more reasonable to attribute the logarithmic law to an inhomogeneous layer growth, as it has been shown that a logarithmic isotherm can be explained by diffusion or nucleation effects in heterogeneous films [2], [7].

(c) Laws of the form

$$a = c_4 \cdot t^{1/3} \tag{4}$$

or

$$a = c_5 \cdot t^{1/4} \tag{5}$$

have been utilized by *Hauffe* and co-workers [7], [13] in order to describe the tarnish oxidation of copper and nickel at medium temperatures. These laws could be verified applying special assumptions about the space charge distribution in the growing layer.

(d) A general theoretical treatment of homogeneous film growth accomplished by *Fromhold* [8] gave evidence for a more general law

$$a = c_6 \cdot t^{1/n} \tag{6}$$

with $n \geq 2$ and passing over to the *square root law* of *Wagner*'s theory after a critical film thickness. Previously, this general law has been experimentally evidenced for the oxidation kinetics of evaporated films of 8 different metals in a wide range of time and temperature (*Weissmantel* [9]). A careful analysis of the present experimental material by means of numerical statistical methods revealed that the best average accuracy (estimated by the correlation coefficients) of all simple terms is obtained for the general law (6) but, in many cases, the direct logarithmic law is equally applicable. Furthermore, the double-logarithmic diagrams revealed in all cases of oxidation at low or medium temperatures a change of the reaction isotherm after a critical layer thickness had been reached, which depends on material and temperature. Figure 1 shows examples of the typical behavior of pure evaporated films during oxidation; the initial *fast* reaction may be described with best accuracy in terms of direct logarithmic expressions, whereas the second region follows the general law (6) with a fair accuracy. The decrease of slope of some of the curves obtained at higher

304

temperatures is to be attributed to a saturation effect, when most of the film is oxidized. Contrary to *Fromhold*, who considered only a homogeneous film growth, we interpreted the kinetic curves by a superposition of different elementary steps. Following the work of *Bénard*, *Evans*, and others, it seemed reasonable to assume that during the initial stage of layer deposition, the formation of nuclei and the growth of separated crystals are rate-controlling. Hence, the logarithmic time dependence observed in this region should be explained by models of a decreasing nucleation rate and/or the descending diffusion rate caused by the blocking of pores and gaps during coalescence of the crystals. During later stages of film formation — after the tilt of the curves — short circuit diffusion along pores is predominated by lattice diffusion via point defects. However, even for this stage, on the basis of the results of electron microscopy investigations we strongly suggested that a superposition of homogeneous film growth and recrystallization within the films occurs. A model leading to a time law consisting of two overlapping stages, an initial part controlled mainly by short circuit diffusion and a consecutive growth with predominating lattice diffusion, was developed by *Schmelzer*, *Hoering* and *Kirkaldy* [10]. The corresponding rate law has the form:

$$a^2 = c_7 \cdot t + k'(1 - e^{-k''t}).\tag{7}$$

This expression consists of a reciprocal logarithmic term and a parabolic term, while, in the generalization, we prefer to use instead the general term $t^{1/n}$.

In conclusion to this brief discussion of the kinetics of tarnish layer formation it can be stated that simple models of continuous film growth are not sufficient to explain the experimental results, and heterogeneous effects as nucleation, recrystallization, etc. must be considered in any satisfactory theory. The intention of our recent experimental work was to obtain more detailed information combining accurate measurements of the reaction kinetics of tarnish layer formation with thorough electron microscopy investigations. So far, we studied mainly the oxidation and sulphidation of evaporated Zn- and Cd-films. The selection of these systems resulted also from the practical interest which CdS- and ZnS-layers deserve in connection with their electronic and optical properties.

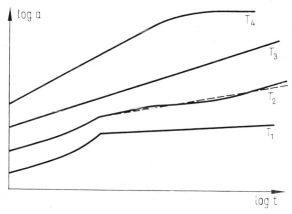

Fig. 1. Typical oxidation isotherms of clean metal films (*a* is the amount of adsorbed oxygen, *t* the reaction time)

Reaction kinetics have been studied in closed high vacuum glass cells. The metal films were produced by thermal evaporation, after pumping down to a vacuum of 10^{-6} torr and baking of the apparatus. In the case of the oxidation experiments, the gas uptake was measured at pressures between 0.1 and 0.3 torr by a special *McLeod* gauge [9]. The sulphidation experiments have been performed at saturation vapor pressures in the temperature region from 50 to 250°C, using radioactive tracer techniques. In Fig. 2 the reaction cell for simple kinetic investigations is shown. Immediately after vapor deposition of the film, the glass cell is separated from the evacuated system by sealing-off, and the cell is heated to the constant temperature of the experiment in an air-thermostat.

A constant vapor pressure may be achieved by the magnetically-operated destruction of a glass ampoule filled with sulphur that was labelled with the radioactive isotope S-35. The metal film was deposited mainly on the inner side of a thin mica window

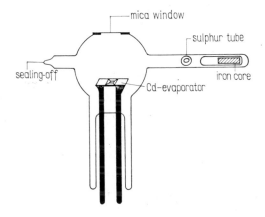

Fig. 2. Reaction cell for the measurement of sulphidation kinetics

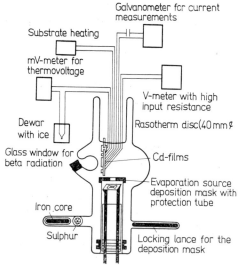

Fig. 3. Reaction cell for simultaneous measurements of sulphidation kinetics and film resistivity

fixed at the top of the glass cell. Through this window the sulphur sorption can be measured continuously by the penetrating β-radiation using a cooled Geiger – Müller tube. For simultaneous measurement of the reaction kinetics and the resistivity of the formed sulphide-films the cells, as displayed in Fig. 3, were constructed. On a glass substrate a metal film consisting of a central circular part and a surrounding open ring is deposited by means of a special mask. Subsequnetly, the substrate is turned by a magnetic device into the position shown in front of a very thin, but still pressure-resistant bulb-window in the cell. The sulphur vapor is then applied and the uptake is measured by the penetrating radiation, whereas the resistivity is determined at the outer film ring.

In order to perform electron microscopy and structure investigations, the experiments were interrupted at certain stages and the films were transferred to the structure laboratory. During this stage of the procedure a short contact with dry air at room temperature was unavoidable; recently, we tried to investigate the initial reaction steps also by an in-situ observation of the reaction in the lens chamber of an electron microscope.

In detail, we studied the morphology and the structure of the films by means of electron microscopy using direct transmission and replica techniques, by observations in a scanning-type electron microscope, and by transmission and reflection electron diffraction studies. Furthermore, X-ray diffraction has been employed for structure investigations at thicker layers and the study of the films with an X-ray microanalyzer proved to be useful, both for analysis of the layer stoichiometry and for a convenient determination of layer thickness even in the region of a few nanometers.

5.2.3.   RESULTS AT THE SYSTEMS Cd–S AND Cd–O [11], [12]

In the following, the most important results of our investigations shall be explained with the help of several series of figures. Some C-Pt-replica micrographs are shown first, informing about the nature of the untreated Cd-films. Figure 4 is the micrograph of a Cd-layer, in the stage of coalescence, deposited on a glass substrate at a magnification of 12 500 ×. Remarkable is the pronounced overgrowth of (0001)-oriented hexagonal crystals. The micrograph (Fig. 5) corresponds to a magnification of 30 000 × and shows the sharp edges and the growth planes of the metal crystallites. During tempering the surface may be smoothened, but the scanning micrograph (Fig. 6), which was taken after heating the deposited film, again reveals a distinct overgrowth structure.

In Fig. 7 the curves of the sulphidation kinetics at 150 and 200°C are depicted in a double-logarithmic scale. In agreement with our earlier results concerning the oxidation behavior of many different metals two stages of film growth must be discerned. On the other hand, also the above-mentioned two-term expression of *Schmelzer* and others yields a satisfying representation of the results (Fig. 8). The following picture (Fig. 9), is a replica micrograph with a 32 000-fold magnification after treatment of a cadmium film during a 5-min sulphur vapor exposure at 150°C. The metal surface is already covered with many CdS-nuclei of 20 to 50 nm diameters, which are distributed almost randomly at a density of about $10^{11} \, cm^{-2}$. During stages of

Fig. 4. Evaporated cadmium film on glass in in the stage of coalescence (C–Pt-replica, 12 500-fold magnification)

layer formation some preference of nucleation sites along grain boundaries and dislocations could sometimes be observed. For instance, the sample shown in Fig. 10 which was treated for 7 min at 150°C, clearly reveals several chains of nuclei giving rise to a chemical decoration effect of certain boundaries and steps. A thorough inspection of the surface at high magnification provides evidence that the surface between the nuclei is covered with a fine granular layer that had not been visible on the untreated metal surface. A comparison of the entire sulphur amount adsorbed at this stage with the approximate fraction bound in the nuclei yields an estimated thickness of 5 nm or something less for this primary layer. In accordance with other authors [14]–[16], we suppose that during the early stage of layer growth a more or less uniform coverage of the metal surface with a highly disordered ultra-thin sulphide film develops from which the nuclei are growing to a height of one hundred nanometers. Neither the primary layer nor the nuclei on the strongly textured (0001)-metal planes are preferentially oriented as evidenced by the reflection electron diffraction patterns.

In the course of further reactions the number of nuclei on the surface is increasing until the entire metal surface is covered with a coalescing sulphide layer. In Fig. 11 the situation after a 90-min treatment is represented at the lower magnification of 12 500×. The corresponding average layer thickness is 18 nm. Finally, in Fig. 12 the situation after a reaction time of 5 000 min is shown. At an average layer thick-

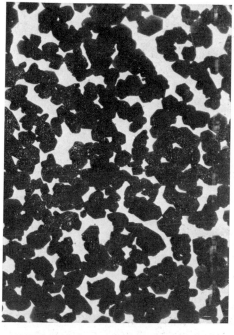

Fig. 5. Evaporated cadmium film on carbon in the stage of coalescence (transmission micrograph, 30 000-fold magnification)

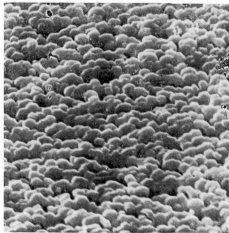

Fig. 6. Cadmium film after thermal treatment (scanning micrograph, 9 000-fold magnification)

ness of 36 nm after coalescence of the nuclei recrystallization with partial uniform orientation has occurred and, in addition, a large number of whiskers have grown with a length of 50 to 200 nm and a diameter of 20 to 30 nm.

The next series of micrographs were taken from Cd-samples exposed to sulphur vapor at a reaction temperature of 200°C. Because of the higher temperature the reaction rate has increased considerably, and much thicker CdS films are formed until the entire metal sample is sulphidized. After a treatment of 5 min on top of the well-formed Cd-crystallites a layer of 76 nm average thickness has grown, from which nuclei with a diameter of 30 to 100 nm are protruding (Fig. 13). Again, the density of these large

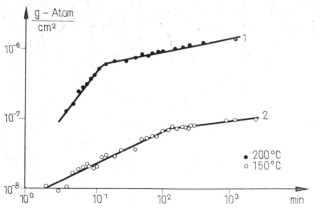

Fig. 7. Sulphidation kinetics of cadmium films at 150 and 200°C, respectively

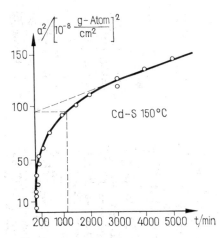

Fig. 8. Parabolic plot of sulphidation kinetics (Cd−S and Zn−S, respectively)

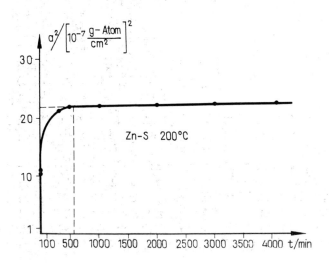

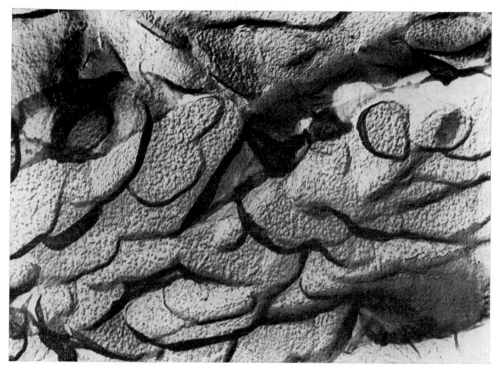

Fig. 9. Cadmium film after a 5 min sulphidation period at 150°C (C — Pt-replica, 32 000-fold magnification)

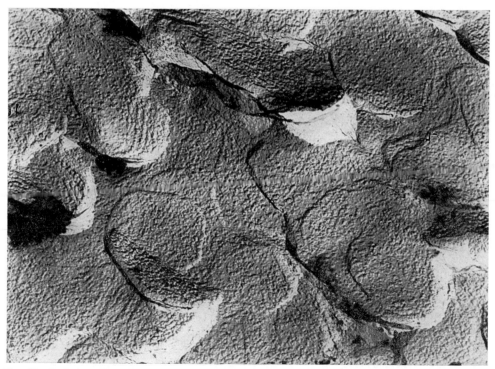

Fig. 10. Cadmium film after a 7 min sulphidation period at 150°C (C — Pt-replica, 32 000-fold magnification)

Fig. 11. Cadmium film after a 90 min sulphidation period at 150°C (C – Pt-replica, 12 500-fold magnification)

Fig. 12. Cadmium film after a 5 000 min sulphidation period at 150°C with some whisker growth (C – Pt-replica, 12 500-fold magnification)

Fig. 13. Cadmium film after a 5 min sulphidation period at 200°C (C—Pt-replica, 12 500-fold magnification)

Fig. 14. Cadmium film after a 30 min sulphidation period at 200°C (C—Pt-replica, 12 500-fold magnification)

nuclei is increasing with proceeding reaction time, and after 30 min the surface is almost completely covered (visible in Fig. 14), whereas the mean nuclei size has only slightly increased (100 to 200 nm). While the layer formed after 5 min did not have any uniform orientation, in the stage of coalescence after 30 min the onset of a (0001)-texture of CdS can be detected. During the rapidly progressing reaction a thick and relatively smooth sulphide layer is formed, but the grain boundaries of the underlying metal crystals are still decorated by chains of nuclei. The following picture shows the morphology after a 200-min reaction on a scanning-micrograph (Fig. 15a, magnification 10 000×) and a C–Pt-replica with a 12 500-fold magnification (Fig. 15b). The mean layer thickness of CdS corresponding to these samples is about 350 nm. During a prolonged treatment, the structure of the layer again changes: Apparently the structure of the fresh layer is not yet stable, and, simultaneously with further thickness growth, a severe recrystallization of the entire reaction product takes place. The result is a pronounced epitaxial orientation with hexagonal CdS in (0001)-texture on top of the metal crystals. Under certain conditions, for instance after a 4 300 min treatment and a layer thickness of more than 500 nm, rather large and well-shaped oriented CdS-crystals are formed, as indicated on a scanning micrograph (Fig. 16, magnification 10 000×) and as very distinctly revealed on a replica micrograph (Fig. 17, 12 500×). During that stage, the reflection electron diffraction pattern displays a strong texture, in full agreement with the morphological impression (Fig. 18). Furthermore, the structure investigation by means of X-ray diffraction indicates a similar orientation of the CdS deposit and the metallic Cd underneath, and, thus, the assumption of genuine epitaxy is verified (Fig. 19). After very long reaction times ($t \approx 4\,000$ min), and favored by increasing temperatures (up to 240°C), a dense growth of whiskers with a length of about 1 μm and a diameter of 10 to 30 nm could be observed on some samples as demonstrated by the scanning micrograph (Fig. 20). The very high density of whiskers in such layers has been demonstrated also by C–Pt-replicas that were deliberately ruptured (Fig. 21). The experiments relating to the oxidation kinetics of cadmium at 100°C and 150°C yielded the results plotted in Fig. 22 with a double-logarithmic scale. There is also a tilt in the curve (upper points), but, compared with the sulphidation, the second region has a steeper slope than the first one. As depicted in Fig. 23 the two-term expression consisting of an asymptotic, and a power, term is also suitable to describe the experimental values.

Within a few minutes of oxidation at 100°C and 0.1 torr a continuous oxide layer of about 10 nm thickness is formed which has no detectable orientation, but is crystalline (Fig. 24). After a longer treatment, rather large "nuclei" (diameter up to 1 μm) of hexagonal shape grow from the primary layer, and, following the coalescence, the entire layer is recrystallizing to a flat surface film in analogy to the case of sulphidation. Neither the large nuclei nor the layer after coalescence displayed a preferential orientation, but after a longer oxidation period at 200°C in air some epitaxial orientation with the (111)-planes of CdO parallel to the (0001)-planes of Cd was observed. However, after only a relatively short reaction time at temperatures in the region of of 150°C and at an oxygen pressure of 0.1 torr a formation of whiskers sometimes occurs (Fig. 25).

Contrary to our expectations (see also *Pfefferkorn* [17]) the oxide and sulphide whiskers have no favored orientations. From the experimental results hitherto obtained,

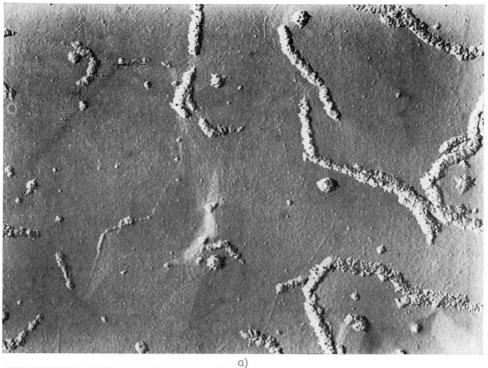

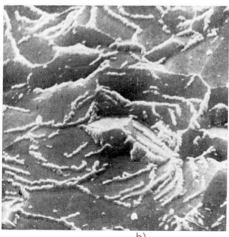

Fig. 15. Cadmium film after a 200-min sulphidation period at 200°C (a) scanning micrograph, 10 000-fold magnification; b) C — Pt-replica, 12 500-fold magnification)

a)

b)

no definite explanation for the mechanism and the structural or morphological reasons of whisker growth can be derived. Because extensive whisker formation is observed mainly after a prolonged exposure to the gaseous reaction component, it seems conceivable to attribute whisker growth to highly disordered film areas with excessive metal atoms (as interstitials or precipitated). It is emphasized that the tarnish films have a concentration gradient of metal atoms perpendicular to their surface. The phase transitions favored by presumably high mechanical stresses in the growing

315

Fig. 16. As Fig. 17 (scanning micrograph, 10 000-fold magnification)

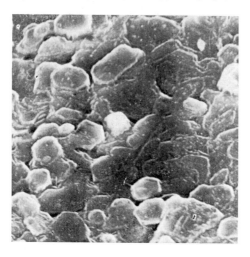

Fig. 17. Cadmium film after a 4 320-min sulphidation period at 200°C (C—Pt-replica, 12 500-fold magnification)

layer lead to the secondary recrystallization resulting in highly ordered epitaxial crystals. In the neighborhood of these crystal sites with a high grade of perturbation, low crystalline order, crevices, screw dislocations and metal precipitates, may remain, and these act as sources of whisker growth. Accordingly, we assume that the whiskers are growing from the solid phase by direct interaction with the gas inside the layer and not at the tip of the needles from the gaseous phase. Although the whiskers may

Fig. 18. Reflection electron diffraction pattern (of the cadmium film, shown in Fig. 16)

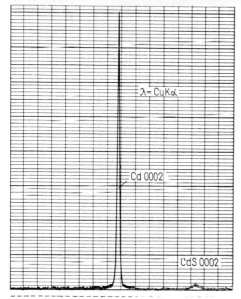

λ = CuKα

Cd 0002

CdS 0002

Fig. 19. X-ray diffraction pattern of the cadmium film shown in Fig. 16

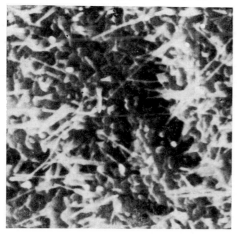

Fig. 20. Cadmium film after a 5 000 min sulphidation period, displaying extended whisker growth (scanning micrograph, 10 000-fold magnification)

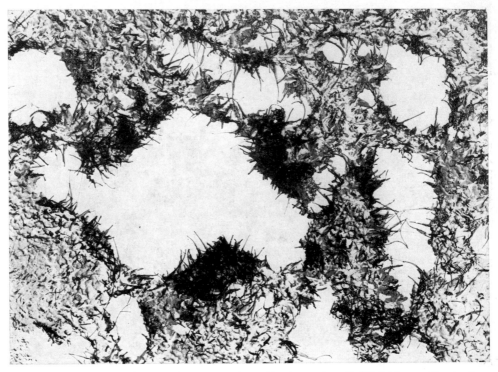

Fig. 21. As Fig. 20, but deliberately ruptured layer (C−Pt-replica, 12 500-fold magnification)

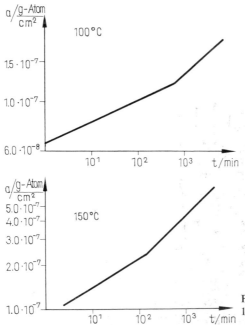

Fig. 22. Oxidation kinetics of cadmium films at 100 and 150°C, respectively

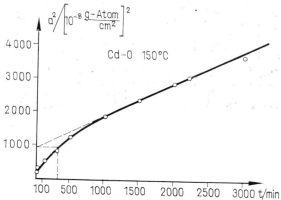

Fig. 23. Parabolic plot of oxidation kinetics (Cd–O)

Cd–O   100°C

Cd–O   150°C

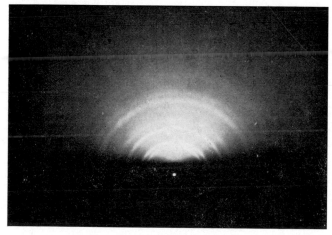

Fig. 24. Reflection electron diffraction pattern of a cadmium film after 3 min oxidation period at 100°C

contain only a very small fraction of the total amount of reaction products of the layer, this model may explain the fact that whisker formation is connected with a reduction of the layer growth rate as has been observed by *Nestler* [18] in the case of oxidation of iron and steel.

5.2.4.   **RESULTS AT THE SYSTEMS Zn–S and Zn–O [11], [12]**

The zinc films, deposited at room temperature on glass substrates, exhibited in the micrographs a morphology of disk-shaped crystals with diameters ranging from 0.5 to 1 μm, and a pronounced overgrowth with a uniform orientation, the (0001) plane being parallel to the substrate surface, as has been found for the Cd-films. During tempering at 150 to 200°C by re-evaporation of metal atoms, the crystal corners and the surface roughness became more or less smooth, and along the crystal walls etching-steps became visible (Fig. 26). At 150°C the reaction with sulphur vapor was very slow and up to 1 000 min no nuclei could be detected. From the measurements of adsorption, we concluded that a continuous layer growth up to a thickness of 100 nm occurred in this phase, but in the micrographs only the granular structure of a smooth layer was observed. Electron diffraction patterns revealed that these primary films consist of very fine crystalline ZnS without a uniform orientation. The kinetic curve in double-logarithmic scale is characterized by a tilt after 10^3 min, and, correspondingly, after some thousand minutes large ZnS "nuclei" (diameters 20 to 100 nm) are growing from the primary film (Fig. 27).

Sulphidation at 200°C produced, after 300 min, a coverage with many randomly distributed nuclei and, later, there occurred coalescence and recrystallization to a rather smooth layer as is known already from the other systems investigated. Figures 28 and 29 are the replica micrograph and the diffraction pattern of a Zn-film sulphidized for 300 min at 250°C. A progressive stage of layer formation is to be seen, but still no texture is obtained. Even after nearly 4 000 minutes of treatment at 250°C, no formation of whiskers and no epitaxial recrystallization occurred (Fig. 30). Finally, after raising the temperature to 350°C and the sulphur pressure to 50 torr (!), both the formation of larger, probably epitaxial ZnS-crystals and some whisker growth started after a long time (Fig. 31).

The investigation of the Zn–O system revealed the same principal effects as evidenced by the kinetic curve with the characteristic tilt shown in Fig. 32. As in the cases of CdS and ZnS and the previously investigated oxides of Fe, Ni, Cr, Pt, Cu, and Al [9], the slope of the first section of the curve is steeper than that of the second region. Therefore, the reversed behavior observed in the system Ce–O is an exception, probably due to the very large size of the nuclei.

Fig. 25. Cadmium film after a 30 min oxidation period at 125°C showing whisker growth (scanning micrograph, 7 000-fold magnification)

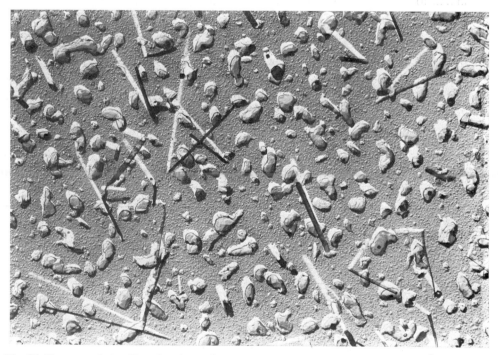

Fig. 26. Evaporated zinc film after thermal treatment at 150°C (C — Pt-replica, 32 000-fold magnification)

Fig. 27. Zinc film after a 4 100 min sulphidation period at 150°C (C—Pt-replica, 32 000-fold magnification)

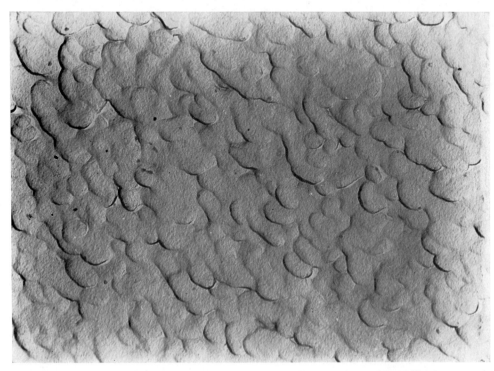

Fig. 28. Zinc film after a 300 min sulphidation period at 250°C (C—Pt-replica, 32 000-fold magnification)

Fig. 29. Reflection electron diffraction pattern of the zinc film shown in Fig. 28

Fig. 30. Zinc film after a 3 900 min sulphidation period at 250°C (C—Pt-replica, 32 000-fold magnification)

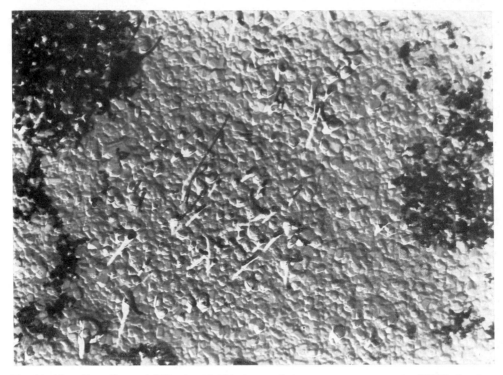

Fig. 31. Zinc film after a 4 260 min sulphidation period at temperatures up to 350°C showing whisker growth (C−Pt-replica, 20 000-fold magnification)

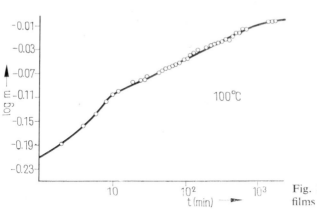

Fig. 32. Oxidation kinetics of zinc films at 100°C

5.2.5. CONCLUSIONS

From the experimental results it can be conclud ed that, in the systems studied, intricate nucleation, coalescence and recrystallization processes have a substantial effect on the kinetics of tarnish layer growth. Taking into account that *Evans, Bénard, Pashley* [19], *Kofstad, Gulbransen* [20], *Pfefferkorn* and other authors report on ana-

324

logous effects obtained for various metal–gas systems and conditions, the conception of a homogeneous tarnish layer growth, still implicated in many publications cannot be sustained. Moreover, theoretical models assuming a rate determining diffusion and field transport through homogeneous crystal layers should not be accepted, even as first-order approximations, for the interpretation of thin tarnish layer growth at low or medium temperatures, although volume diffusion certainly is one of the elementary processes which must be considered. The experimentally obtained curves of reaction kinetics, therefore, always represent a process caused by several overlapping and consecutive elementary mechanisms and, usually, neither of these processes can be considered to be the rate determining one. This is the reason why, in most cases, the analytical functions of the time-dependence of layer growth do not exactly correspond to the different terms derived from the basic concept of one distinct rate controlling effect.

Now the question arises, whether the experimental facts, obtained mainly from the investigation of some special systems, are sufficient to propose at least a qualitative model for the general mechanism of tarnish layer growth on metals. Our attempt to develop the general principles of such a general model is supported by the results of the other authors mentioned above, who have studied nucleation and crystal growth during tarnish layer formation. Moreover, very recently, *Nestler* [18] investigated the oxidation of iron and steel in the temperature range from 260 to 520°C by means of electron microscopy and found the same principal features as those we have observed in other systems.

Summarizing, we propose a model for the tarnish layer formation that takes into account the following stages (Fig. 33):

(a) Chemisorption of the gas particles on the metal surface and relatively rapid formation of a primary layer, the thickness of which may range — depending on the particular system and the reaction parameters — from a few, to some hundred, nm. This initial layer seems, in fact, to be rather homogeneous, and it consists of fine grains, or "primary nuclei", characterized by a high degree of disorder and a considerable interstitial space. Hence, both the transport of gas to the metal interface and the migration of metal atoms towards the outer surface are possible with low activation energy, a behavior which is denoted in more recent publications as *short-circuit diffusion*.

(b) Because of the high degree of disorder and probably favored by increasing internal stresses during the growth process in connection with the misfit between the metal and the compound a "secondary nucleation" — at first observed by *Bénard* —

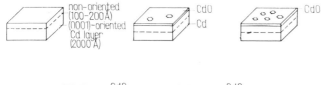

Fig. 33. Elementary steps and phase transitions of tarnish layer formation

occurs. During the course of this process, larger and better ordered crystals will grow from the primary layer. Our hypothesis is that these nuclei are formed both by a relatively rapid coalescence and growth of suitably orientated primary nuclei within the layer and, additionally, by epitaxial arrangement and growth of nuclei attached at the metal surface. Presumably the mechanism investigated by *Kern* and co-workers [21] of the orientation and movement of small nuclei during epitaxy may be applicable in a modified form to this process. During the subsequent reaction stages the secondary nuclei are only slowly growing, because growth can only be accomplished by a slower diffusion of the respective components, but the number of nuclei is continuously increasing until the entire surface is covered.

(c) After coalescence of the secondary nuclei, the thickness of the entire layer is increasing, but at a smaller rate than during the first stage, and the metal becomes covered by a tarnish layer with improving smoothness.

(d) At sufficiently high temperatures and after a sufficiently long time a new recrystallization process is starting, during the course of which larger crystals of the reactant with epitaxial orientation in respect to the surface of the metalcrystallites are formed. Probably this effect consists mainly of a favored growth of the suitably orientated secondary nuclei at the expense of the others.

(e) Simultaneously with the last mentioned process, but evidently favored by high gas pressures, an extensive growth of whiskers occurs, the latter being randomly spread about the surface. In all the systems investigated and during the above-mentioned experiments of *Nestler* on iron and steel, whisker growth is a very common phenomenon. We assume that whisker growth occurs at interstices or imperfections caused by mechanical stresses as a result of the secondary recrystallization, and that the interstices act as sources of the material transport for whisker growth. From several facts, we conclude that whisker growth in these systems starts from the metal phase, the whisker being squeezed out of the layer.

(f) Of course, the model outlined here must be developed further, and many details, especially the growth and structure of the primary layer and the precise mechanism of whisker growth, have to be established both by experimental and theoretical work.

5.2.6.   **REFERENCES**

[1] *Evans, U. R.:* The Corrosion and Oxidation of Metals. Arnold, London (1960).
[2] *Bénard, J.:* L'Oxydation des Métaux. Gauthier Villars, Paris (1962/64).
[3] *Kofstad, P.:* High Temperature Oxidation of Metals. J. Wiley (1966).
[4] *Kubaschewski, O., Hopkins, B. E.:* Oxidation of Metals and Alloys. Butterworths, London (1962).
[5] *Wagner, C.:* Atom Movements. Cleveland (1951).
[6] *Cabrera, N., Mott, N. P.:* Rep. Progr. Physics 12 (1948/49) 163.
[7] *Hauffe, K.:* Oxidation of Metals. Plenum Press, New York (1965).
[8] *Fromhold, A. T.:* J. Phys. Chem. Solids 24 (1963) 1081, 1309; 25 (1964) 1129.
[9] *Weissmantel, Chr.:* Habilitation-Thesis, Techn. Universität Dresden (1963).
[10] *Schmelzer, W. W., Hoering, R. R., Kirkaldy, J. S.:* Acta Met. 9 (1961) 880.

[11] *Weissmantel, Chr., Hecht, G., Herberger, J., Glauche, R., Wagner, I.:* Symposium: Oberflächenkristallisation — Grundlagen der Epitaxie, Berlin (1969. Kristall und Technik 5 (1970) 299.

[12] *Hecht, G., Herberger, J., Weissmantel, Chr.:* Internl. Conf. Thin Films, Boston 1969. J. Vac. Sci. Technol. 7 (1969) 547.

[13] *Engell, H., Hauffe, K., Ilschner, B.:* Z. Elektrochem. 58, (1954) 478.

[14] *Bénard, J.:* Met. Rev. 9 (1964) 473.

[15] *Bénard, J.:* J. Electrochem. Soc. 114 (1967) 139 C.

[16] *Seyboldt, A. U.:* Adv. Phys. 12 (1963) 2.

[17] *Pfefferkorn, G., Dachselt, W. D.:* Z. Naturforsch. 15a (1960) 412.

[18] *Nestler, C.—G.:* Dissertation, TH Karl-Marx-Stadt. (1971).

[19] *Pashley, D. W.:* Advances in Physics 14 (1965) 361.

[20] *Gulbransen, E. A., Copan, T. P.:* Nature 186 (1960) 959.

[21] *Kern, R., Masson, A., Metois, J. J.:* Advances in Epitaxy and Endotaxy (Eds.: H. G. Schneider and V. Ruth) Chapter 2.2. Leipzig (1971).

334

7. SUBJECT INDEX

Ch. 5.2

344